Masonry
Level Two

Trainee Guide
Fourth Edition

PEARSON

Boston Columbus Indianapolis New York San Francisco Upper Saddle River
Amsterdam Cape Town Dubai London Madrid Milan Munich Paris Montreal Toronto
Delhi Mexico City São Paulo Sydney Hong Kong Seoul Singapore Taipei Tokyo

NCCER

President: Don Whyte
Director of Product Development: Daniele Dixon
Masonry Project Manager: Rob Richardson, Tim Davis
Senior Manager of Product Development: Tim Davis
Quality Assurance Coordinator: Debie Hicks

Desktop Publishing Coordinator: James McKay
Permissions Specialist: Megan Casey
Production Specialist: Megan Casey
Editor: Chris Wilson

Writing and development services provided by S4Carlisle Publishing Services, Dubuque, IA

Lead Writer/Project Manager: Michael B. Kopf
Writer: Paul Lagassse, Jack Klasey
Art Development: S4Carlisle Publishing Services

Permissions Specialist: Kim Schmidt, Karyn Morrison, Katherine Benzer
Media Specialist: Genevieve Brand
Copy Editor: Michael H. Toporek

Pearson Education, Inc.

Director, Global Employability Solutions: Jonell Sanchez
Head of Associations: Andrew Taylor
Editorial Assistant: Douglas Greive
Program Manager: Alexandrina B. Wolf
Project Manager: Janet Portisch
Operations Supervisor: Deidra M. Skahill
Art Director: Diane Ernsberger
Directors of Marketing: David Gesell, Margaret Waples
Field Marketers: Brian Hoehl, Stacey Martinez

Composition: NCCER
Printer/Binder: LSC Communications
Cover Printer: LSC Communications
Text Fonts: Palatino and Univers

Credits and acknowledgments for content borrowed from other sources and reproduced, with permission, in this textbook appear at the end of each module.

PEARSON

Perfect bound   ISBN-13: 978-0-13-377970-7
                   ISBN-10:   0-13-377970-X

9 2020

Preface

To the Trainee

Masons are recognized as premier craftworkers on any construction site. Although masonry is one of the world's oldest crafts, masons also use 21st-century technology on the job. Using brick, block, or stone, and bound with mortar, masons build durable structures with optimized energy performance.

With the support of the Mason Contractor Association of America (MCAA), NCCER's program has been designed and revised by subject matter experts from across the nation and industry to update the curriculum with modern techniques. Our three levels present an apprentice approach to the masonry field and will help to keep you knowledgeable, safe, and effective on the job.

We wish you the best as you continue your training for an exciting and promising career. This newly revised masonry curriculum will help you enter the workforce with the knowledge and skills needed to perform productively in either the residential or commercial market.

New with *Masonry Level Two*

NCCER is proud to release the newest edition of *Masonry Level Two* in full color with updates to the curriculum that will engage you and give you the best training possible. In this edition, you will find that the layout has changed to better align with the learning objectives. There are also new end-of-section review questions to compliment the module review. The text, graphics, and special features have been enhanced to reflect advancements in masonry technology and techniques.

We invite you to visit the NCCER website at **www.nccer.org** for information on the latest product releases and training, as well as online versions of the *Cornerstone* magazine and Pearson's NCCER product catalog.

Your feedback is welcome. You may email your comments to **curriculum@nccer.org** or send general comments and inquiries to **info@nccer.org**.

NCCER Standardized Curricula

NCCER is a not-for-profit 501(c)(3) education foundation established in 1996 by the world's largest and most progressive construction companies and national construction associations. It was founded to address the severe workforce shortage facing the industry and to develop a standardized training process and curricula. Today, NCCER is supported by hundreds of leading construction and maintenance companies, manufacturers, and national associations. The NCCER Standardized Curricula was developed by NCCER in partnership with Pearson, the world's largest educational publisher.

Some features of the NCCER Standardized Curricula are as follows:

- An industry-proven record of success
- Curricula developed by the industry for the industry
- National standardization providing portability of learned job skills and educational credits
- Compliance with the Office of Apprenticeship requirements for related classroom training (*CFR 29:29*)
- Well-illustrated, up-to-date, and practical information

NCCER also maintains a National Registry that provides transcripts, certificates, and wallet cards to individuals who have successfully completed a level of training within a craft in NCCER's Curricula. *Training programs must be delivered by an NCCER Accredited Training Sponsor in order to receive these credentials.*

Special Features

In an effort to provide a comprehensive, user-friendly training resource, we have incorporated many different features for your use. Whether you are a visual or hands-on learner, this book will provide you with the proper tools and information to orient you to the important skills and techniques of the masonry trade.

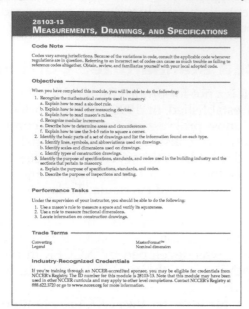

Color Illustrations and Photographs

Full-color illustrations and photographs are used throughout each module to provide vivid detail. These figures highlight important concepts from the text and provide clarity for complex instructions. Each figure reference is denoted in the text in *italic type* for easy reference.

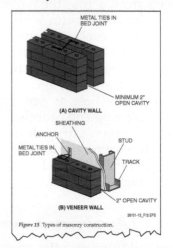

Figure 15 Types of masonry construction.

Introduction

This page is found at the beginning of each module and lists the Objectives, Performance Tasks, Trade Terms, and Required Trainee Materials for that module. The Objectives list the skills and knowledge you will need in order to complete the module successfully. The Performance Tasks give you an opportunity to apply your knowledge to real-world tasks you will undertake as a mason. The list of Trade Terms identifies important terms you will need to know by the end of the module. Required Trainee Materials list the materials and supplies needed for the module.

Special Features

Features provide a head start for those learning masonry by presenting technical tips and professional practices. These features often include real-life scenarios similar to those you might encounter on the job site.

Fall Protection

Most workers who die from falls were wearing harnesses but had failed to tie off properly. Always follow the manufacturer's instructions when wearing a harness. Know and follow your company's safety procedures when working on roofs, ladders, and other elevated locations.

Notes, Cautions, and Warnings

Safety features are set off from the main text in highlighted boxes and are organized into three categories based on the potential danger of the issue being addressed. Notes simply provide additional information on the topic area. Cautions alert you of a danger that does not present potential injury but may cause damage to equipment. Warnings stress a potentially dangerous situation that may cause injury to you or a co-worker.

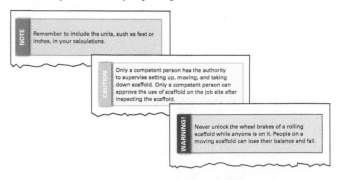

Going Green

Going Green looks at ways to preserve the environment, save energy, and make good choices regarding the health of the planet. Through the introduction of new construction practices and products, you will see how the greening of America has already taken root.

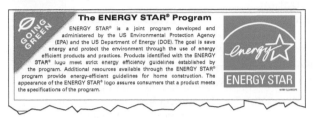

Did You Know?

The *Did You Know?* features offer hints, tips, and other helpful bits of information.

Did you know?
Safety Training

Safe working conditions on mine sites fall under the Mine Safety and Health Administration (MSHA), and every other job site is regulated by the Occupational Safety and Health Administration (OSHA). However, except for a few industry-specific requirements, the regulations are the same. Safety training is required for all activities. Never operate tools, machinery, or equipment without prior training.

Step-by-Step Instructions

Step-by-step instructions are used throughout to guide you through technical procedures and tasks from start to finish. These steps show you not only how to perform a task but also how to do it safely and efficiently.

The mason needs to determine whether the brick is too dry for a good bond with the mortar. The following test can be used to measure the absorption rate of brick:

Step 1 Draw a circle about the size of a quarter on the surface of the brick with a crayon or wax marker.

Step 2 With a medicine dropper, place 20 drops of water inside the circle.

Step 3 Using a watch with a second hand, note the time required for the water to be absorbed.

If the time for absorption exceeds 1½ minutes,

Trade Terms

Each module presents a list of Trade Terms that are discussed within the text and defined in the Glossary at the end of the module. These terms are denoted in the text with **bold, blue type** upon their first occurrence. To make searches for key information easier, a comprehensive Glossary of Trade Terms from all modules is located at the back of this book.

Masons are recognized as premier craftworkers at any construction site. Although masonry is one of the world's oldest crafts, masons also use 21st-century technology on the job. Masons build structures out of masonry units. The two main types of masonry units manufactured today are made of clay and concrete. Clay products are commonly known as brick and tile; concrete products are commonly known as concrete masonry units (CMUs) or block. Masonry units are also made from ashlar, glass, adobe, and other materials. In the most common forms of masonry, a mason assembles walls and other structures of clay brick or CMUs using mortar to bond the units together.

Review Questions

Review Questions are provided to reinforce the knowledge you have gained. This makes them a useful tool for measuring what you have learned.

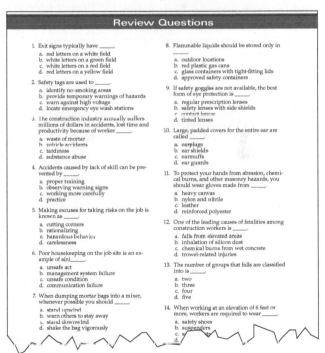

NCCER Standardized Curricula

NCCER's training programs comprise more than 80 construction, maintenance, pipeline, and utility areas and include skills assessments, safety training, and management education.

Boilermaking
Cabinetmaking
Carpentry
Concrete Finishing
Construction Craft Laborer
Construction Technology
Core Curriculum:
    Introductory Craft Skills
Drywall
Electrical
Electronic Systems Technician
Heating, Ventilating, and
    Air Conditioning
Heavy Equipment Operations
Highway/Heavy Construction
Hydroblasting
Industrial Coating and Lining
    Application Specialist
Industrial Maintenance Electrical
    and Instrumentation Technician
Industrial Maintenance
    Mechanic
Instrumentation
Insulating
Ironworking
Masonry
Millwright
Mobile Crane Operations
Painting
Painting, Industrial
Pipefitting
Pipelayer
Plumbing
Reinforcing Ironwork
Rigging
Scaffolding
Sheet Metal
Signal Person
Site Layout
Sprinkler Fitting
Tower Crane Operator
Welding

Maritime

Maritime Industry Fundamentals
Maritime Pipefitting
Structural Fitter

Green/Sustainable Construction

Building Auditor
Fundamentals of Weatherization
Introduction to Weatherization
Sustainable Construction
    Supervisor
Weatherization Crew Chief
Weatherization Technician
Your Role in the Green
    Environment

Energy

Alternative Energy
Introduction to the Power Industry
Introduction to Solar Photovoltaics
Introduction to Wind Energy
Power Industry Fundamentals
Power Generation Maintenance
    Electrician
Power Generation I&C
    Maintenance Technician
Power Generation Maintenance
    Mechanic
Power Line Worker
Power Line Worker: Distribution
Power Line Worker: Substation
Power Line Worker: Transmission
Solar Photovoltaic Systems
    Installer
Wind Turbine Maintenance
    Technician

Pipeline

Control Center Operations, Liquid
Corrosion Control
Electrical and Instrumentation
Field Operations, Liquid
Field Operations, Gas
Maintenance
Mechanical

Safety

Field Safety
Safety Orientation
Safety Technology

Supplemental Titles

Applied Construction Math
Careers in Construction
Tools for Success

Management

Fundamentals of Crew Leadership
Project Management
Project Supervision

Spanish Titles

Acabado de concreto: nivel uno,
    nivel dos
Aislamiento: nivel uno, nivel dos
Albañilería: nivel uno
Andamios
Aparejamiento básico
Aparajamiento intermedio
Aparajamiento avanzado
Carpintería:
    Introducción a la carpintería,
    nivel uno; Formas para
    carpintería, nivel tres
Currículo básico: habilidades
    introductorias del oficio
Electricidad: nivel uno, nivel dos,
    nivel tres, nivel cuatro
Encargado de señales
Especialista en aplicación de
    revestimientos industriales:
    nivel uno, nivel dos
Herrería: nivel uno, nivel dos,
    nivel tres
Herrería) de refuerzo: nivel uno
Instalación de rociadores: nivel
    uno
Instalación de tuberías: nivel uno,
    nivel dos, nivel tres, nivel cuatro
Instrumentación: nivel uno, nivel
    dos, nivel tres, nivel cuatro
Orientación de seguridad
Mecánico industrial: nivel uno,
    nivel dos, nivel tres, nivel
    cuatro, nivel cinco
Paneles de yeso: nivel uno
Seguridad de campo
Soldadura: nivel uno, nivel dos,
    nivel tres

Portuguese Titles

Currículo essencial: Habilidades
    básicas para o trabalho
Instalação de encanamento
    industrial: nível um, nível dois,
    nível três, nível quatro

Acknowledgments

This curriculum was revised as a result of the farsightedness and leadership of the following sponsors:

Arizona Masonry Contractors Association
Brick Industry Association
Central Cabarrus High School
Florida Masonry Apprentice & Educational Foundation, Inc.
Mason Contractors Association of America
Mortar Net Solutions

Mortenson Construction
Pyramid Masonry
Rhino Masonry, Inc.
Rocky Mountain Masonry Institute
Samuell High School
Skyline High School

This curriculum would not exist were it not for the dedication and unselfish energy of those volunteers who served on the Authoring Team. A sincere thanks is extended to the following:

Jeff Buczkiewicz
Kenneth Cook
Steve Fechino

John Foley
Todd Hartsell
Lawrence Johnson

Bryan Light
Moroni Mejia
Dennis Neal

NCCER Partners

American Fire Sprinkler Association
Associated Builders and Contractors, Inc.
Associated General Contractors of America
Association for Career and Technical Education
Association for Skilled and Technical Sciences
Carolinas AGC, Inc.
Carolinas Electrical Contractors Association
Center for the Improvement of Construction Management and Processes
Construction Industry Institute
Construction Users Roundtable
Construction Workforce Development Center
Design Build Institute of America
GSSC – Gulf States Shipbuilders Consortium
Manufacturing Institute
Mason Contractors Association of America
Merit Contractors Association of Canada
NACE International
National Association of Minority Contractors
National Association of Women in Construction
National Insulation Association
National Ready Mixed Concrete Association
National Technical Honor Society
National Utility Contractors Association

NAWIC Education Foundation
North American Technician Excellence
Painting & Decorating Contractors of America
Portland Cement Association
SkillsUSA®
Steel Erectors Association of America
U.S. Army Corps of Engineers
University of Florida, M. E. Rinker School of Building Construction
Women Construction Owners & Executives, USA

NCCER Business Partners

Contents

Module Seven

Construction Inspection and Quality Control

Introduces the quality control requirements for masonry construction. Presents procedures for inspection and testing of masonry materials and finished masonry construction. (Module ID 28207-14; 15 Hours)

Glossary

Index

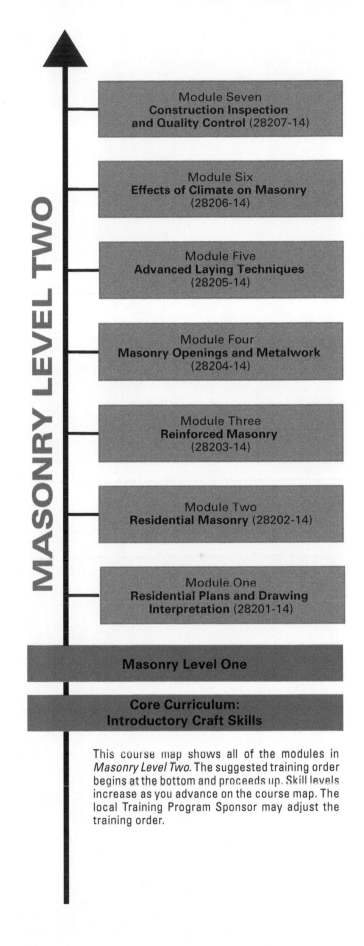

This course map shows all of the modules in *Masonry Level Two*. The suggested training order begins at the bottom and proceeds up. Skill levels increase as you advance on the course map. The local Training Program Sponsor may adjust the training order.

28201-14

Residential Plans and Drawing Interpretation

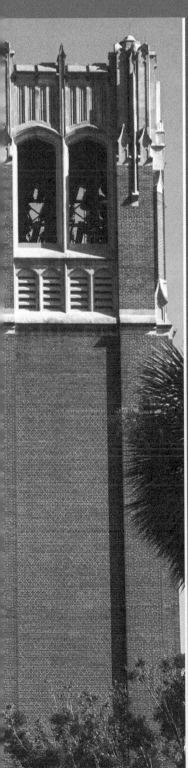

All masons must be able to read and interpret plans and drawings. Reading plans is important whether you are working by yourself or supervising others. In *Masonry Level One*, you learned the basics of masonry math, drawing interpretation, and specifications. This module covers the information you will need in order to work with residential plans and construction drawings and convert that information into action on the job.

Module One

Trainees with successful module completions may be eligible for credentialing through NCCER's National Registry. To learn more, go to **www.nccer.org** or contact us at **1.888.622.3720**. Our website has information on the latest product releases and training, as well as online versions of our *Cornerstone* magazine and Pearson's product catalog.

Your feedback is welcome. You may email your comments to **curriculum@nccer.org**, send general comments and inquiries to **info@nccer.org**, or fill in the User Update form at the back of this module.

This information is general in nature and intended for training purposes only. Actual performance of activities described in this manual requires compliance with all applicable operating, service, maintenance, and safety procedures under the direction of qualified personnel. References in this manual to patented or proprietary devices do not constitute a recommendation of their use.

Objectives

When you have completed this module, you will be able to do the following:

1. Describe the basic parts of a set of residential drawings and list the information found on each type of drawing.
   a. Identify keys and legends, as well as selected lines, architectural terms, abbreviations, and symbols on residential drawings.
   b. Explain how to use scales and dimensions in residential drawings.
   c. Explain how to interpret the various types of residential drawings.
2. Explain how to estimate material quantities from residential drawings.
   a. Explain how to use the rule-of-thumb method.
   b. Explain how to use estimating aids.

Performance Tasks

Under the supervision of your instructor, you should be able to do the following:

1. From a plan, calculate the square footage of one elevation, including openings.
2. Estimate the amount of brick and mortar from that same elevation.
3. Estimate the size and number of lintel block for that same elevation.

Trade Terms

Change order
Legend

Sectional drawing
Shop drawing

Industry-Recognized Credentials

If you're training through an NCCER-accredited sponsor, you may be eligible for credentials from NCCER's Registry. The ID number for this module is 28201-14. Note that this module may have been used in other NCCER curricula and may apply to other level completions. Contact NCCER's Registry at 888.622.3720 or go to **www.nccer.org** for more information.

Code Note

Codes vary among jurisdictions. Because of the variations in code, consult the applicable code whenever regulations are in question. Referring to an incorrect set of codes can cause as much trouble as failing to reference codes altogether. Obtain, review, and familiarize yourself with your local adopted code.

Contents

Topics to be presented in this module include:

Figures and Tables

1.0.0 RESIDENTIAL DRAWINGS AND THEIR ELEMENTS

Objective

Describe the basic parts of a set of residential drawings and list the information found on each type of drawing.

    a. Identify keys and legends, as well as selected lines, architectural terms, abbreviations, and symbols on residential drawings.

    b. Explain how to use scales and dimensions in residential drawings.

    c. Explain how to interpret the various types of residential drawings.

Trade Terms

Change order: A document or form used during the construction process to document a change in the construction requirements from the original plans or specifications.

Legend: A listing that explains or defines symbols or special marks placed on plans or drawings. Usually the legend is on the front sheet or index of the plan set.

Sectional drawing: A drawing that shows the inside of a component or structure. The view would be as if you cut the item into two pieces and looked at the end of the cut.

Shop drawing: A drawing that is usually developed by manufacturers, fabricators, or contractors to show specific dimensions and other pertinent information concerning a particular piece of equipment and its installation methods.

A set of plans is needed for every residence, from a tract-built house to a custom-designed mansion. The plans provide a road map. They guide the contractor and all the craftspeople. Plans enable each craftsperson to work in cooperation with other trades.

Plans provide guidance before, during, and after construction:

- Before the job begins, plans are used for planning the sequence of work and estimating the required materials and other resources.
- During the construction, plans are used to convey information about dimensions, assemblies, materials use, and architectural finish.

- After the construction, plans are used to check the accuracy and quality of the construction work.

Residential drawing sets contain several different kinds of drawings (*Figure 1*). A complete set of plans and working drawings for a structure usually includes the following sections:

- Title sheet(s)
- Site or plot plan
- Architectural drawings
- Specialty plans
- Schedules

Specialty plans include plumbing, mechanical, and electrical plans. Larger, more-complex projects may include a full set of plans for each specialty. This helps each craftsperson locate specific information.

The drawings in the set illustrate the structure using different views. The types of written information and views normally contained in a drawing set are:

- Plot plan
- Foundation plan
- Floor plan
- Elevation drawings
- Sectional drawings
- Detail drawings

This section will focus on using the information included in foundation drawings, elevation drawings, and floor plans. These are the drawings that masons use most frequently when working on residential projects. You should be familiar with the other types of drawings as well, in case you need to refer to them for a specific project. Keep in mind that the need to consult specific drawings will vary from project to project.

When studying plans, be sure to read all notes included in the drawings. Notes supply important information that supplements the drawing. It is also important to consult the specifications for any information not shown on the plans.

Sometimes changes are made by the owner or architect after the contract has been signed. These changes, called revisions, must accompany the plans. This way, the contractor is informed of them before work is started. Revisions should be held to a minimum, but are a part of most plans.

As an example of a necessary revision, consider a door to the main entrance of a house. The size of the masonry opening is given on the floor plan, but the architect decides to change the size because of the way it looks. To change the door size so the right proportions are obtained, a revision to the plans must be made.

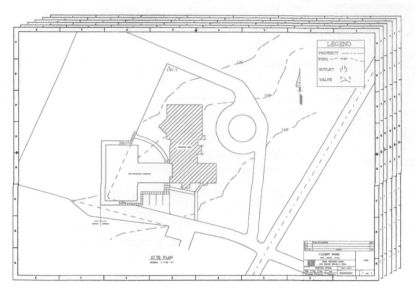

TITLE SHEET(S)
ARCHITECTURAL DRAWINGS
- SITE (PLOT) PLAN
- FOUNDATION PLAN
- FLOOR PLAN
- INTERIOR/EXTERIOR ELEVATIONS
- SECTIONS
- DETAILS
- SCHEDULES

STRUCTURAL DRAWINGS
MECHANICAL PLANS
ELECTRICAL PLANS
PLUMBING PLANS

28201-14_F01.EPS

Figure 1 Typical format of a set of working drawings.

If the work has already been laid out by the mason, a change order is then issued by the architect or construction manager. The main point to remember is that before there is any departure from the plans as drawn, approval must be made by the proper authority.

Plans that are used on the job site are typically called construction drawings. Prints can be duplicated with conventional copiers. The plans are white with black lines, though you may see older prints that are blue with white lines. These types of drawings are often referred to as blueprints. The care of the plans will depend on which type you are using, as some types are more susceptible to wear and tear than other types.

Proper care must be taken of all drawings if they are going to last the entire job. The drawings should not be allowed to get wet. They should not be left in the sun for any length of time, as they will fade and become unreadable. The plans should be kept away from mortar or concrete mixtures, and caution should be used when handling drawings where there is any welding going on because of the danger of fire. They should be collected at the close of the workday and returned to the job office or other secured area.

1.1.0 Identifying Keys and Legends, Lines, Terms, Abbreviations, and Symbols

In this section, you will learn how to identify and use the various elements that are commonly used on residential drawings. These include keys, legends, lines, abbreviations, terms, and symbols. Be

Presentation Drawings

Presentation drawings are three-dimensional (3-D) drawings that show a building from a desirable vantage point in order to show its most interesting features. Presentation drawings do not provide detailed information for construction purposes.

These drawings are used by the architect or a contractor to sell the design of the building to a prospective customer. As a mason, you should be able to visualize and draw a rough three-dimensional sketch of a building after interpreting the architectural plans for the building.

28201-14_SA01.EPS

Computer-Aided Design

Architectural plans have been used for centuries to pictorially describe buildings and structures before they are built. In the past, drafters would draw these plans by hand.

Today, most drawings are generated using computers. This process is called computer-aided design (CAD). The drafter creates the drawings for the building on the computer using architectural software. When they are completed, the electronic drawings are sent to a printer or plotter and printed on paper.

aware that these elements may vary among different sets of drawings. Always refer to the keys and legends before using a drawing, to ensure that you are interpreting it correctly.

1.1.1 Keys and Legends

Keys and legends (*Figure 2*) inform the reader of special symbols or abbreviations used on a plan. It can be as basic as the scale of the drawing or as complicated as a set of symbols used for architectural finishes. This information is usually placed on the cover sheet. It may appear on the index sheet if the plans are large enough to require one.

1.1.2 Lines

Plans and drawings use lines, symbols, and numbers to convey information about the structure's shape, size, dimensions, and finish. Lines used on plans and drawings usually follow a standard pattern. *Figure 3* contains commonly used standard line symbols for drawing plans. If the architect or designer uses nonstandard lines, then the legend should explain what they mean.

The standard definitions for the most commonly used lines are as follows:

- *Object lines* – Heavy line used to show the main outline of a structure, including exterior walls, interior partitions, porches, patios, sidewalks, parking lots, and driveways.
- *Dimension and extension lines* – Lightweight lines used to provide the dimensions of an object. An extension line is "extended" from an object at both ends of the object to be measured. Extension lines should not touch the object lines, to avoid confusion with object lines. A dimension line is drawn at right angles between the extension lines and a number placed above, below, or to the side of it to indicate the length of the object being measured. Sometimes a gap is provided in the dimension line and the dimension is shown in the gap.
- *Center line* – Used to designate the center of an area or object and to provide a reference point for dimensioning. Center lines are typically used to indicate the centers of objects such as columns, posts, footings, and door openings.
- *Cutting plane (section line)* – Used to indicate an area that is being cut away and shown in a section view so that the interior features can be seen. The arrows at the ends of the cutting-plane line indicate the direction in which the section is viewed. Letters identify the section view of that specific part of the structure. More elaborate methods of labeling cutting-plane lines are used in larger sets of plans (*Figure 4*) where many sections are being used. The section view may be on the same page as the cutting-plane line or on another page.
- *Break line* – Used to indicate that an object or area is not being shown in its entirety. Break lines are common where there are long continuous extensions of a structure that would require a larger plan sheet if completely drawn

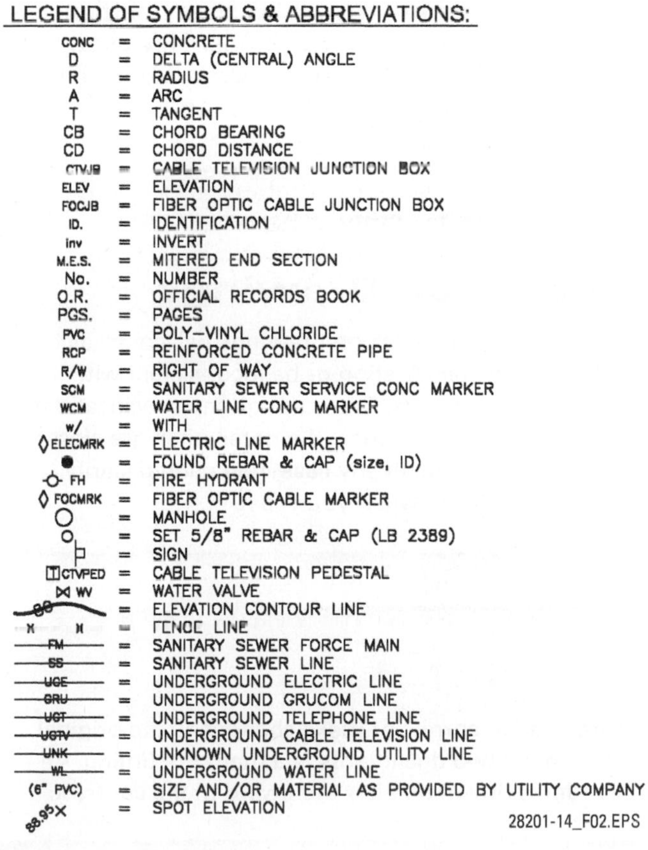

LEGEND OF SYMBOLS & ABBREVIATIONS:

CONC	=	CONCRETE
D	=	DELTA (CENTRAL) ANGLE
R	=	RADIUS
A	=	ARC
T	=	TANGENT
CB	=	CHORD BEARING
CD	=	CHORD DISTANCE
CTVJB	=	CABLE TELEVISION JUNCTION BOX
ELEV	=	ELEVATION
FOCJB	=	FIBER OPTIC CABLE JUNCTION BOX
ID.	=	IDENTIFICATION
inv	=	INVERT
M.E.S.	=	MITERED END SECTION
No.	=	NUMBER
O.R.	=	OFFICIAL RECORDS BOOK
PGS.	=	PAGES
PVC	=	POLY–VINYL CHLORIDE
RCP	=	REINFORCED CONCRETE PIPE
R/W	=	RIGHT OF WAY
SCM	=	SANITARY SEWER SERVICE CONC MARKER
WCM	=	WATER LINE CONC MARKER
w/	=	WITH
◊ELECMRK	=	ELECTRIC LINE MARKER
●	=	FOUND REBAR & CAP (size, ID)
-◊- FH	=	FIRE HYDRANT
◊ FOCMRK	=	FIBER OPTIC CABLE MARKER
○	=	MANHOLE
○	=	SET 5/8" REBAR & CAP (LB 2389)
	=	SIGN
CTVPED	=	CABLE TELEVISION PEDESTAL
WV	=	WATER VALVE
	=	ELEVATION CONTOUR LINE
X X	=	FENCE LINE
FM	=	SANITARY SEWER FORCE MAIN
SS	=	SANITARY SEWER LINE
UGE	=	UNDERGROUND ELECTRIC LINE
GRU	=	UNDERGROUND GRUCOM LINE
UGT	=	UNDERGROUND TELEPHONE LINE
UGTV	=	UNDERGROUND CABLE TELEVISION LINE
UNK	=	UNKNOWN UNDERGROUND UTILITY LINE
WL	=	UNDERGROUND WATER LINE
(6" PVC)	=	SIZE AND/OR MATERIAL AS PROVIDED BY UTILITY COMPANY
X	=	SPOT ELEVATION

28201-14_F02.EPS

Figure 2 Sample legend.

Blueprints

The term *blueprint* is derived from a method of reproduction used in the past. A true blueprint shows the details of the structure as white lines on a blue background.

The method involved coating a paper with a specific chemical. After the coating dried, an original hand drawing was placed on top of the paper. Both papers were then covered with a piece of glass and set in the sunlight for about an hour. The coated paper was developed much like a photograph. After a cold-water wash, the coated paper turned blue and the lines from the drawing remained white.

Today, most drawing reproduction methods produce a black, blue, or brownish line on a white background. These copies or prints are typically made using an engineering copier or a diazo, or similar copying machine. However, the term blueprint is still occasionally used to refer to construction drawings.

out. There can be long breaks and short breaks. Long break lines are thin lines drawn with a straightedge (refer to *Figure 3*). Short break lines are heavier and are drawn freehand.

- *Hidden line* – Used to indicate an outline that is invisible to an observer because it is covered by another surface or object that is closer to the observer.
- *Stair indicator line* – A short line with an arrowhead that shows the ascent or descent of a stairway on a floor plan.

1.1.3 Architectural Terms

There are many standard terms used in the construction industry. Some of them started out as slang terms for particular objects. Some have been adopted over time as the standard terms. *Appendix A* contains a list of architectural terms commonly found on plans.

When dealing with a set of construction plans, information will be written on the plans to identify specific objects and materials. You need to understand the terminology used in these explanations.

There may be more terms that are used in your area, or some common terms may have a slightly different meaning based on regional differences.

If you are not sure how a term is used, ask your supervisor for clarification. If the term is unfamiliar to everyone on the job, consult the architect for an explanation before proceeding with the work. Making an assumption about what is meant can result in very costly rework.

1.1.4 Abbreviations

Abbreviations are commonly used on plans. Often, there is not enough space to write out the whole word for a specific item. It is necessary to abbreviate where possible. The problem is that there are no standard abbreviations.

In most cases, architects and engineers try to follow some commonly accepted abbreviations for well-known items. But this is not always the case. More specialized items may not be universally known. Abbreviations may differ from person to person. For example, a closet can be abbreviated as C, CL, or CLO. Length can be abbreviated as L, LG, or LNG.

A number of abbreviations commonly used on plans are listed in *Appendix B*. Many of them also appear on written specifications. However, there is less standardization on specifications. The abbreviations may contain periods and may be written in lowercase.

Notes on Drawings

Sometimes the notes on drawings will contradict the written specification or be inconsistent with other requirements. Even though the specifications usually take precedence, the notes often are closer to the true intent. In any case, you should clarify any discrepancy when you see it.

Addenda and Change Orders

Addenda and change orders are contractual documents. They are used to correct or make changes to the original construction drawings and/or specifications. The difference between the two documents is a matter of timing. An addendum is written before the contract is awarded. A change order is drawn up after the award of the contract.

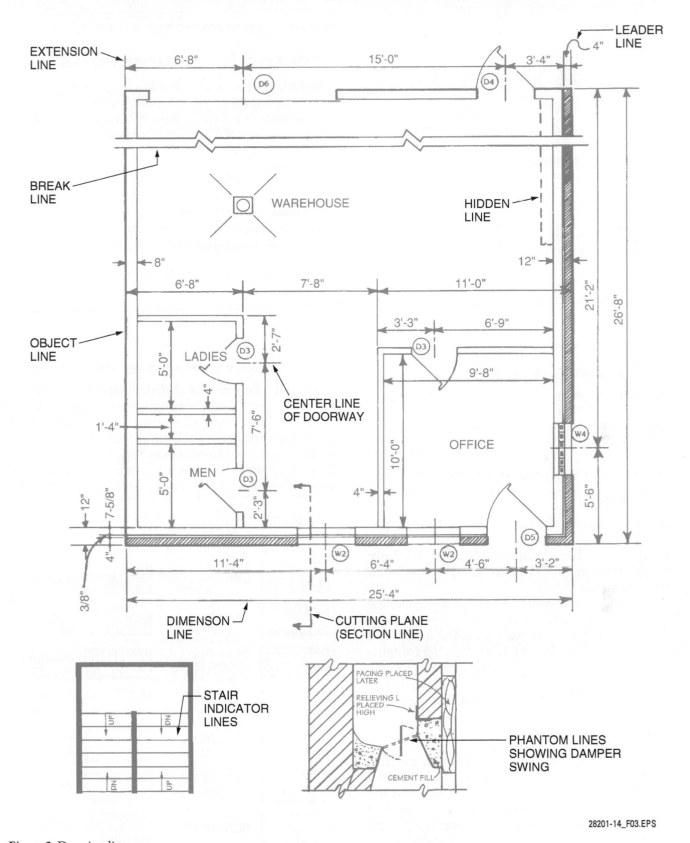

Figure 3 Drawing lines.

28201-14_F03.EPS

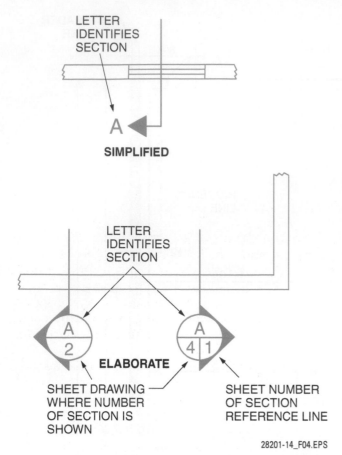

SIMPLIFIED

ELABORATE

28201-14_F04.EPS

Figure 4 Methods of labeling section reference lines.

1.1.5 Symbols

Symbols are another form of abbreviation for different types of materials used in construction. There are a number of commonly used symbols; however, there are no standardized symbols for specific materials.

The key block or legend on the front page or the index page of the plans usually has a list of all the symbols used throughout the plans. For a small set of plans this may not be practical. *Figure 5* contains a selection of symbols that have achieved some standardization over time. There are many variations of the material within the basic category.

1.2.0 Using Scales and Dimensions

Dimensions and scales show critical measurements to the tradespersons carrying out the plans. The mason must understand scales and the various methods of dimensioning so the project can be laid out and constructed correctly. A plan without accurate measurements is a useless picture. When studying working drawings, notice that most dimensions (commonly called working measurements) appear on the floor plans. Dimensions indicate the size of all important parts of the structure. They are drawn to a preset scale.

Metric Measurements

In the United States, dimensions on plans are typically shown in feet and inches. In Canada and other parts of the world, the metric system is used. The meter and the millimeter are two common length measurements used in architectural drawings. The millimeter is $\frac{1}{1000}$ of a meter.

On drawings drawn to scales between 1:200 and 1:2,000, the meter is generally used. Again, the meter symbol (m) will not be shown and the drawing should have a note indicating that all dimensions are given in meters unless otherwise noted.

Land distances on plot or site plans expressed in metric units are typically given in meters or kilometers (1,000 meters). Conversion factors that can be used to convert between English and metric units are provided in *Appendix C*.

Do not assume that all information on a drawing is correct. If you are suspicious of any dimension or value, check before proceeding. Individual measurements on a wall can be checked. Add all of the separate dimensions together and check this value with the overall dimension. If there is any difference in the two figures, have someone

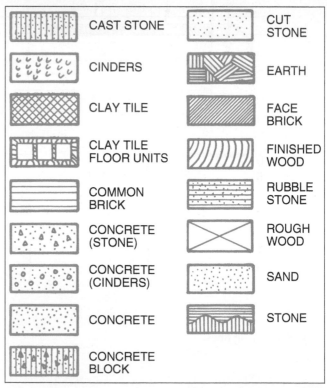

28201-14_F05.EPS

Figure 5 Typical materials symbols.

NCCER – *Masonry Level Two* 28201-14

Leader and Phantom Lines

Leader lines and phantom lines are commonly used on residential drawings, but masons will need to refer to them less often than other common types of lines. Leader lines are used to connect a note or dimension to a related part of the drawing. They are usually curved or at an angle from the feature being distinguished, to avoid confusion with dimension and other lines. Phantom lines are used to indicate alternative positions of moving parts, such as a damper's swing or adjacent positions of related parts. They may also be used to represent repeated details.

else check your calculations. If there is still a difference, ask your supervisor for guidance.

You should never change the original drawing without first consulting your supervisor or the project engineer. If there is any discrepancy between the specifications and the plans, the specifications typically take precedence.

Dimensions serve two purposes. The first is to indicate the location of a construction feature. The second is to indicate the size of this object or feature. Dimensions given on drawings show actual sizes, distances, and heights of the objects and spaces being represented. They may be from outside to center, center to center, wall to wall, or outside to outside (*Figure 6*).

There are two commonly used methods of dimensioning exterior walls and foundations. One consists of measuring the overall dimension of the structure. The other method is to measure the dimensions just to the outside of the unfinished

wall, allowing space for sheathing. These two methods are shown in *Figure 7*.

The foundation is 30 feet 0 inches. This is the overall dimension. The exterior face of the structure is covered with ¾-inch sheathing or plywood. The sheathing is flush with the foundation wall. The dimension to the outside face of the wood studs is 29 feet 10½ inches (the overall dimension of 30 feet 0 inches minus 1½ inches for two sides with ¾-inch plywood).

Some architects prefer to set the outside studs even with the outside face of the foundation wall. The sheathing would, in this case, overlap on the face of the foundation wall. This method simplifies the procedure since the overall dimensions and the framing measurements are the same. Be-

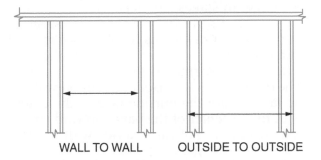

OUTSIDE TO CENTER   CENTER TO CENTER

WALL TO WALL   OUTSIDE TO OUTSIDE

28201-14_F06.EPS

Figure 6 Common methods of indicating dimensions on drawings.

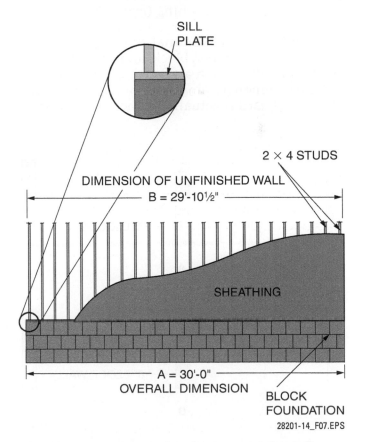

SILL PLATE

2 × 4 STUDS

DIMENSION OF UNFINISHED WALL
B = 29'-10½"

SHEATHING

A = 30'-0"
OVERALL DIMENSION

BLOCK FOUNDATION

28201-14_F07.EPS

Figure 7 Dimension measurements on an exterior wall.

fore laying out any work, be sure that you understand the dimension measurements.

1.2.1 Dimensioning Interior Walls

Masonry walls can be used to divide a space into smaller spaces or rooms. They may or may not be loadbearing. There are two general methods of laying out masonry partitions.

- Measure from the outside of the exterior wall to the center of the interior wall (*Figure 8A*).
- Measure from the outside of the exterior masonry wall to the outside face of the interior wall, as shown on the left side of *Figure 8B*. The dimensions of the interior of the room are given on the right side of *Figure 8B*.

The most accurate method is to measure from the outside face of the wall to the center of the partition wall because you are measuring the overall dimension and not assuming the thickness of the block wall.

1.2.2 Dimensioning Openings in Masonry Walls

There are two ways to dimension openings in a masonry wall. The measurements may be taken from either the center of the opening (*Figure 9*), or from the side of the opening (*Figure 10*).

Sectional and detail views may be labeled using nominal sizes. Nominal sizes are the rough sizes by which lumber, block, and other materials are commonly known and sold. For example, a 2 × 4 is a common nominal size for lumber. Ten feet of a 2 × 4 board is actually 1½ inches × 3½ inches × 10 feet.

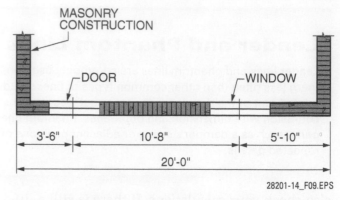

Figure 9 Measurements taken from the center of the opening.

All drawings are not alike. There are differences in size, scale, and detail. This can be confusing because what is being used is different from what you are used to. When studying dimensions on a drawing you should keep the following points in mind; however, the practices in your area may be different.

- Architectural dimension lines are unbroken lines, with the dimensions placed above and near the center of the line.
- Generally, measurements are taken from the outside wall to the edge of an opening. Measure from the center of the opening to the edge of the jambs. Double-check all measurements of openings before proceeding with the work.
- Overall measurements are taken from one extreme end of the wall or object to the other extreme.
- Dimensions should be shown on drawings according to their importance. The overall dimension is always given first, with the dimension to the center of openings next. The sizes of the openings are then shown. All of the smaller dimensions are shown inside the complete overall measurement.
- The overall dimensions of a certain length and the total of all segmented portions of the same length must be equal. If the lengths do not add up, larger mistakes will happen.
- An overall dimension must include the sheathing thickness as well as the distance from the outside of the studding to the outside of the studding in the framework.
- The dimensions of masonry partition walls can be taken from the outside face of the exterior wall to the center of the partition wall, or from the outside face of the exterior wall to the inside face of the partition wall.
- Rooms are sometimes dimensioned from the center lines of partition walls, but wall-to-wall dimensions are more common.

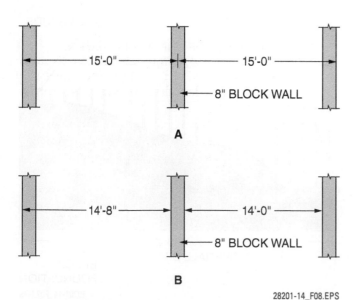

Figure 8 Measuring partition walls.

NCCER – *Masonry Level Two* 28201-14

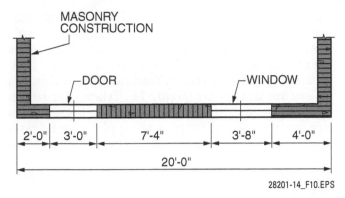

Figure 10 Measurements taken from the side of the opening.

1.2.3 Scales

Working drawings are reduced to scale. This means that all dimensions are reduced proportionately. The ratio between the size of the drawing and the actual size of the finished structure is known as the scale.

In the United States, plans and drawings are usually produced on a standard size of paper. This is dictated by the scale required for the particular drawing. Standard paper sizes for the English system of measures are shown in *Figure 11*. Each paper size is double the area of the previous sheet.

1.2.4 Scales on a Drawing

In order to obtain dimensions of a structure from a plan, you must know the scale of the drawing. The scale for the individual sheet is usually shown in the title box located in one of the corners. It is possible that the scale will change several times on the same sheet. This is especially true for detail drawings and sections. Be sure to check all views for a possible change of scale.

Floor plans for different floors of the same structure are usually drawn to the same scale on each sheet. Plan sets that have many sheets may have an index in the front following the title page. Even though a scale is given on the title page or index page, be sure to check each page as you review it. The scale for that page may be different.

There are 10 basic scales used in dimensioning plans and drawings besides the full scale. They are as follows:

- $\frac{3}{32}$" = 1'-0"
- $\frac{1}{8}$" = 1'-0"
- $\frac{3}{16}$" = 1'-0"
- $\frac{1}{4}$" = 1'-0"
- $\frac{3}{8}$" = 1'-0"
- $\frac{1}{2}$" = 1'-0"
- $\frac{3}{4}$" = 1'-0"

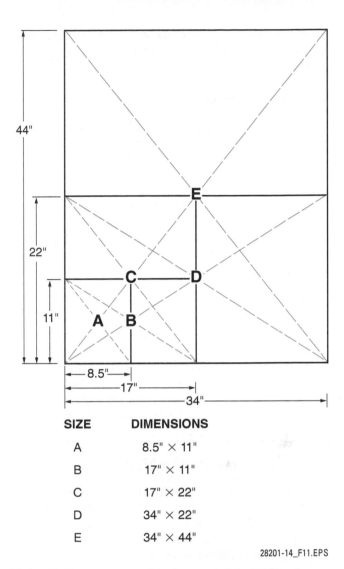

SIZE	DIMENSIONS
A	8.5" × 11"
B	17" × 11"
C	17" × 22"
D	34" × 22"
E	34" × 44"

Figure 11 Common sizes for plans and drawings.

- 1" = 1'-0"
- $1\frac{1}{2}$" = 1'-0"
- 3" = 1'-0"

These scales are referred to as architect's scales and are printed on the architect's ruler. A commonly used scale is $\frac{1}{4}$" = 1'-0". This indicates that every $\frac{1}{4}$ inch on the plan sheet represents 1 foot of actual structure area.

For example, 1 inch would represent 4 feet on the job, and 5 inches would represent 20 feet. This $\frac{1}{4}$-inch scale is commonly used on residential plans because all of the information can be drawn on a small sheet of paper (C or D size) and still be seen clearly.

For larger structures, a scale of $\frac{1}{8}$" = 1'-0" is most often used. Scales smaller than $\frac{1}{8}$ inch would be very difficult to read, with an increased possibility of error. The smaller the scale, the more careful you must be when dimensioning. For example, a mistake of $\frac{1}{16}$ inch on the $\frac{1}{8}$-inch scale would represent 6 inches on the job. A 2½-inch line drawn to the ⅛-inch scale would actually measure 20 feet on the job.

1.2.5 Use of Rules

Sometimes it is necessary to determine dimensions not shown on a drawing. This can be done by measuring them with an architect's scale. Methods for using the architect's rule to scale a drawing were covered earlier in the Core Curriculum module *Introduction to Construction Drawings*.

Plan drawings are created using a specified scale. Inches or fractions of an inch on the drawing are used to represent feet in the actual measurement of a building. For example, in a plan drawn to ¼" scale, ¼ inch on the drawing represents 1 foot of the building. The scale of a drawing is usually shown directly below the drawing. The same scale may not be used for all the drawings in a set of plans.

In order to accurately measure lines on a set of plans, you need good measuring instruments. The two most commonly used measuring instruments are the architect's ruler and the engineer's ruler. The architect's ruler is widely used for measuring lines on residential plans. Engineer's rulers are more commonly used for measuring lines on civil plans. In this section, you will learn how to use an architect's ruler. Traditionally, architect's rulers are triangular (*Figure 12*). Triangular rulers show up to six different scales. While these rulers are normally used for preparing plans and drawings, they are not practical on the job because of their shape and length. A straight, flat architect's ruler (see *Figure 12*) is easy to use and will fit into your shirt pocket.

> **NOTE**
>
> When they are available, always use the dimensions shown on a drawing rather than the ones obtained by scaling the drawing. This is because reproduction methods used to make copies of drawings can introduce errors in the reproduced image.

A flat architect's ruler will usually have two or four scales that are part of an architect's scale. Typically, the right and left edges are laid out using two different scales with a common denomi-

On-Screen Estimators

Today, most construction drawings are created electronically using computer-aided design (CAD) technology and then printed out on paper. Material takeoffs and estimates can now be prepared using the same electronic plans. On-screen estimators, such as those developed by Tradesmen's Software, Inc., have the ability to turn digital blueprints into 3-D renderings that automatically calculate the amount of materials required to complete the structure. This allows you to spot mistakes before they happen, saving your company time and money from the start of a project.

On-screen estimators are growing in popularity among the masonry trade because of their speed and ease of use. It takes just a few seconds for the software to estimate the number of masonry units and the amount of mortar or grout for a wall, pier, or other masonry structure. Material estimates are automatically updated whenever a dimensional change is made in the plans. It can even calculate labor rates, mortar yields, and lay rates based on information that you enter into the system. The information can also be exported into BIM (building information management) files and used to create 3-D proposal drawings for clients and contractors.

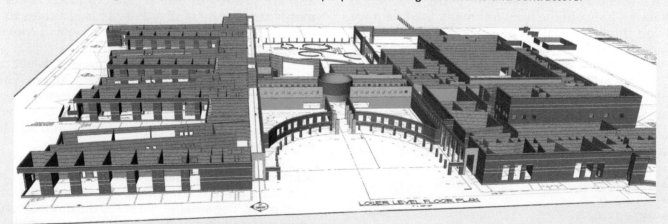

28201-14_SA02.EPS

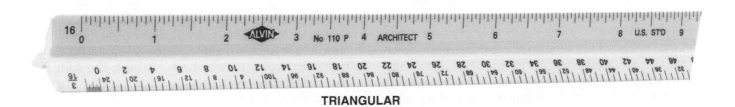

TRIANGULAR

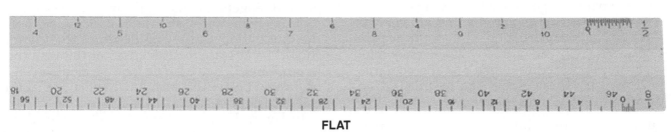

FLAT

28201-14_F12.EPS

Figure 12 Architect's rulers.

nator. On the left end, the ⅛" scale is marked off with short lines, each representing 1 foot. This scale is read from left to right. On the right end, the ¼" scale is marked with longer lines, each representing 1 foot, and is read from right to left. Each long line represents 1 foot on the ¼" scale or 2 feet on the ⅛" scale. Each short line represents 1 foot on the ⅛" scale and 6 inches on the ¼" scale.

The scale has a short section with several lines at the end. These fine lines represent inches or fractions of inches. Using the measurement 10 feet 6 inches as an example, the 10-foot measurement is read on the scale and the remaining 6 inches are found in the finely calibrated lines.

On a ¾" scale, there are twelve ¹⁄₁₆-inch marks in ¾ of an inch. Since ¾ inch equals 1 foot according to the scale, each of the 12 marks represents 1 inch because there are 12 inches in a foot. Therefore, ¹⁄₁₆ inch in a ¾" scale always represents 1 inch on the job.

To determine what fraction is equal to 1 inch on the job, divide the fraction that equals 1 foot on the job by 12 (12 inches per foot). For example, with a scale of ½" = 1'-0", divide the ½ inch by 12; ¹⁄₂₄ inch equals 1 inch on the job.

1.2.6 Dimensions

Understanding the dimensions of an object is an essential aspect of reading the drawings. Dimensions are written in standard format to avoid confusion. Dimensions over one foot are shown in feet and inches, not decimals, for example, 5'-4", 1'-4", 6'-6", 10'-5½". Full foot measurements are shown with a hyphen and a 0 with inch marks (") following the foot measurement; for example, 4'-0". The 0" is used to avoid any confusion. Dimensions less than 1 foot are shown in inches only.

The exceptions to this practice are dimensions that are recognized standards in the modular system of construction work. The common measurement of 16 inches center-to-center for framing studs would be shown as 16" C to C.

This measurement is also referred to as 16" on center and written 16" OC.

Generally, most the of dimensions needed by the mason are located on the floor plans. Finish details are also shown on elevation and section drawings.

1.3.0 Interpreting Residential Drawings

Residential construction drawings or architectural plans show the designer's ideas in graphical form. The builder must follow the plans. The resulting structure will only be acceptable to the owner if the plans are clearly presented and carefully followed.

To avoid confusion, construction drawings are presented in a somewhat standardized form. As a mason, you must learn the basics of this graphical language. Once you know what to look for in a set of residential construction drawings, you will have few problems in converting the ideas presented into an actual structure.

Residential drawings contain all of the elements found in a standard set of construction drawings. These were discussed in the *Masonry Level One* module *Measurements, Drawings, and Specifications*. They include the site plan, floor plans, elevations, sections, details, schedules, structural plans, and specialty plans, which show the electrical, plumbing, and mechanical systems. **Shop drawings**

may also be included as part of the installation requirements.

Some of these elements are combined or shown in an abbreviated form in residential drawings. For example, the specialty plans are often shown on the architectural floor plans and not as separate drawings. A complete set of residential plans may consist of only two or three pages. There may not be a separate page for each drawing type.

The keys to accurate plan reading are very basic. Yet errors in construction often occur because the plans are not read correctly. If the mason follows a simple procedure when reading construction drawings, these errors can be minimized. This procedure includes the following steps:

Step 1 Review all architectural symbols to make sure that you recognize them. If a symbol is unfamiliar, check with your supervisor for an interpretation.

Step 2 Read all notes on the plans. The architect will often give unusual features a label so that there is no chance for error.

Step 3 Review all line symbols so that you can accurately read all dimensions.

Step 4 Review the basic construction procedures that are required to build the structure. By doing this, items omitted from the drawing may be discovered.

Step 5 Check and recheck the plans. Before the construction begins, make sure you fully understand the plans.

1.3.1 Foundation Plans

All structures must have an adequate foundation. Foundations serve several purposes. They provide a level surface to build on. They distribute the weight of the structure to the surrounding earth. They also help to prevent water or air from entering under a structure. It is very important for a mason to thoroughly understand the foundation plan.

There are several types of foundations used for residential construction. These include continuous footings, isolated or spread footings, slab or mat foundations, and pole- or pile-supported foundations. The type of foundation used depends on the size and location of the structure and the local soil and weather conditions. In order to accurately read the foundation plans, masons must be able to recognize basic materials symbols and understand the basic techniques used in constructing a foundation.

Foundation drawings include plan views, sections, and details. In addition, the drawings should include symbols, dimensions, and notations that reveal information about the footing, the foundation wall, and the foundation surface or flooring (*Figure 13*).

Footings – Footings distribute the weight of the structure to the soil and should be on firm ground. A general rule of thumb for residential footings is that they should be twice as wide as the foundation walls. The depth of the footing should be at least the thickness of the wall (between 8 and 12 inches). The footing should always be located below the frost line to ensure that the soil does not freeze under the footing.

In *Figure 13* the footings are shown in three sectional views marked A-A, B-B, and C-C on the foundation plan. Section view A-A shows the details of the slab-on-grade for the garage. Section B-B shows the details of the footings and crawl space under the house. Section C-C shows the detail of the exterior foundation.

Foundation plans will provide the following footing information:

- The location of footings, often shown by dashed lines (hidden lines) on a plan view.
- The dimensions of footings or of thickened sections of a slab. This information is often shown graphically with a footing or slab detail, or it may be noted in a footing schedule.
- Sizes and locations of reinforcing requirements. This is described in a note or in a schedule.

Foundation walls – Foundation walls are any walls between the loadbearing structure walls and the footing. They act as the base for the sill. Foundation walls transfer the building weight to the footing. They can be made of concrete, stone, brick, or block.

Foundation plans will show foundation wall type, location, and dimensions. Section and detail drawings will give the wall type; sill placement, anchoring procedures; sheathing applications; and stud, pier, or column placement. In the case of masonry walls, the wall sections may also include details about wall ties, special jointing procedures, and waterproofing requirements (*Figure 14*).

Foundation surfaces and flooring – A concrete floor slab is often a part of a building's foundation system. The slab may be designed to carry the weight of the structure or to provide a durable moisture-resistant flooring system for the building. The two most common types of slabs are shown in *Figure 15*.

The foundation plans will incorporate slab details such as the slab's position, thickness, slope,

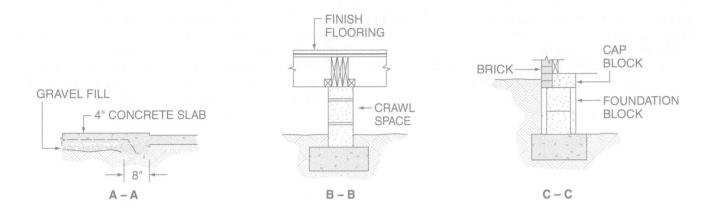

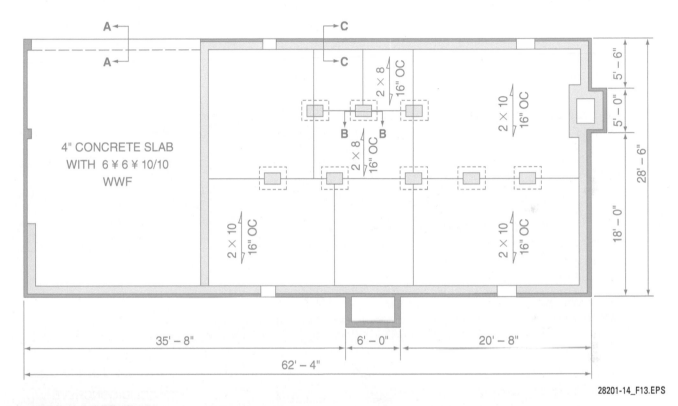

Figure 13 Foundation plan.

and elevation. This information will be supplemented with notes if required. Areas where the slab must be thickened to support loadbearing walls or depressed to receive a finished flooring should be clearly indicated in a plan view.

Section and detail drawings will include the proper reinforcement for slab construction, the fill on which the slab should rest, and the appropriate vapor barrier application and insulation. Slab foundations are often reinforced with wire mesh. The designation for such embedded mesh gives the mesh pattern size and the wire size. For example, 6 × 6 × 10/10 WWF refers to the commonly used welded wire fabric that has a 6-inch grid and is made from 10-gauge wires in both directions.

Crawl space foundations will include information on the floor framing requirements along with the foundation plans. This would include the type of lumber required; the size, location, and support conditions for all beams or girders; and the size and spacing for all floor joists. A note about rough flooring requirements may also be included. With frame or slab construction techniques, the foundation plans describe the part of the structure below the finished flooring.

In reading plans, be sure to identify the location of all section and detail notations. To fully understand the working drawings, the mason must be able to tie all the parts of the drawing together. Remember that the plan views will have section marks noting where a section or detail will be taken and the direction of the view.

These should be clearly indicated by a solid line with a number or letter reference and an arrow indicating the direction of the view.

1.3.2 Floor Plans

The floor plan shows the interior layout of the building. It includes the location of all walls, windows, doors, and other permanent fixtures. In a typical set of residential working drawings, it may also contain information about the location of items supplied by the electrical and mechanical contractors. The floor plan is drawn as if the structure were cut horizontally about 4 feet above floor level and viewed from above. This horizontal cutting plane will show all of the windows and doors (see *Figure 16*).

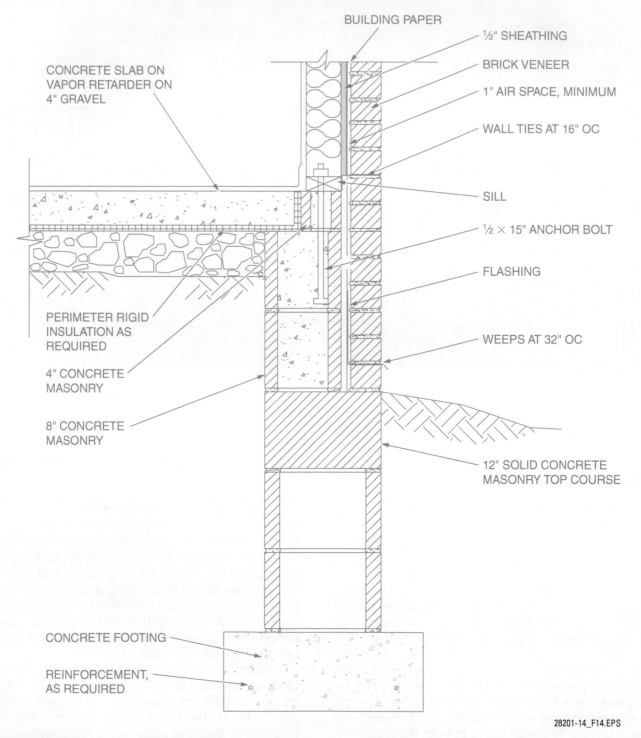

Figure 14 Detail drawing for foundation wall.

28201-14_F14.EPS

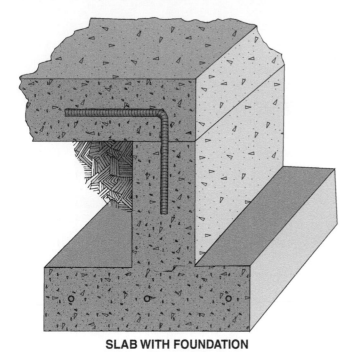

SLAB WITH FOUNDATION

SLAB WITH THICKENED EDGE

28201-14_F15.EPS

Figure 15 Types of slabs.

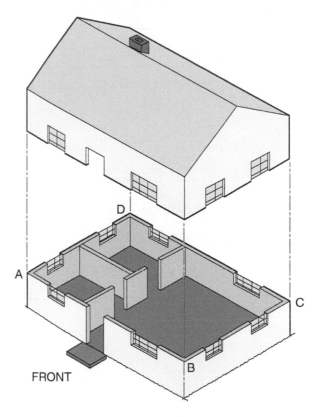

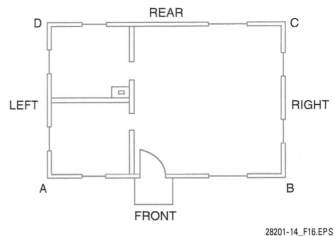

28201-14_F16.EPS

Figure 16 The hypothetical horizontal cutting plane for a floor plan.

Floor plans contain detailed dimensions and materials information. Floor plans are drawn to scales of ⅛" = 1'-0" or ¼" = 1'-0". They contain as precise a description of the structure as is possible on a single plan (*Figure 17*). However, because some details are too small to be drawn to these scales, separate drawings are often included to describe the smaller details.

Once you understand the symbols used on a floor plan, you should have little trouble finding the following information:

• An architectural scale, such as ¼" = 1'-0", as shown in *Figure 17*.

• Layout dimensions for exterior walls and interior partitions, including the overall dimensions for the building.

• A floor plan for each story level in the building. A two-story house will have two floor plans in the working drawings.

• Layout dimensions for all permanent parts of the structure. This includes door and window location, size, and type; fireplace location and size; and stair location and construction information.

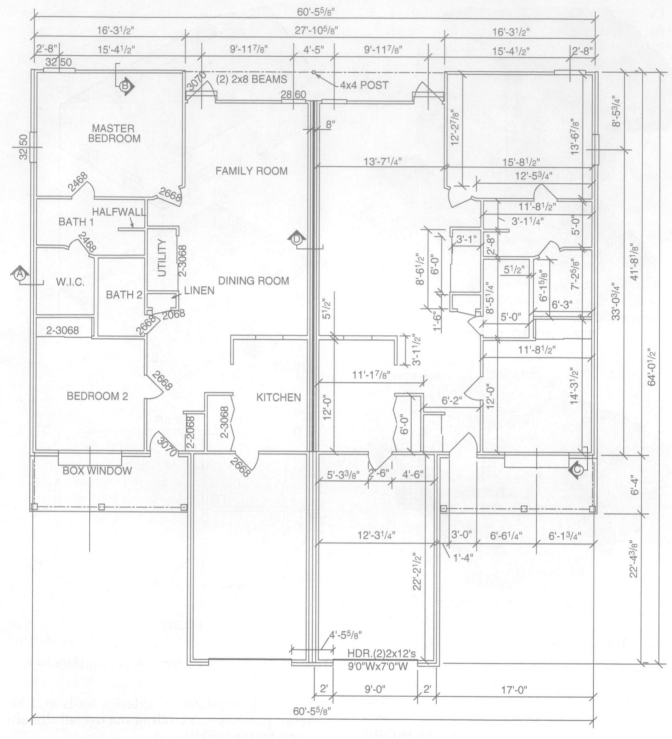

Figure 17 Floor plan for a residence.

- Designations for the location of built-in equipment such as bathroom fixtures, kitchen fixtures or appliances, and closets or cabinets. Hidden (dashed) lines are used to locate different items that occur at the same floor location. For example, wall cabinets will appear as hidden lines while a solid line will be used to represent the base cabinet below.
- Symbols indicating the materials used. For example, the symbols used in the wall lines will show if the wall is a wood frame, concrete, concrete block, brick, or veneer wall.
- Notes for all features of the structure wherever the drawing is not self-explanatory. For example, a plan view can easily show the width of a cased opening, but the height is most easily indicated with a note.
- The location of lighting fixtures and/or HVAC ducts. If the floor plan becomes too cluttered, this information can be shown on a separate plan.
- Schedule marks for doors and windows. These marks serve to reference the plan to a door or window schedule for a detailed description of the item. Schedule marks are numbers or letters placed in a small circle, square, triangle, or other geometric shape.
- Section reference lines. Since vertical information cannot be clearly indicated in a plan view, it must be shown elsewhere in the plans. The plan view is cut by section lines to reveal this information. The relationship between the plan views and the section views must be clearly understood to convert the two-dimensional drawings into a three-dimensional visualization.

1.3.3 Elevation Drawings

Elevation drawings provide a picture-like view of one side of the structure. If you were standing on the ground, what you would see would be a two-dimensional above-ground view from a specific external point.

The elevation drawings supply the most information as to the look of the finished structure. They are very helpful in visualizing the relationship between the floor plans and the wall sections. However, elevation drawings provide very little additional construction information that is not found on other drawings.

The typical elevation drawing is of the structure exterior. It normally includes the foundation as a hidden line. Sometimes, all hidden lines are omitted and landscaping features are included.

This type of elevation is called a presentation elevation. It is of more interest to the owner or prospective client than the builder. Interior elevations are used to show interior walls or details which are very important to the cabinetmaker and the mason constructing a fireplace.

In most cases, four elevations are sufficient to show the structure. But when a floor plan has more than four sides, additional elevation views, called auxiliary elevations, may be used. The standard elevation views are usually labeled north, south, east, and west in relationship to the north arrow indication on the plot plan. When the elevation reads "north elevation," it refers to the elevation that faces northward. Alternatively, they are labeled front, rear, right, and left. If two sides of a building are very similar in appearance, one of the elevation views may be omitted.

Typical elevation drawings show the following:

- Grade lines.
- Floor heights.
- Window and door types.
- Roof lines and slope, roofing material, vents, gravel stops, and projection of eaves.
- Exterior finish a and trim.

1.3.4 Exterior Elevations

Exterior elevations show the outside of the structure. The drawings contain information on the grade line and foundation line, siding materials, chimney placement and proportion, roof slope and overhang, and door and window style and placement. Examples of some of these items are shown in *Figure 18*. It is important to completely understand the relationship between the elevations and floor plans in order to visualize the final structure.

In viewing elevation drawings, you should consider the following:

- Compare the elevations with the floor plans. This helps you to understand the relationship of the structure's interior to the exterior.
- Review and read the materials symbols. The materials symbols for elevation drawings appear much like the materials they represent, but most of these symbols are simplified. For instance, brick is often shown by simple horizontal lines rather than individual brick and mortar joints.
- Review the foundation plan. Hidden lines on elevation drawings are used to show footing locations and exterior below-grade features.

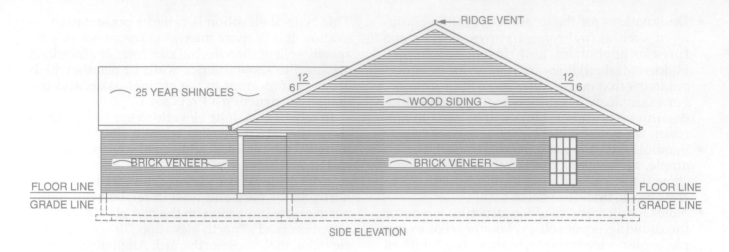

SIDE ELEVATION

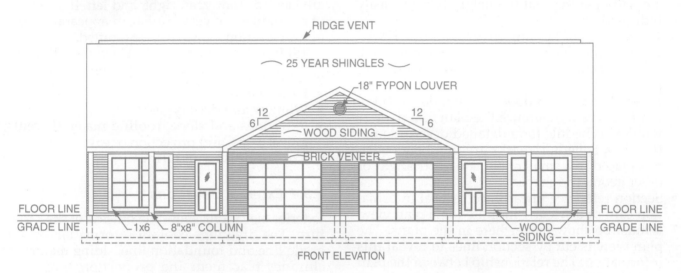

FRONT ELEVATION

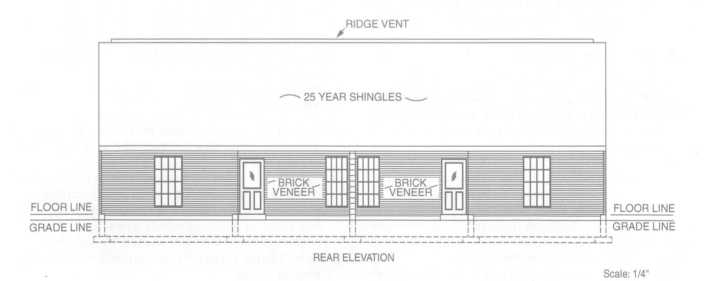

REAR ELEVATION

Scale: 1/4"

28201-14_F18.EPS

Figure 18 Elevation drawings.

Interior Elevations

Interior elevations are included in a set of construction plans to show special features, particularly in areas such as bathrooms and kitchens. These drawings show interior wall designs. They include appliance positioning, cabinet types and location, built-in carpentry items such as bookcases and china cabinets, and fireplace elevations.

Interior elevations are not drawn in any particular sequence or location. They are, however, always referenced by a detail number and arrow shown on a plan view. The detail number may be found on the floor plan for the entire building, or a floor plan of the particular room may be redrawn at a larger scale to show more detail, like the one shown.

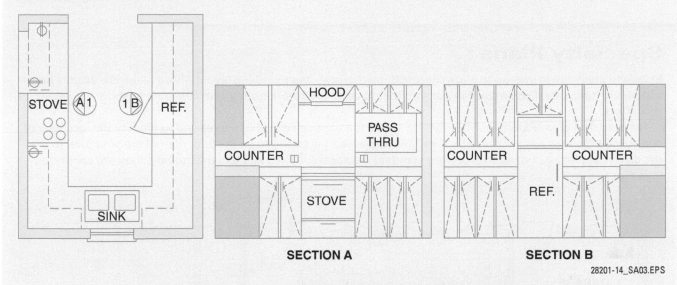

28201-14_SA03.EPS

Wall and Detail Sections

Sectional drawings are created by passing a vertical cutting plane through an object. This may be drawn through the entire building, a single wall, or through some small part of the building. Once this imaginary plane is cut, interior structural details become apparent. It reveals vertical dimensions, material requirements, and structural support conditions. These elements, to a large degree, dictate the construction procedures required for the structure. Two commonly used sectional drawings are wall sections and detail sections.

Full wall sections are usually shown for each major type of loadbearing wall in a structure. A wall section is often divided into three different details that align vertically to form the complete wall. The foundation detail includes footing and foundation wall requirements. The floor-sill detail shows the junction of the foundation wall and lower building wall. The cornice detail provides detailed information on the upper wall and roof framing construction requirements.

Detail sections cover specific construction procedures. These details are enlarged views of any part of the structure that cannot be adequately described elsewhere. Most sets of plans will contain several detail sections.

Because sectional drawings are used to show very precise information, they are often drawn in a larger scale. Notes with each drawing advise the reader about sizes, shapes, and special placement of materials.

- Review and read the window and door schedules. Because these features significantly affect the structure's appearance, they are usually drawn with as much detail as possible; however, the most detailed information will appear on the schedules and specifications. If a discrepancy occurs between the specifications and the schedules, the specifications will most likely be correct.

- Review dimensions on all elevation drawings. Such dimensions are usually limited to vertical dimensions and labeled features. The dimensions that may appear are depth of footings, floor or story heights, roof height, chimney height, window and door heights, and roof slope indication. Most of these dimensions can be verified by checking the section views.

Specialty Plans

Specialty plans are drawn up to show electrical, mechanical, and plumbing layout. These plans are usually drawn separately for clarity. Each trade can perform their takeoffs and work from a plan designed to show the details they need.

The electrical plan (*Figure SA04*) is used to locate the various electrical fixtures and to identify the locations of control switches for each of these fixtures. This information is provided on the floor plan of the structure. To read the electrical plan, you need to recognize standard electrical symbols and understand how they are connected (see *Figure SA05*).

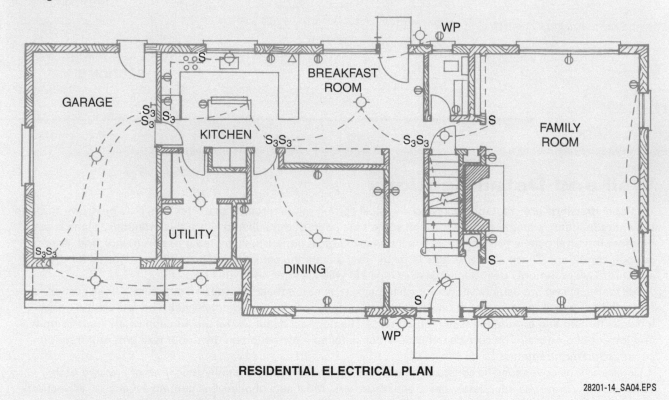

RESIDENTIAL ELECTRICAL PLAN

28201-14_SA04.EPS

Because HVAC equipment is the most expensive single system in a residential structure, an engineer usually consults with the architect or owner before a system is drawn up. The mason needs to understand the basics of the HVAC system in order to read the HVAC plans and avoid conflicts. In most cases, the mason will only have to read abbreviated or basic plans using the symbols shown in *Figure SA06*. These designate system type; thermostat positioning; heat unit types; and delivery system position, flow direction, and duct size.

Plumbing drawings usually appear as plan-view drawings and as isometric drawings called riser diagrams. The plan views show the horizontal distances or piping runs, and the riser diagram is used to show vertical pipes in the walls.

S	SWITCH
S_2	TWO-WAY SWITCH
S_3	THREE-WAY SWITCH
S_4	FOUR-WAY SWITCH
	INCANDESCENT LIGHT FIXTURE (CEILING)
	INCANDESCENT LIGHT FIXTURE (WALL MOUNTED)
	INDIVIDUAL FLUORESCENT FIXTURE (SURFACE OR PENDANT)
	TELEPHONE
	BUZZER
	BELL
	HEAVY-DUTY OUTLET (DW) Dishwasher (RA) Range
	DUPLEX RECEPTACLE OUTLET
	SINGLE RECEPTACLE OUTLET
s	SWITCH AND DUPLEX RECEPTACLE

ELECTRICAL WIRING SYMBOLS

28201-14_SA05.EPS

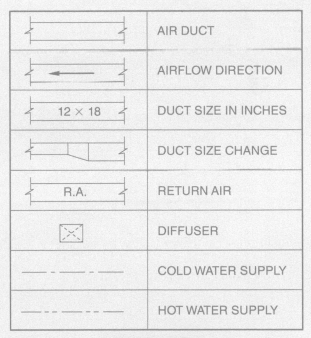

	AIR DUCT
	AIRFLOW DIRECTION
12 × 18	DUCT SIZE IN INCHES
	DUCT SIZE CHANGE
R.A.	RETURN AIR
	DIFFUSER
	COLD WATER SUPPLY
	HOT WATER SUPPLY

HVAC SYMBOLS

28201-14_SA06.EPS

Additional Resources

Bricklaying: Brick and Block Masonry. 1988. Brick Industry Association. Orlando, FL: Harcourt Brace & Company.

Concrete Masonry Handbook for Architects, Engineers, Builders, Fifth edition. 1991. W. C. Panerese, S. K. Kosmatka, and F. A. Randall, Jr. Skokie, IL: Portland Cement Association.

1.0.0 Section Review

1. The heavy line used to show the main outline of a structure on a construction drawing is called the _____.

   a. object line
   b. center line
   c. section line
   d. dimension line

2. Besides the full scale, the number of basic scales used in dimensioning plans and drawings is _____.

   a. 2
   b. 5
   c. 10
   d. 12

3. In residential construction, continuous footing, isolated footings, and pole- or pile-supported are examples of _____.

   a. walls
   b. foundations
   c. roof supports
   d. door frames

2.0.0 ESTIMATING MATERIAL QUANTITIES

Objective

Explain how to estimate material quantities from residential drawings.

   a. Explain how to use the rule-of-thumb method.

   b. Explain how to use estimating aids.

Performance Tasks

From a plan, calculate the square footage of one elevation, including openings.

Estimate the amount of brick and mortar from that same elevation.

Estimate the size and number of lintel block for that same elevation.

Proper estimates for materials are perhaps the most critical factor in preconstruction planning. Material quantities will, in part, determine the cost of a structure. Therefore, the task of estimating quantities can affect whether a profit is made or a bid is accepted.

Clients will select a particular masonry contractor based on a number of factors, primarily the quality of work and the price. The price is largely determined by the quantity of materials required, as determined by a materials, or quantity, takeoff. Only through an accurate quantity takeoff can a mason determine the price of a job with any degree of accuracy.

The mason must recognize the importance of accurate materials estimates and be able to apply the mathematical skills necessary for such estimates. The mason must also be able to estimate the number of masonry units required for a job as well as the quantity of each mortar ingredient.

Errors in estimating can be very costly. An overestimate will most likely result in losing the job to a more competitive contractor. At best, it will lead to the additional costs of loading and transporting unused materials from the job site. Underestimates will result in lost productivity and work stoppages while additional materials are ordered and delivered to the job site. An underestimate can be even more serious if special materials are being used, because the mason may have trouble matching the materials.

The mason can generally make a quantity takeoff by carefully applying basic math, and by referring to charts and tables that have been developed for this purpose. For any estimate the mason must consider the following:

- The size of the walls to be constructed.
- The size and type of masonry units to be used.
- The size of the mortar joints to be used.
- The number of openings such as doors and windows.
- The type of wall and the bond type.
- The type of mortar to be used.

Most masonry materials used today are based on the modular system. This greatly simplifies estimating requirements. The approach to estimating based on the modular system is by wall area or square foot. It consists of multiplying the length and the height of the wall minus any openings. In fact, experience has shown that this system of estimating is the most accurate and reliable and is known as the rule-of-thumb method. Remember to follow the local applicable code and standards when estimating masonry materials.

2.1.0 Using the Rule-of-Thumb Method

Estimating by rule of thumb is not intended to be a mathematically perfect method of estimation. It is workable, however, when estimating materials for the construction of small- and average-sized jobs. Some examples of these types of jobs are homes, garages, retaining walls, or chimneys. As an apprentice mason, you are not expected to estimate materials for a large or complicated job, but you should be able to figure the necessary materials for daily work projects.

Estimating by rule of thumb involves finding the area of each wall, dividing that area by the number of units per square foot, and subtracting the number of block for all openings. A percentage for breakage and waste is also added, based on the experience of the mason. These figures may range from 2–5 percent for block and brick, and 2–5 percent for mortar. Charts and tables may be consulted for unusual or special bonds and for determining the quantity of mortar materials.

The rule-of-thumb method of estimation was developed through years of practical experience. It is designed to allow for some waste on the job. Most small contractors estimate their materials by this method. Rule-of-thumb estimating can best be explained by using a problem as an example.

2.1.1 Estimating Block by Wall Area

To determine the number of masonry units needed by the rule-of-thumb method, you must determine the following:

- The total square feet of wall area in the structure
- The square feet of wall area occupied by a single block
- The total number of block required

The wall-area method is simple to use. Most block has the same nominal face dimensions: 8 inches by 16 inches (*Figure 19*). A common concrete masonry unit has a face area of 128 square inches (8 × 16 = 128), and therefore 1.125 block are required for each square foot (144 square inches per square foot ÷ 128 square inches per block).

The following steps demonstrate this procedure using an 8-foot-high rectangular structure with outside dimensions of 40 feet wide by 64 feet long with no openings, as shown in *Figure 20*.

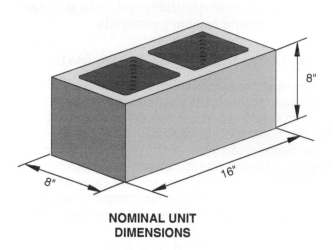

NOMINAL UNIT DIMENSIONS

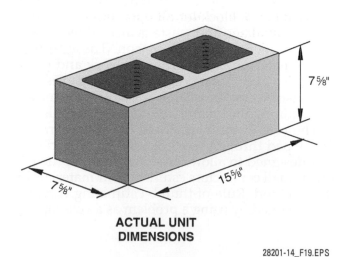

ACTUAL UNIT DIMENSIONS

28201-14_F19.EPS

Figure 19 Surface area of a standard block.

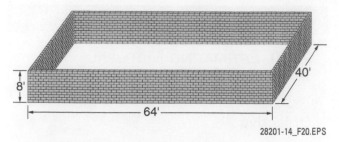

28201-14_F20.EPS

Figure 20 Rectangular block structure.

Step 1 Find the total square feet of wall area by multiplying the total length by the height. Remember there are two walls, back and front.

$$\text{Total length} = 2\,(a + b)$$

Where:

$$a = \text{length of side } a$$
$$b = \text{length of side } b$$

Total length = 2(64 + 40) = 2(104) = 208
Total square feet = total length × height
Total square feet = 208 × 8
Total square feet = 1,664 square feet

Step 2 Determine the number of block required for each square foot of wall area. A standard block is used; therefore, figure 1.125 block per square foot.

Step 3 Determine the number of block required by multiplying the total area by the number of block per square foot.

$$1,664 \times 1.125 = 1,872 \text{ block}$$

Step 4 Add a percentage (5 percent) for wastage and breakage.

$$1,872 \times 1.05 = 1,965.6 \text{ rounded to } 1,966 \text{ block required}$$

Step 5 Recheck the accuracy of all figures.

Estimating block, allowing for openings – The step-by-step process for these calculations can be seen using the same foundation as in the first example. The structure, instead of being a solid enclosure, will have a single door measuring 32 inches by 6 feet 8 inches, and two windows that measure 4 feet by 24 inches, as shown in *Figure 21*.

To accurately estimate the number of block required, a deduction must be made for any openings that occur in the wall area. This deduction should be made from the total square feet of wall area before converting to block and before any allowance is made for waste and breakage. Small

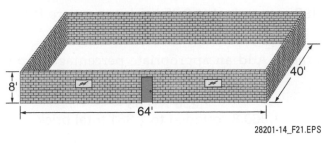

Figure 21 Rectangular building with a door and windows.

28201-14_F21.EPS

openings (less than window size) are generally disregarded because the extra cutting required around such openings tends to offset any savings in block.

Step 1 Determine the total square feet of wall area without considering the openings. From the previous example: 1,664 square feet.

Step 2 Determine the number of square feet in each opening and add them together.

For the door
Convert 32 inches to decimal
equivalent in feet:
32 ÷ 12 = 2.67 feet
Convert 6 feet 8 inches to its decimal
equivalent in feet:
6 + (8 ÷ 12) = 6.67 feet
Determine the number of square feet
for the door:
2.67 feet (width) × 6.67 feet (height) =
17.8 square feet

For each window
Convert 24 inches to decimal equivalent
in feet:
24 ÷ 12 = 2.00 feet
Determine the number of square feet
for each window:
4.00 feet (width) × 2.00 feet (height) =
8.00 square feet
Determine the total square feet
of all openings:
17.8 + 8.0 + 8.0 = 33.8 square feet,
rounded to 34 square feet

Step 3 Deduct the square feet of the openings from the total square feet of the wall area.

1,664 − 34 = 1,630 square feet

Step 4 Determine the total number of concrete block (as before).

1,630 square feet × 1.125 block/square foot =
1,833.75 rounded to 1,834 block

Step 5 Add a percentage (5 percent) to the total number of block required to allow for wastage and damage.

1,834 (total block required) × 0.05
5 percent wastage =
91.7 (or 92 block for wastage and breakage)
92 (block wastage) + 1,834 (block required) =
1,926 (total block required)

Step 6 Determine the number of lintel block required. For this example, use standard lintel block over all openings, extending 8 inches on each side of an opening.

Determine the total length of lintel block:
32 inches (door width) + 16 inches =
48 inches
24 inches (window width) + 16 inches =
40 inches
24 inches (window width) + 16 inches =
40 inches
Total length =
48 inches + 40 inches + 40 inches =
128 inches

Determine the required number of lintel block:
128 inches ÷ 16 inches per block =
8 lintel block required

Step 7 Recheck the accuracy of all figures and the reliability of all mathematical calculations.

> NOTE
> A lintel made of steel, concrete, or masonry will be required over most openings. Complete Step 6 only if necessary.

2.1.2 *Estimating Materials for Brick Construction*

Because of the widely used modular grid system of brick construction, the mason can generally estimate quantities just as easily with brick construction as with block construction. However, brick estimates depend on the mason's knowledge of the 8-inch modular grid system.

Most brick in the United States is manufactured so that the actual size plus the appropriate mortar joint thickness of ¼, ⅜, or ½ inch will equal a nominal size that will fit in the 8-inch modular grid. However, nonmodular standard brick is still in use and will require some special measuring steps before estimates can be made efficiently.

The nonmodular brick still manufactured are the standard brick (3¾ inches × 2¼ inches × 8 inches) and the oversized brick (3¾ inches × 2¾ inches × 8 inches). However, even standard brick are being manufactured in a modular size with nominal dimensions of 4 inches × 2⅔ inches × 8 inches. In actual practice, even the modular units may vary slightly in size. This size difference should make no difference in the brick estimate. You can simply vary the joint thickness of your mortar slightly to make up for any size differences in units on the job.

Other factors that may make brick estimates somewhat more difficult than block estimates are the pattern bond specified, the wall type specified, and the number of specialized brick types required (such as headers and face brick). Most charts or tables will make allowances for such variations, but you should be very careful in how these figures are used.

Bonds that require full headers must be estimated by using correction factors to account for the differences. If a multiwythe wall is specified, the mortar estimates must be increased to allow for full collar joints. For these particular variations you will find that tables will usually provide adequate information. These correction factors are used only after the initial estimates are made. As in the case of block estimates, tables such as the one shown in *Table 1* can be used to help with the estimates.

In order to better understand the estimating procedure, follow this step-by-step procedure for estimating quantities for the following construction task: Job specifications call for the construction of a 4-inch-thick wall with ⅜-inch joints and standard modular brick. The wall is to be built 24 feet long and 8 feet high in a running bond.

Step 1 Determine the total wall area of the construction.

24 feet (specified length) ×
8 feet (specified height) =
192 square feet (wall area)

Step 2 Consult *Table 1* to determine the number of brick needed for each 100 square feet of wall. Since the specifications are for 4 inch × 2⅔ inch × 8 inch brick, the figure on the first line in the Number of Brick column is appropriate.

675 brick per 100 square feet or
6.75 brick per square foot

Step 3 Determine the number of brick required.

192 square feet × 6.75 brick per square foot =
1,296 total number of brick

Step 4 Add an appropriate percentage (5 percent) for breakage and wastage.

1,296 × 1.05 =
1,360.8 rounded to 1,361 total brick

Step 5 Recheck the accuracy of all figures and the reliability of all mathematical procedures.

When estimating for openings, a certain number of brick should be deducted from the exact number of brick required for the structure. This can be done by deducting the total square feet of openings from the total wall area. To illustrate this, consider the last example with one rectangular opening 36 inches wide and 6 feet high:

Step 1 Determine the total wall area. From the last example: 192 square feet.

Step 2 Determine the square feet of the area in each opening.

36 inches = 3 feet
3 feet × 6 feet = 18 square feet

Step 3 Deduct the square feet of the openings from the square feet of the wall area.

192 square feet − 18 square feet =
174 square feet

Continue the estimating process as before (Steps 2 through 4 in the previous example). If there are numerous small openings in the masonry structure, the percentage for wastage should be increased to account for the units that must be cut around the openings.

2.1.3 Estimating Mortar

Estimating mortar and mortar materials is done in the same way for both block and brick construction. If the mason chooses to use masonry cement rather than portland cement–based mortar, five rules of thumb can be used. For brick, this assumes the use of a ⅜-inch joint and standard modular brick. These rules of thumb will be sufficient for the estimating work you will perform in this module; estimating is covered in more detail in the *Masonry Level Three* module titled *Estimating*.

- One 80-pound bag of preblended mortar is adequate for 12 block or 35 brick.
- Three bags of masonry cement are adequate for approximately 100 concrete block.

Table 1 Modular Brick and Mortar Required for Single-Wythe Walls in Running Bond

Brick Designation	Nominal Dimensions, in.			Joint Thickness, in.	Number of Brick per 100 sq ft	Cubic Feet of Mortar	
	W	H	L			Per 100 sq ft	Per 1,000 Brick
Modular	4	2⅔	8	⅜	675	5.5	8.1
				½		6.9	10.3
Engineer Modular	4	3⅕	8	⅜	563	4.8	8.5
				½		6.1	10.8
Closure Modular	4	4	8	⅜	450	4.1	9.1
				½		5.2	11.6
—	4	6	8	⅜	300	3.2	10.7
				½		4.1	13.7
—	4	8	8	⅜	225	2.8	12.3
				½		3.5	15.7
Roman	4	2	12	⅜	600	6.4	10.7
				½		8.2	13.7
Norman	4	2⅔	12	⅜	450	5.1	11.2
				½		6.5	14.3
Engineer Norman	4	3⅕	12	⅜	375	4.4	11.7
				½		5.6	14.9
Utility	4	4	12	⅜	300	3.7	12.3
				½		4.7	15.7
—	6	3⅕	12	⅜	375	6.8	18.1
				½		8.8	23.4
—	6	4	12	⅜	300	5.7	19.1
				½		7.4	24.7
—	8	4	12	⅜	300	7.8	25.9
				½		10.1	33.6
—	4	2⅔	16	⅜	338	4.9	14.5
				½		6.5	19.2
Meridian	4	4	16	⅜	225	3.5	15.4
				½		4.4	19.7
Double Meridian	4	8	16	⅜	113	2.1	18.6
				½		2.7	23.8
6-in. Through-Wall Meridian	6	4	16	⅜	225	5.4	24.0
				½		7.0	31.0
8-in. Through-Wall Meridian	8	4	16	⅜	225	7.3	32.5
				½		9.5	42.3

- Seven bags of masonry cement are adequate for approximately 1,000 brick.
- One ton of sand is required for 1,000 brick.
- One cubic yard of sand is required for approximately nine bags of masonry cement.

For the 24-foot × 8-foot wall from the example in the section titled *Estimating Materials for Brick Construction*, the step-by-step process is as follows:

Step 1 Determine the number of bags of masonry cement using the third rule of thumb.

1,000 brick ÷ 7 bags of masonry cement = 142.8 brick per bag, rounded up to 143 brick per bag
1,361 ÷ 143 = 9.5, round up to 10
10 bags of masonry cement are needed to lay up the structure

Step 2 Determine the appropriate ratio of materials in the specified American Society for Testing and Materials (ASTM) mortar type (as in the block example), or use the fourth rule of thumb.

(1,361 brick ÷ 1,000 brick per ton of sand) =
1.361 tons of sand
1.361 × 1 = 1.4 tons of sand
needed for the proposed structure

As a result of these calculations you can now accurately report the estimates for the construction of the wall in question:

- 1,361 standard modular brick
- 10 bags of masonry cement
- 2 tons of sand (quantities are typically rounded up to the next ton)

If all calculations have been carefully performed and checked, you can be certain that the estimate will be accurate. This important step of preplanning will hopefully result in a cost-effective job without any loss of time, money, efficiency, or quality. On big jobs, sand is ordered by the truckload; for example, 9 tons on an 18-ton truck.

2.2.0 Using Estimating Aids

There are a number of estimating aids that are available to help determine quantities of block and brick for different types of wall (see *Figure 22*). Three organizations that provide various types of charts for estimating quantities are the Masonry Institute of America, the Brick Industry Association, and the National Concrete Masonry Institute. Other organizations, such as the Portland Cement Association, also publish manuals and reference guides that use the charts and graphs developed by other organizations.

An example of a chart that can be used for establishing the quantity of concrete masonry units for a single-wythe concrete masonry wall is shown in *Table 2*. This chart gives the number of units required per 100-square-foot wall area, assuming the mortar joints to be ⅜ inch. The number of units varies with the nominal size of the concrete masonry unit.

Figure 23 lists material quantities for constructing various types of composite walls bonded with metal ties or masonry headers. The mortar quantities include the customary allowance for waste that occurs during construction. Below the table, a section view of the six types of walls is shown so there is no mistake in the type of bond pattern being applied.

Computer-Aided Estimating

There are several calculators and computer programs on the market that can assist you with estimating. These calculators are programmed to automatically perform the calculations and conversions explained in this module.

There are several handheld calculators that compute the number of block or brick needed when the wall area or length and width dimension are entered. They will also convert between weight and volume for standard construction materials. Some will perform dimensional math and estimate material volumes and cost.

28201-14_SA07.EPS

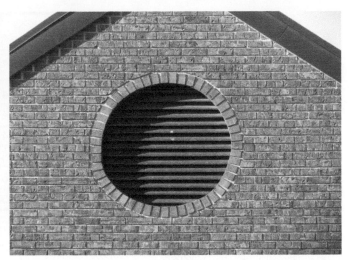

28201-14_F22.EPS

Figure 22 Different types of walls have their own estimating requirements.

Table 2 Material Quantities for Single-Wythe Concrete Masonry Construction

Nominal Wall Thickness (inches)	Nominal Size (width × height × length) of Concrete Masonry Units (Inches)	Average Weight of 100 sq ft Wall Area (in lbs)		Material Quantities for 100 sq ft Wall Area		
		Units Made with Sand-Gravel Aggregate	Units Made with Lightweight Aggregate	# of units	Mortar (cu ft)	Mortar for 100 Units (cu ft)
4	4 × 4 × 16	4,550	3,550	225	13.5	6.0
6	6 × 4 × 16	5,100	3,900	225	13.5	6.0
8	8 × 4 × 16	6,000	4,450	225	13.5	6.0
4	4 × 8 × 16	4,050	3,000	112.5	8.5	7.5
6	6 × 8 × 16	4,600	3,350	112.5	8.5	7.5
8	8 × 8 × 16	5,550	3,950	112.5	8.5	7.5
12	12 × 8 × 16	7,550	5,200	112.5	8.5	7.5

| Wall Thickness (inches) | Type of Bonding | No. and Size of Block | | # of Units | Mortar (cu ft) |
		Stretchers	Headers		
8	A—metal ties	112.5—4 × 8 × 16	—	675	20.0
	B—7th course headers	97—4 × 8 × 16	—	770	12.2
	C—7th course headers	197—4 × 5 × 12	—	770	13.1
12	D—metal ties	112.5—8 × 8 × 16	—	675	20.0
	E—7th course headers	97—8 × 8 × 16	—	868	13.5
	F—course headers	57—8 × 8 × 16	57—8 × 8 × 16	788	13.6

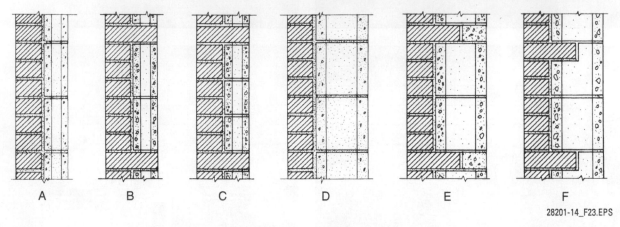

| A | B | C | D | E | F |

28201-14_F23.EPS

Figure 23 Material quantities for 100-square-foot area of composite walls.

A general estimate of other materials used for mortar and grout can also be obtained from tabular references. For example, once the volume of mortar is known, the individual quantities of materials that go into the specified mix can be estimated. A breakdown of materials by volume for various types of cementitious materials is given in *Table 3*.

Table 4 shows the volume of grout required for hollow single-wythe walls made from block. The table shows the volume of grout, in cubic feet per 100 square feet of wall, required for various widths of wall. Tables similar to this one are available for estimating grout in double-wythe walls as well as walls made with reinforced concrete masonry. Always refer to the estimating aids that are approved for use in your region.

NOTE

Some of the sample mixes fit the classification of two types of mortar. This is because mortar specifications permit a range in the amounts of lime and sand used for any type of mortar.

Table 3 Proportion Specifications for Cementitious Materials

Mortar	Type	Portland or Blended Cement	Mortar Cement M	Mortar Cement S	Mortar Cement N	Masonry Cement M	Masonry Cement S	Masonry Cement N	Hydrated Lime or Lime Putty	Aggregate Ratio (Measured in Damp, Loose Conditions)
Cement–Lime	M	1							¼	Not less than 2¼ and not more than 3 times the sum of the separate volumes of cementitious materials
	S	1							over ¼ to ½	
	N	1							over ½ to 1¼	
	O	1							over 1¼ to 2½	
Mortar Cement	M	1			1					
	M		1							
	S	½			1					
	S			1						
	N				1					
	O				1					
Masonry Cement	M	1						1		
	M					1				
	S	½						1		
	S						1			
	N							1		
	O							1		

Table 4 Volume of Grout in Hollow Single-Wythe Concrete Masonry Walls[a]

Grout Spacing, in.	6 in.	8 in.	10 in.	12 in.	14 in.
8	25.6	36.1	47.0	58.9	74.5
16	12.8	18.1	23.5	29.5	37.3
24	8.6	12.1	15.7	19.7	24.8
32	6.4	9.1	11.8	14.8	18.6
40	5.2	7.3	9.4	11.8	14.9
48	4.3	6.1	7.9	9.9	12.4
56	3.7	5.2	6.8	8.5	10.6
64	3.2	4.6	5.9	7.4	9.3
72	2.9	4.1	5.3	6.6	8.3
80	2.6	3.7	4.7	5.9	7.5
88	2.4	3.3	4.3	5.4	6.8
96	2.2	3.1	4.0	5.0	6.2
104	2.0	2.8	3.7	4.6	5.7
112	1.9	2.6	3.4	4.3	5.3
120	1.8	2.5	3.2	4.0	4.9

[a] Assumes two-core hollow concrete masonry units and 3% waste.

Additional Resources

Technical Note TN10, *Dimensioning and Estimating Brick Masonry.* 2009. Reston, VA: The Brick Industry Association. **www.gobrick.com**

TEK 4-2A, Estimating Concrete Masonry Materials. 2004. Herndon, VA: National Concrete Masonry Association. **www.ncma.org**

2.0.0 Section Review

1. When estimating block by wall area using the rule-of-thumb method, you do *not* need to determine _____.

    a. the total square feet of wall area in the structure
    b. the square feet of wall area occupied by a single block
    c. the total number of bags of mortar
    d. the total number of block required

2. In *Table 2*, the number of units varies with the _____.

    a. number of concrete masonry units used
    b. height of the concrete masonry unit
    c. nominal size of the concrete masonry unit
    d. actual size of the concrete masonry unit

SUMMARY

Masons must be able to read and interpret plans and drawings if they are to work on their own or supervise other masons. Plans and drawings provide a "road map" that guides each craft in doing its job without having to interfere with other workers.

Typical residential plans contain architectural, structural, mechanical, plumbing, and electrical drawings. Within each of these sections are floor plans and foundation plans, as well as sectional, elevation, and detail drawings. Masons are mostly concerned with foundation and elevation drawings, and with floor plans.

To read a set of residential plans, you must learn the shorthand used to describe the construction process. This includes knowing what different types of lines and symbols represent and what the abbreviations used in the notes and callouts mean.

Being able to accurately scale and measure out dimensions from the plans is very important. Using the wrong scale or making an error in reading a measurement can lead to major problems. Always double-check your measurements. If something still does not look right, have someone else check your work.

Proper estimates for materials are very important because their cost will have a direct impact on the overall cost of the structure. You can generally make estimates by carefully applying basic math and by referring to charts and tables that have been developed for this purpose. Estimating by rule of thumb is not intended to be mathematically perfect, but can be used for the construction of a small or average-sized job, such as a home, garage, retaining wall, or chimney.

Review Questions

1. Plumbing, mechanical, and electrical drawings for a residential project are included in the _____.
   a. detail drawings
   b. trades drawings
   c. specialty plans
   d. sectional drawings

2. To document a revision in the original construction drawings, a change order must be issued by the construction manager or the _____.
   a. building owner
   b. architect
   c. building inspector
   d. site supervisor

3. Keys and legends are typically placed on the _____.
   a. cover sheet
   b. plot plan
   c. floor plan
   d. elevations

4. On plans and drawings, dimension lines are shown as _____.
   a. dashed lines
   b. lightweight lines
   c. intermittent lines
   d. heavy lines

5. The abbreviation commonly used on plans to indicate a footing is _____.
   a. FOOT
   b. FG
   c. FTNG
   d. FTG

6. The dimensions shown on the floor plan are commonly called _____.
   a. scaled measurements
   b. plan measurements
   c. working measurements
   d. detail measurements

7. One method of dimensioning an exterior wall is to measure just to the outside of the unfinished wall. This allows space for _____.
   a. sheathing
   b. expansion
   c. insulation
   d. exterior finish materials

8. Measurements taken from one extreme end of a wall or object to the other extreme are referred to as _____.
   a. general measurements
   b. overall measurements
   c. total measurements
   d. extreme measurements

9. On a flat architect's ruler, the ⅛" scale is read _____.
   a. from right to left
   b. from left to right
   c. from the middle
   d. from either side

10. As a general rule of thumb, if a residential foundation wall is 8 inches thick, it should rest on a footing that is _____.
    a. 8 inches wide
    b. 12 inches wide
    c. 16 inches wide
    d. 24 inches wide

11. Residential floor plans are often drawn to a scale of _____.
    a. ¼" = 1'
    b. ½" = 1'
    c. ¾" = 1'
    d. 1" = 1'

12. In order to visualize the final structure, it is important to completely understand the relationship between the elevations and _____.
    a. foundation plans
    b. floor plans
    c. plot plans
    d. mechanical plans

13. Estimating quantities is made easier because most masonry materials today are based on the _____.

    a. nominal system
    b. metric system
    c. modular system
    d. standard system

14. When estimating quantities of mortar needs, the waste allowance may be as high as _____.

    a. 2 percent
    b. 3 percent
    c. 4 percent
    d. 5 percent

15. Refer to the *Review Question Figure 1*. Brick with nominal dimensions of 4 inches × 2 inches × 12 inches is designated as _____.

    a. Roman brick
    b. meridian brick
    c. Norman brick
    d. utility brick

Brick Designation	Nominal Dimensions, in.			Joint Thickness, in.	Number of Brick per 100 sq ft	Cubic Feet of Mortar	
	W	H	L			Per 100 sq ft	Per 1,000 Brick
Modular	4	2⅔	8	⅜ ½	675	5.5 6.9	8.1 10.3
Engineer Modular	4	3⅕	8	⅜ ½	563	4.8 6.1	8.5 10.8
Closure Modular	4	4	8	⅜ ½	450	4.1 5.2	9.1 11.6
—	4	6	8	⅜ ½	300	3.2 4.1	10.7 13.7
—	4	8	8	⅜ ½	225	2.8 3.5	12.3 15.7
Roman	4	2	12	⅜ ½	600	6.4 8.2	10.7 13.7
Norman	4	2⅔	12	⅜ ½	450	5.1 6.5	11.2 14.3
Engineer Norman	4	3⅕	12	⅜ ½	375	4.4 5.6	11.7 14.9
Utility	4	4	12	⅜ ½	300	3.7 4.7	12.3 15.7
—	6	3⅕	12	⅜ ½	375	6.8 8.8	18.1 23.4
—	6	4	12	⅜ ½	300	5.7 7.4	19.1 24.7
—	8	4	12	⅜ ½	300	7.8 10.1	25.9 33.6
—	4	2⅔	16	⅜ ½	338	4.9 6.5	14.5 19.2
Meridian	4	4	16	⅜ ½	225	3.5 4.4	15.4 19.7
Double Meridian	4	8	16	⅜ ½	113	2.1 2.7	18.6 23.8
6-in. Through-Wall Meridian	6	4	16	⅜ ½	225	5.4 7.0	24.0 31.0
8-in. Through-Wall Meridian	8	4	16	⅜ ½	225	7.3 9.5	32.5 42.3
Double Through-Wall Meridian	8	8	16	⅜ ½	113	4.4 5.7	39.1 51.0

Figure 1

28201-14_RQ01.EPS

Trade Terms Quiz

Fill in the blank with the correct term that you learned from your study of this module.

1. A drawing that shows the inside of a component or structure is called a(n) _____.

2. A(n) _____ is a listing that explains or defines symbols or special marks placed on plans or drawings.

3. A drawing that is usually developed by manufacturers, fabricators, or contractors to show specific dimensions and other pertinent information concerning a particular piece of equipment and its installation methods is called a(n) _____.

4. A(n) _____ is a document or form used during the construction process to document a change in the construction requirements from the original plans or specifications.

Trade Terms:

Change order
Legend

Sectional drawing
Shop drawing

Trade Terms Introduced in This Module

Change order: A document or form used during the construction process to document a change in the construction requirements from the original plans or specifications.

Legend: A listing that explains or defines symbols or special marks placed on plans or drawings. Usually the legend is on the front sheet or index of the plan set.

Sectional drawing: A drawing that shows the inside of a component or structure. The view would be as if you cut the item into two pieces and looked at the end of the cut.

Shop drawing: A drawing that is usually developed by manufacturers, fabricators, or contractors to show specific dimensions and other pertinent information concerning a particular piece of equipment and its installation methods.

ARCHITECTURAL TERMS COMMONLY FOUND ON PLANS

WINDOW TERMS

Apron – A plain or molded piece of finish below the stool of a window that is put on to cover the rough edge of the plastering.

Drip cap – A projection of masonry or wood on the outside top of a window to protect the window from rain.

Head jamb – Horizontal top post used in the framing of a window or doorway.

Light – A pane of glass.

Lintel – Horizontal structural member supporting a wall over a window or other opening.

Meeting rail – The horizontal center rail of a sash in a double-hung window.

Mullion – A large, vertical division of a window opening.

Muntin – A strip of wood or metal that separates and supports the panes of glass in a window sash.

Sash – The part of a window in which panes of glass are set; it is generally movable, as in double-hung windows. The two side pieces are called stiles and the upper and lower pieces are called rails.

Side jambs – Vertical side posts used in the framing of a window or doorway.

Sill – Horizontal member at the bottom of a window or doorway.

Stool – A flat, narrow shelf forming the top member of the interior trim at the bottom of a window.

Stop bead – The strip on a window frame against which the sash slides.

PITCHED ROOF TERMS

Flashing – Sheet metal, copper, lead, or tin that is used to cover open joints to make them waterproof.

Gable – The end of a ridged roof as distinguished from the front or rear side.

Louver – An opening for ventilation that is covered by sloping slats to exclude rain.

Ridge – The top edge of a roof where the two slopes meet.

Ridgeboard – A board that is placed on edge at the ridge of a roof to support the upper ends of rafters.

Saddle – A tent-shaped portion of a roof between a chimney and the main part of the roof; built to support flashing and to direct water away from the chimney.

Valley – The intersection of the bottom two inclined sides of a roof.

CORNICE TERMS

Cornice – The part of a roof that projects beyond a wall.

Cornice return – The short portion of a cornice that is carried around the corner of a structure.

Crown molding – The molding at the top of the cornice and just under the roof.

Fascia – The outside flat member of a cornice.

Frieze – A trim member used just below the cornice.

Soffit – The underside of a cornice.

STAIR TERMS

Head room – The distance between flights of steps or between the steps and the ceiling above.

Landing – The horizontal platform in a stairway.

Nosing – The overhanging edge of a stair tread.

Rise – The vertical distance from the top of a tread to the top of the next highest tread.

Riser – The vertical portion of a step.

Run – The horizontal distance that is covered by a flight of steps. Also, the net width of a step.

Stringer – The supporting timber at the sides of a staircase.

Tread – The horizontal part of a step on which the foot is placed.

STRUCTURAL TERMS

Anchor bolt – A bolt with the threaded portion projecting from a structure; generally used to hold the frame of a building secure against wind load. Anchor bolts may also be referred to as hold-down bolts, foundation bolts, and sill bolts.

Batt – A type of insulation designed to be installed between framing members.

Battens – Narrow strips of wood or metal used to cover vertical joints between boards and panels.

Beam – One of the principal horizontal members of a building.

Bridging – The process of bracing floor joists by fixing lateral members between them.

Camber – The concave or convex curvature of a surface.

Expansion joint – The separation between adjoining parts to allow for small relative movements such as those caused by temperature changes.

Footing – The foundation for a column or the enlargement at the bottom of a wall to distribute the weight of the superstructure over a greater area to prevent settling.

Furring – Strips of wood or metal applied to a wall or other surface to make it level, form an airspace, or provide a fastening surface for a finish covering.

Girder – The main supporting beam (either timber or steel) that is used for supporting a superstructure.

Header – A wood beam that is set at a right angle to a joist to provide a seat or support.

Joist – A heavy piece of horizontal timber to which the boards of a floor or a ceiling are nailed. Joists are laid edgewise to form a floor support; they rest on the wall or on girders.

Monolithic concrete – A continuous mass of concrete that is cast as a single unit.

Plate – A structural member with a depth that is substantially smaller than its length or width.

Rake – Trim members that run parallel to the roof slope and form the finish between the roof and the wall at the gable end.

Reinforced concrete – Concrete containing metal rods, wires, or other slender members. It is designed in such a manner that the concrete and metal act together to resist forces.

Sill – A horizontal member that is supported by a foundation wall or piers, and which in turn bears the upright members of a frame.

Slab – A poured concrete floor.

Studs – The vertical, slender wood or metal members that are used to support the elements in walls and partitions.

Vapor barrier – A material that is used to retard the flow of vapor or moisture into the walls or floors and thus prevent condensation within them.

Veneer – The covering layer of material for a wall or facing materials applied to the external surface of steel, reinforced concrete, or frame walls.

COMMON ABBREVIATIONS

Aluminum	AL	Foundation	FND
Asbestos	ASB	Full size	FS
Asphalt	ASPH	Galvanized	GALV
Basement	BSMT	Galvanized iron	GI
Beveled	BEV	Gauge	GA
Brick	BRK	Glass	GL
Building	BLDG	Glass block	GL BL
Cast iron	CI	Grade	GR
Ceiling	CLG	Grade line	GL
Cement	CEM	Height	HGT, H, or HT
Center	CTR	High point	H PT
Center line	C or CL	Horizontal	HOR
Clear	CLR	Hose bibb	HB
Column	COL	Inch or inches	" or IN
Concrete	CONC	Insulating (insulated)	INS
Concrete block	CONC B	Length	LGTH, LG, or L
Copper	COP	Length overall	LOA
Corner	COR	Level	LEV
Detail	DET	Light	LT
Diameter	DIA	Line	L
Dimension	DIM.	Lining	LN
Ditto	DO.	Long	LG
Divided	DIV	Louver	LV
Door	DR	Low point	LP
Double-hung window	DHW	Masonry opening	MO
Down	DN or D	Metal	MET. or M
Downspout	DS	Molding	MLDG
Drawing	DWG	Mullion	MULL
Drip cap	DC	North	N
Each	EA	Number	NO. or #
East	E	Opening	OPNG
Elevation	EL	Outlet	OUT
Entrance	ENT	Outside diameter	OD
Excavate	EXC	Overhead	OVHD
Exterior	EXT	Panel	PNL
Finish	FIN.	Perpendicular	PERP
Flashing	FL	Plate glass	PL GL
Floor	FL	Plate height	PL HT
Foot or feet	' or FT		

Radius	R	Square	SQ
Revision	REV	Square inch	SQ. IN.
Riser	R	Stainless steel	SST
Roof	RF	Steel	STL
Roof drain	RD	Stone	STN
Roofing	RFG	Terra-cotta	TC
Rough	RGH	Thick or thickness	THK or T
Saddle	SDL or S	Typical	TYP
Scale	SC	Vertical	VERT
Schedule	SCH	Waterproofing	WP
Section	SECT	West	W
Sheathing	SHTHG	Width	W or WTH
Sheet	SH	Window	WDW
Shiplap	SHLP	Wire glass	W GL
Siding	SDG	Wood	WD
South	S	Wrought iron	WI
Specifications	SPEC		

COMMON ABBREVIATIONS USED ON PLAN VIEWS

Access panel	AP	Blueprint	BP
Acoustic	ACST	Boiler	BLR
Acoustical tile	AT	Bookshelves	BK SH
Aggregate	AGGR	Brass	BRS
Air conditioning	AIR COND	Brick	BRK
Aluminum	AL	Bronze	BRZ
Anchor bolt	AB	Broom closet	BC
Angle	AN	Building	BLDG
Apartment	APT	Building line	BL
Approximate	APPROX	Cabinet	CAB.
Architectural	ARCH	Caulking	CLKG
Area	A	Casing	CSG
Area drain	AD	Cast iron	CI
Asbestos	ASB	Cast stone	CS
Asbestos board	AB	Catch basin	CB
Asphalt	ASPH	Cellar	CEL
Asphalt tile	AT	Cement	CEM
Basement	BSMT	Cement asbestos board	CEM AB
Bathroom	B	Cement floor	CEM FL
Bathtub	BT	Cement mortar	CEM MORT
Beam	BM	Center	CTR
Bearing plate	BRG PL	Center to center	C to C
Bedroom	BR	Center line	C or CL
Blocking	BLKG	Center matched	CM

Ceramic	CER	Fireproof	FPRF
Channel	CHAN	Fixture	FIX.
Cinder block	CIN BL	Flashing	FL
Circuit breaker	CIR BKR	Floor	FL
Cleanout	CO	Floor drain	FD
Cleanout door	COD	Flooring	FLG
Clear glass	CL GL	Fluorescent	FLUOR
Closet	C, CL, or CLO	Flush	FL
Cold air	CA	Footing	FTG
Cold water	CW	Foundation	FND
Collar beam	COL B	Frame	FR
Concrete	CONC	Full size	FS
Concrete block	CONC B	Furring	FUR
Concrete floor	CONC FL	Galvanized iron	GI
Conduit	CND	Garage	GAR
Construction	CONST	Gas	G
Contract	CONT	Glass	GL
Copper	COP	Glass block	GL BL
Counter	CTR	Grille	G
Cubic feet	CU FT	Gypsum	GYP
Cutout	CO	Hardware	HDW
Detail	DET	Hollow metal door	HMD
Diagram	DIAG	Hose bibb	HB
Dimension	DIM.	Hot air	HA
Dining room	DR	Hot water	HW
Dishwasher	DW	Hot-water heater	HWH
Ditto	DO.	I-beam	I
Double acting	DA	Inside diameter	ID
Double-strength glass	DSG	Insulation	INS
Down	DN	Interior	INT
Downspout	DS	Iron	I
Drain	D or DR	Jamb	JB
Drawing	DWG	Kitchen	K
Dressed and matched	D & M	Landing	LDG
Dryer	D	Lath	LTH
Electric panel	EP	Laundry	LAU
End to end	E to E	Laundry tray	LT
Excavate	EXC	Lavatory	LAV
Expansion joint	EXP JT	Leader	L
Exterior	EXT	Length	L, LG, or LNG
Finish	FIN.	Library	LIB
Finished floor	FIN. FL	Light	LT
Firebrick	FBRK	Limestone	LS
Fireplace	FP	Linen closet	L CL
		Lining	LN

Linoleum	LINO	Room	RM or R
Living room	LR	Rough	RGH
Louver	LV	Rough opening	RGH OPNG
Main	MN	Rubber tile	RTILE
Marble	MR	Scale	SC
Masonry opening	MO	Schedule	SCH
Material	MATL	Screen	SCR
Maximum	MAX	Scuttle	S
Medicine cabinet	MC	Section	SECT.
Minimum	MIN	Select	SEL
Miscellaneous	MISC	Service	SERV
Mixture	MIX	Sewer	SEW.
Modular	MOD	Sheathing	SHTHG
Mortar	MOR	Sheet	SH
Molding	MLDG	Shelf and rod	SH & RD
Nosing	NOS	Shelving	SHELV
Obscure glass	OBSC GL	Shower	SH
On center	OC	Sill cock	SC
Opening	OPNG	Single-strength glass	SSG
Outlet	OUT	Sink	SK or S
Overall	OA	Soil pipe	SP
Overhead	OVHD	Specifications	SPEC
Pantry	PAN	Square feet	SQ FT
Partition	PTN	Stained	STN
Plaster	PL or PLAS	Stairs	ST
Plastered opening	PO	Stairway	STWY
Plate	PL	Standard	STD
Plate glass	PL GL	Steel	ST or STL
Platform	PLAT	Steel sash	SS
Plumbing	PLBG	Storage	STG
Porch	P	Switch	SW or S
Precast	PRCST	Telephone	TEL
Prefabricated	PREFAB	Terra-cotta	TC
Pull switch	PS	Terrazzo	TER
Quarry tile floor	QTF	Thermostat	THERMO
Radiator	RAD	Threshold	TH
Random	RDM	Toilet	T
Range	R	Tongue-and-groove	T & G
Recessed	REC	Tread	TR or T
Refrigerator	REF	Typical	TYP
Register	REG	Unexcavated	UNEXC
Reinforce or reinforcing	REINF	Unfinished	UNF
Revision	REV	Utility room	URM
Riser	R	Vent	V
Roof	RF		
Roof drain	RD		

Vent stock	VS
Vinyl tile	VTILE
Warm air	WA
Washing machine	WM
Water	W
Water closet	WC
Water heater	WH
Waterproof	WP
Weatherstripping	WS
Weep hole	WH
White pine	WP
Wide flange	WF
Wood	WD
Wood frame	WF
Welded wire fabric	WWF
Yellow pine	YP

CONVERSION CHARTS

ENGLISH TO METRIC CONVERSIONS		
WEIGHTS		
1 ounce	=	28.349523 grams
1 pound	=	53,5924 grams or
		0.4536 kilograms
1 (short) ton	=	907.2 kilograms
LENGTHS		
1 inch	=	2.540 centimeters
1 foot	=	30.48 centimeters
1 yard	=	91.44 centimeters or
		0.9144 meters
1 mile	=	1.6093 kilometers
AREAS		
1 square inch	=	6.4516 square centimeters
1 square foot	=	929.0 square centimeters
		or 0.0929 square meters
1 square yard	=	0.8361 square meters
VOLUMES		
1 cubic inch	=	16.387 cubic centimeters
1 cubic foot	=	0.02832 cubic meter
1 cubic yard	=	0.7646 cubic meter
LIQUID MEASUREMENTS		
1 (fluid) ounce	=	0.02957 liter or
		28.349523 grams
1 pint	=	473.1765 cubic centimeters
1 quart	=	0.94635 liter
1 (US) gallon	=	3,785.412 cubic centimeters
		or 3.785 liters

TEMPERATURE MEASUREMENTS

To convert degrees Fahrenheit to degrees Celsius, use the following formula: $C = 5/9 \times (F - 32)$.

METRIC TO ENGLISH CONVERSIONS		
WEIGHTS		
1 gram (G)	=	0.035273962 ounces
1 kilogram (kg)	=	2.2046 pounds
1 metric ton	=	2,205 pounds
LENGTHS		
1 millimeter (mm)	=	0.03937 inches
1 centimeter (cm)	=	0.3937 inches
1 meter (m)	=	3.2808 feet or 1.0936 yards
1 kilometer (km)	=	0.6214 miles
AREAS		
1 square millimeter	=	0.00155 square inches
1 square centimeter	=	0.155 square inches
1 square meter	=	10.7640 square feet or
		1.196 square yards
VOLUMES		
1 cubic centimeter	=	0.06100 cubic inches
1 cubic meter	=	35.3160 cubic feet or
		1.3080 cubic yards
LIQUID MEASUREMENTS		
1 cubic centimeter (cm^3) =		0.06102 cubic inches
1 liter (1,000 cm^3)	=	1.057 quarts, 2.113 pints, or
		61.02 cubic inches

TEMPERATURE MEASUREMENTS

To convert degrees Celsius to degrees Fahrenheit, use the following formula: $F = (9/5 \times C) + 32$.

28201-14_A01.EPS

Additional Resources

This module presents thorough resources for task training. The following resource material is suggested for further study.

Bricklaying: Brick and Block Masonry. 1988. Brick Industry Association. Orlando, FL: Harcourt Brace & Company.

Concrete Masonry Handbook for Architects, Engineers, Builders, Fifth Edition. 1991. W. C. Panerese, S. K. Kosmatka, and F. A. Randall, Jr. Skokie, IL: Portland Cement Association.

Technical Note TN10, *Dimensioning and Estimating Brick Masonry.* 2009. Reston, VA: The Brick Industry Association. **www.gobrick.com**

TEK 4-2A, *Estimating Concrete Masonry Materials.* 2004. Herndon, VA: National Concrete Masonry Association. **www.ncma.org**

Figure Credits

Section Review Answers

Answer	Section Reference	Objective
Section One		
1.a	1.1.2	1a
2.c	1.2.4	1b
3.b	1.3.1	1c
Section Two		
1.c	2.1.1	2a
2.c	2.2.0	2b

NCCER CURRICULA — USER UPDATE

NCCER makes every effort to keep its textbooks up-to-date and free of technical errors. We appreciate your help in this process. If you find an error, a typographical mistake, or an inaccuracy in NCCER's curricula, please fill out this form (or a photocopy), or complete the online form at **www.nccer.org/olf**. Be sure to include the exact module ID number, page number, a detailed description, and your recommended correction. Your input will be brought to the attention of the Authoring Team. Thank you for your assistance.

Instructors – If you have an idea for improving this textbook, or have found that additional materials were necessary to teach this module effectively, please let us know so that we may present your suggestions to the Authoring Team.

NCCER Product Development and Revision
13614 Progress Blvd., Alachua, FL 32615

Email: curriculum@nccer.org
Online: www.nccer.org/olf

❏ Trainee Guide   ❏ Lesson Plans   ❏ Exam   ❏ PowerPoints   Other _____

Craft / Level: _____   Copyright Date: _____

Module ID Number / Title: _____

Section Number(s): _____

Description: _____

_____

_____

_____

_____

Recommended Correction: _____

_____

_____

_____

_____

Your Name: _____

Address: _____

Email: _____   Phone: _____

28202-14

Residential Masonry

Masonry work continues to be in high demand in residential construction, which includes single-family houses and town-house complexes. Residential masonry projects have their own specific construction processes and masonry techniques, as well as nonmasonry work that may be part of the overall masonry contract. This can include concrete work for footings and foundations, compaction of base materials, and subfloor construction. This module introduces the mason to construction techniques for residential and small-structure foundations, steps, patios, decks, chimneys, and fireplaces, which are all part of residential construction.

Module Two

Trainees with successful module completions may be eligible for credentialing through NCCER's National Registry. To learn more, go to **www.nccer.org** or contact us at **1.888.622.3720**. Our website has information on the latest product releases and training, as well as online versions of our *Cornerstone* magazine and Pearson's product catalog.

Your feedback is welcome. You may email your comments to **curriculum@nccer.org**, send general comments and inquiries to **info@nccer.org**, or fill in the User Update form at the back of this module.

This information is general in nature and intended for training purposes only. Actual performance of activities described in this manual requires compliance with all applicable operating, service, maintenance, and safety procedures under the direction of qualified personnel. References in this manual to patented or proprietary devices do not constitute a recommendation of their use.

RESIDENTIAL MASONRY

Objectives

When you have completed this module, you will be able to do the following:

1. Explain the requirements for construction of various types of residential foundations.
   a. Explain what spread foundations are.
   b. Explain what raft and mat foundations are.
   c. Explain what foundation walls are.
2. Identify and explain the characteristics, uses, and installation techniques for clay brick and concrete pavers.
   a. Describe the various types of clay brick pavers.
   b. Explain how to install clay brick pavers.
   c. Describe the various types of concrete and interlocking pavers.
   d. Explain how to install concrete and interlocking pavers.
3. Lay out and build steps, patios, and decks made from masonry units.
   a. Describe the various types of steps.
   b. Explain how to recognize patterns and tread designs.
   c. Explain how to build a concrete base.
   d. Explain how to set clay brick in steps.
   e. Explain how patios are constructed.
   f. Explain how decks are constructed.
4. Explain how to lay out and build fireplaces and chimneys.
   a. Explain the basic theory of the fireplace.
   b. Describe the parts of a fireplace.
   c. Explain the key points of workmanship.
   d. Explain how to lay out chimneys and fireplaces.
   e. Explain how to begin the fireplace.
   f. Explain how to finish the fireplace.
   g. Describe a multi-opening fireplace.

Performance Tasks

Under the supervision of your instructor, you should be able to do the following:

1. Lay out and construct a set of steps with three risers.
2. Lay out and construct a 5-foot by 7-foot clay brick patio section.

Trade Terms

Bearing pressure	Downdraft	Parge
Bedrock	Footing	Pavers
Chamfered	Hydrostatic pressure	Pilaster
Corbel	Infiltration	Slushing
Cove	Loadbearing	
Damper	Mortarless paving	

Industry-Recognized Credentials

If you're training through an NCCER-accredited sponsor, you may be eligible for credentials from NCCER's Registry. The ID number for this module is 28202-14. Note that this module may have been used in other NCCER curricula and may apply to other level completions. Contact NCCER's Registry at 888.622.3720 or go to **www.nccer.org** for more information.

Code Note

Codes vary among jurisdictions. Because of the variations in code, consult the applicable code whenever regulations are in question. Referring to an incorrect set of codes can cause as much trouble as failing to reference codes altogether. Obtain, review, and familiarize yourself with your local adopted code.

Contents

Topics to be presented in this module include:

Contents (continued)

Figures and Tables

1.0.0 REQUIREMENTS FOR RESIDENTIAL FOUNDATIONS

Objective

Explain the requirements for construction of various types of residential foundations.

   a. Explain what spread foundations are.
   b. Explain what raft and mat foundations are.
   c. Explain what foundation walls are.

Trade Terms

Bearing pressure: The load on a bearing surface divided by its area, expressed in pounds per square inch.

Bedrock: Solid rock that cannot be easily dislodged or removed from the soil. Typically, the rock that forms the outer crust of the earth.

Footing: An enlargement at the bottom of a wall that distributes the weight of the superstructure over a greater area to prevent settling.

Hydrostatic pressure: The pressure at any point in a liquid at rest, equal to the depth of the liquid multiplied by its density.

Loadbearing: Any structure that supports any vertical load in addition to its own weight.

Parge: To apply a thin coat of mortar or grout on the outside surface of a masonry surface to prepare it for the attachment of veneer or tile, or to waterproof it.

Pilaster: A wall portion projecting from a wall face and serving as a vertical column and/or beam.

The foundation supports the weight of a structure and resists settling. Footings transmit the weight from the foundation walls to the soil beneath them. The most common types of foundation are the spread foundation, the raft foundation, the mat foundation, and the pile foundation. Masons rarely work with pile foundations, however.

Spread foundations, like the example shown in *Figure 1*, are usually wider than the foundation wall. This provides extra support that will stop or reduce uneven settling. As a result, the structure should remain sound. The load applied at the top of the foundation wall is distributed down through the wall to the footing at the base of the wall. The load is then transmitted down through the footing and distributed to the soil supporting the footing. Footings are typically made of concrete. The concrete is poured in place or precast and placed on undisturbed native soil, unless the soil is unsuitable, weak, or soft. In this case, the soil should be removed and replaced with compacted soil, gravel, or concrete. Foundation walls made of concrete masonry units (CMUs), commonly called block, or other material are then constructed on top of the footing.

Raft foundations and mat foundations are usually made with concrete and reinforced with steel. The entire foundation acts as a single unit. These types of foundation are commonly called floating foundations because the pressure from the soil resists the downward pressure from the foundation and makes it float in the soil.

Pile foundations, as their name suggests, rely on piles that are driven into the soil to a depth that is sufficient to support the load. If the pile is driven far enough into the ground, it will hit bedrock. The bedrock will support the pile and the load above it. Sometimes the pile cannot be driven deep enough to hit bedrock. In such cases, the friction between the soil and the sides of the pile supports the pile. When pile foundations are used, beams or other footings are placed on top of the piles to support the foundation walls. A poured concrete footing is not needed. The foundation wall is built on the beam footings.

1.1.0 Understanding Spread Foundations

Spread foundations consist of walls, pilasters, columns, or piers that rest on a wider base called a footing. Spread foundations distribute the structure loads over a wider area of soil. Raft and mat foundations are usually classified as spread foundations. They are typically used on soils that do not have the ability to carry a large amount of weight. Raft and mat foundations are discussed in the section *Understanding Raft and Mat Foundations*.

All foundation walls constructed of masonry units bonded with mortar should rest on footings, which are usually made from concrete and classified as plain, reinforced, continuous, stepped, and isolated. Footings made with poured concrete can be either formed or unformed.

Unformed footings are made by digging a trench in the soil and pouring the concrete into the trench. The shape of the bottom and walls of the trench define the shape of the sides and bottom of the footing. The top is usually smoothed to some specified level of flatness and correct height.

Formed footings are constructed by pouring the concrete into a form with specified dimensions. Forms can be made of wood, metal, or Styrofoam™.

Figure 2 shows a typical job-constructed wooden form used for making an on-ground foundation. The form has sides made of wood. It is braced with wooden or metal stakes driven into the soil and nailed to the back side of the form. Reinforcement can also be placed inside the form. This gives added strength to the finished concrete footing and ties the foundation wall to the footing. Notice that the rebar must be positioned so that it is completely enclosed by the concrete once it is poured. The rebar must be supported to keep it from resting on the soil.

1.1.1 Continuous Footings

Continuous footings are made so that there is no break in the support provided to the foundation above (*Figure 3*). A continuous footing would uniformly support an entire wall or several columns in a row without any break. In residential construction, the footing is typically designed to run completely around the perimeter of the structure.

Footing size should be related to the local soil conditions and the weight of the structure. Local building codes will specify the minimum area required for footings based on the bearing pressure of the soil. However, the rule of thumb for residential footings placed on average soil is as follows:

- The width of the footing should be at least twice the thickness of the wall.
- The depth of the footing should be at least equal to the thickness of the wall, or a minimum of 8 inches, and below the frost line.

28202-14_F02.EPS

Figure 2 Wooden form for footing.

Figure 4 shows the proportions for a standard footing used in residential construction. The design of the foundation is the job of the architect

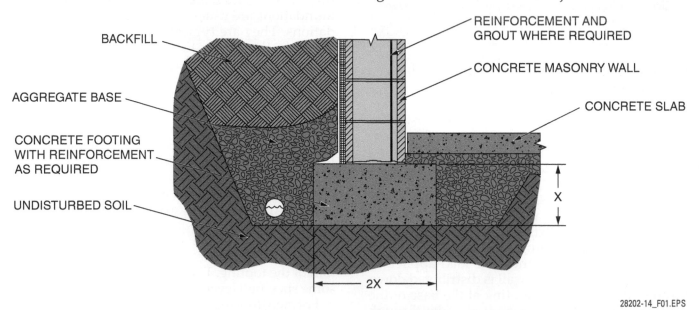

28202-14_F01.EPS

Figure 1 Basic spread footing.

or engineer. In most cases, the mason implements the design and constructs the foundation using the plans and drawings.

1.1.2 Combined Footings

A combined footing is a single footing that supports more than one column. The column transfers the weight to the footing at specific points along the length of the footing. The space between the columns can be enclosed by a wall (sometimes called a curtain wall) that can rest on the footing, but the wall should not be loadbearing. This type of arrangement can be used for aboveground decks that need several columns for support. The underneath portion would be enclosed with screening or wooden walls.

1.1.3 Stepped Footings

Stepped footings are designed to support a foundation wall on a sloping grade. There are two types of stepped footings. One widens the base to provide added support. The other changes levels to accommodate a sloping grade (*Figure 5*). In modern construction, reinforced footings have largely replaced the widened footing.

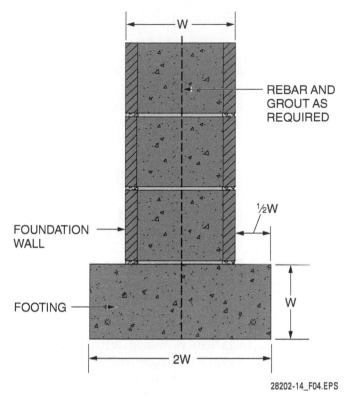

28202-14_F04.EPS

Figure 4 Standard footing proportions.

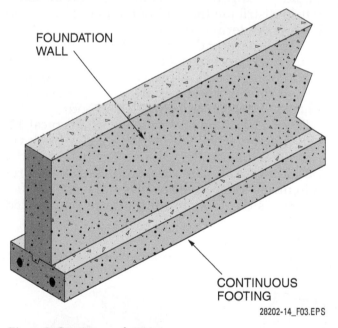

28202-14_F03.EPS

Figure 3 Continuous footing.

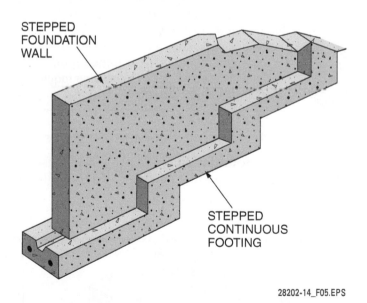

28202-14_F05.EPS

Figure 5 Stepped footing.

Reinforced Footings

Steel reinforcing rods, commonly called rebar, are often embedded in footings to make them stronger. This provides greater resistance to bending and tension. Reinforcing should be used over weak spots in the soil and over locations where excavation under the footing has been performed for sewer, gas, or other connections.

1.1.4 Isolated Footings

Isolated footings (*Figure 6*) are not part of the foundation. They receive the loads of free-standing columns or piers. This type of footing is used where minimum force is applied through the column. In residential construction, they are typically used to support deck or porch columns that do not carry a large amount of weight.

1.2.0 Understanding Raft and Mat Foundations

A mat foundation is a thickened slab that transmits the structure's load over the entire slab and soil area. It ties the slab or floor and the footings into one unit. The foundation is usually poured as one continuous element. This type of foundation may be used for residential structures where there is no basement. Walls and partitions can be built on the floor slab as required.

A raft foundation is made with reinforced walls and floor cast in a single unit. The slab is on the bottom of the foundation, and the walls are tied into the slab on the top side. Raft foundations can be used for residential structures where basements are required and soil conditions dictate the use of a floating foundation.

> **WARNING!**
>
> Federal law requires that you contact the designated authority at least 72 hours before the start of excavation. Local utility companies will come and mark the locations of underground utility lines and cables. Some areas also require you to obtain a digging permit.

1.3.0 Understanding Foundation Walls

Like footings, foundation walls are load-transferring elements. They must be able to do the following:

- Support the weight of the structure above
- Resist pressures from the surrounding soil
- Provide anchorage for the superstructure of the building
- Resist moisture penetration and remain durable

Foundation walls can be made from various materials. Local conditions and construction practices usually determine the type of foundation walls to be used.

> **NOTE**
>
> In some areas, the foundation must withstand frost heaving or earthquakes. Check local codes for the specific requirements for foundations.

The most common type of foundation wall is the T foundation. The name comes from the shape of the foundation and the footings, which look like an inverted T-shape. The footing and the foundation wall are generally two separate parts, but may be cast as a single unit if using reinforced concrete.

Hollow concrete block is a common material for building foundation walls for houses and other small structures. If concrete block is used, it should be capped with a course of solid masonry to help distribute the load of the floor beams and to serve as a termite barrier.

In some parts of the United States, solid block of 4-inch thickness is used. If stretcher block is used, a strip of metal lath is placed under the cores of the top course. The cores are then filled in with concrete or mortar and troweled smooth.

Figure 7 shows a section view of a typical 12-inch concrete block foundation for clay brick veneer on frame. In this design, the height of the foundation wall can vary according to the requirements for space below ground level. The figure shows the 18-inch minimum recommended

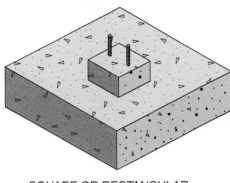

SQUARE OR RECTANGULAR
STEPPED PIER

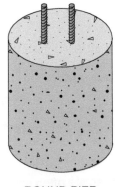

ROUND PIER

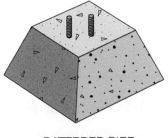

BATTERED PIER

28202-14_F06.EPS

Figure 6 Typical isolated footings.

crawl space for houses without basements. For a house with a basement, the foundation wall could be 10 feet or higher.

1.3.1 Lateral Pressure

Foundation walls must be built to withstand some lateral pressure so they are not overturned or displaced. The magnitude and direction of soil pressure on the wall depends on the height and shape of the surface. It also depends on the nature and physical properties of the backfill. Additional lateral pressure may be created by operating cars, trucks, or construction equipment on the soil surface behind the foundation wall.

The overall stability of a below-grade wall may be improved by increasing the stiffness in either direction or by reducing the length of the horizontal span. Horizontal stiffness can be increased by incorporating bond beams into the design. Placing prefabricated joint reinforcement in the horizontal mortar joints at vertical intervals of not more than 16 inches will also improve stiffness. Vertical stiffness may be increased in one of two ways:

- Steel reinforcement may be grouted into hollow cells (*Figure 8*).
- Pilasters may be added at intervals determined by the engineer.

1.3.2 Drainage

The purpose of drainage systems is to direct water away from the structure. Proper drainage and water resistance of below-grade foundation walls not only prevent the buildup of hydrostatic pressure, but also maintain dry conditions and eliminate dampness in interior spaces. Note the drainage and moisture control features in a typical foundation in *Figure 7*.

Provisions must be made to prevent accumulation of water behind a wall. Water accumulation causes increased soil pressure, seepage, and, in areas subject to frost, powerful expansive forces near the top of the wall. Drain tiles are generally placed around footings to carry away groundwater. They are connected to a dry well or storm sewer or directed to a lower elevation on the plot of land.

Footings

Continuous footing forms are typically made from construction lumber. The side boards are placed on edge and held in place by stakes, braces, and cross spreaders. Rebar is placed and concrete is poured into the form creating the footing. An example of concrete placed in a continuous footing form is shown here.

28202-14_SA01.EPS

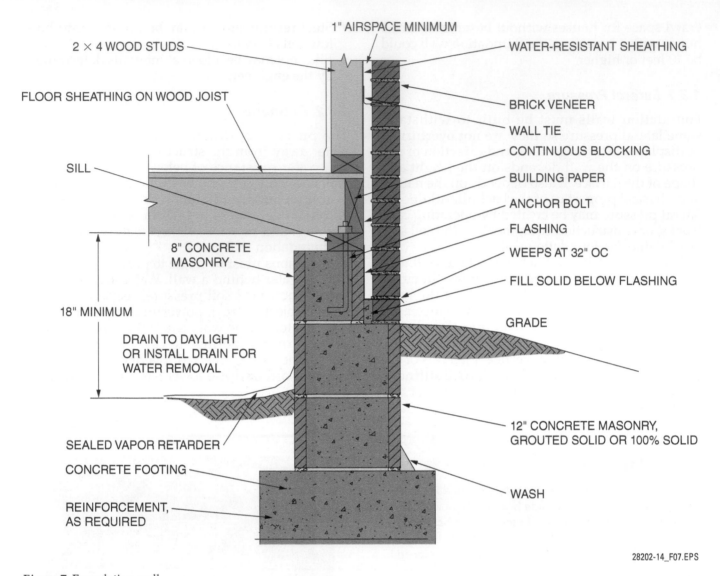

Figure 7 Foundation wall.

28202-14_F07.EPS

Load Capacity of Different Soils

The ground below a foundation must be able to uniformly support the weight carried by the slab. Under certain circumstances, a loadbearing-capacity test of the soil is required in order to determine the soil's actual loadbearing capacity.

Some general figures for the loadbearing capacity of different soils, expressed in tons per square inch, are as follows:

Soil Type	Loadbearing Capacity (tons/sq in)
Soft clay	1
Wet sand or firm clay	2
Fine or dry sand	3
Hard dry clay or coarse sand	4
Gravel	6

28202-14_F08.EPS

Figure 8 Steel- and grout-reinforced foundation walls.

The concrete and clay tiles are usually 4 inches in diameter and are laid with open joints. Plastic pipe is perforated with holes that allow the water to enter. These pipes are laid in a bed of gravel that surrounds the pipe by at least 12 inches.

Some damp-proofing techniques call for the outside of masonry basement walls to be parged with a ½-inch coating of portland cement plaster or mortar. It should be applied in two coats, each ¼-inch thick. The wall surface should be clean and damp before applying the plaster. Do not soak the wall; the plaster will not adhere if the wall is too wet. The first coat should be troweled firmly over the wall.

When the first coat has partially hardened, it should be roughened with a notched flat trowel. This technique provides a good bond for the next coat. Keep the first coat moist, and allow it to harden for a minimum of 24 hours before applying the second coat. Cover the wall from the footing to 6 inches above the finished grade line. Form a wash at the intersection of the wall and the footing to prevent water from collecting at this point.

Some areas of the United States with certain soil conditions require additional damp-proofing. A heavy coat of tar, two coats of cement-based paint, or a covering of thin plastic film may be used in addition to the parge coat.

Additional Resources

Concrete Masonry Handbook, Fifth edition. W. C. Panerese, S. K. Kosmatka, and F. A. Randall, Jr. Skokie, IL: Portland Cement Association.

Bricklaying: Brick and Block Masonry. 1988. Brick Industry Association. Orlando, FL: Harcourt Brace & Company.

1.0.0 Section Review

1. For residential footings placed on average soil, the rule of thumb for the width of the footing is that it should be at least _____.

   a. as wide as the depth of the wall
   b. three times the thickness of the wall
   c. twice the thickness of the depth of the wall
   d. twice the thickness of the wall

2. A foundation in the form of a thickened slab that transmits the structure's load over the entire slab and soil area is called a _____.

   a. mat foundation
   b. slab foundation
   c. raft foundation
   d. spread foundation

3. The most common type of foundation wall is the _____.

   a. L foundation
   b. T foundation
   c. mat foundation
   d. isolated foundation

SECTION TWO

2.0.0 CLAY BRICK AND CONCRETE PAVERS

Objective

Identify and explain the characteristics, uses, and installation techniques for clay brick and concrete pavers.

    a. Describe the various types of clay brick pavers.
    b. Explain how to install clay brick pavers.
    c. Describe the various types of concrete and interlocking pavers..
    d. Explain how to install concrete and interlocking pavers.

Trade Terms

Chamfered: Stone block or brick with beveled edges that do not go all the way across the edge or end of the unit.

Infiltration: The drainage of storm water and other runoff into the ground beneath a paved surface.

Mortarless paving: The placement of paving brick or block on a horizontal surface in some pattern to form a smooth, flat surface. No mortar is used to bond the units to the surface underneath or to each other.

Pavers: Brick, solid concrete block, or patterned concrete block that are used to build smooth, horizontal surfaces. Pavers are manufactured in many different thicknesses and shapes.

Paving materials include clay brick paver units, concrete pavers, and interlocking concrete pavers. Among these choices, some materials are cut to more precise sizes and shapes than others, and some materials are better suited to particular climates. However, for most pavers, the construction techniques and required bases used are similar. This section of the module describes the use of clay brick and concrete pavers for residential applications.

2.1.0 Identifying Clay Brick Pavers

Because of the various colors, patterns, and bonds available, clay brick has long been popular as a paving material. Clay brick is often used in and around residences and commercial structures because of its natural appeal (*Figure 9*). The warm colors and interesting textures and patterns of clay brick make it a popular choice for walks, terraces, patios, and interior floors.

A wide variety of clay brick is suitable for paving, but three main factors should be considered before making a selection: grade, size, and pattern. Other necessary considerations are the amount and type of traffic, freeze-thaw cycles, site drainage, and the design and appearance of the unit once it is laid. If clay brick is used on an exterior floor, it should be solid enough so that weather conditions cannot penetrate the unit. Also since these surfaces must bear foot traffic, the units should not have a rough texture. Clay brick pavers used for residential applications should meet the requirements of *ASTM* (American Society for Testing and Materials) *C902, Standard Specification for Pedestrian and Light Traffic Paving Brick*. Clay brick pavers used for vehicular traffic should meet the requirements of *ASTM C1272, Standard Specification for Heavy Vehicular Paving Brick*.

Salvaged or used clay brick should not be used for outdoor structures. The light-pink clay brick found in many older homes is generally too soft and should not be re-used. When removed from these homes and rebuilt into exterior structures, they quickly start flaking and break into pieces.

28202-14_F09.EPS

Figure 9 Clay brick pavers on a pedestrian plaza.

2.1.1 Sizes and Shapes

There are many sizes and shapes of clay brick for paving and floors. Shapes range from rectangular or square to hexagonal. Some of the most frequently used sizes and shapes are shown in *Figure 10*.

Clay brick pavers are made in thicknesses ranging from ½ inch to 2½ inches. Standard sizes from ½-inch- to 3½-inch-thick clay brick pavers are the most common. The ½-inch thickness is popular because it is easy to install and relatively inexpensive. Widths range from 3⅜ inches to 4 inches; lengths range from 7½ inches to 11¾ inches. There are also square shapes measuring 4, 12, and 16 inches on a side, and 6-, 8-, and 12-inch hexagonal units.

Clay brick sizes may be given as nominal or actual. Nominal sizes allow for a ½-inch mortar thickness between clay brick. Actual sizes follow the 4-inch modular system; both units may be used in mortarless paving.

Pattern bonds can be laid with or without mortar joints. Clay brick pavers with length dimensions exactly twice that of their width, such as 4 inches × 8 inches, should be laid without mor-

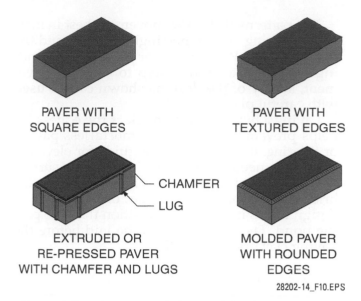

PAVER WITH
SQUARE EDGES

PAVER WITH
TEXTURED EDGES

CHAMFER

LUG

EXTRUDED OR
RE-PRESSED PAVER
WITH CHAMFER AND LUGS

MOLDED PAVER
WITH ROUNDED
EDGES

28202-14_F10.EPS

Figure 10 Popular sizes and shapes of clay brick pavers.

tar. Some examples of these patterns are basket weave, herringbone, running, Spanish, and stack mixed (*Figure 11*). The patterns shown are developed with clay brick having an exposed face of 4 inches × 8 inches actual size.

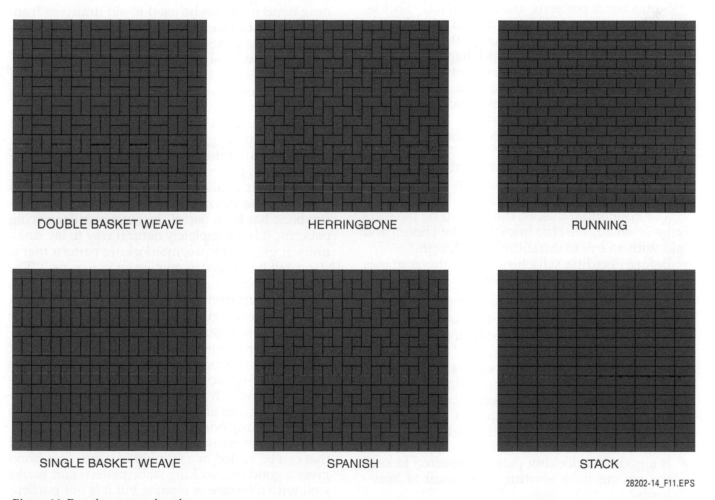

DOUBLE BASKET WEAVE

HERRINGBONE

RUNNING

SINGLE BASKET WEAVE

SPANISH

STACK

28202-14_F11.EPS

Figure 11 Popular pattern bonds.

The patterns that resist movement best in mortarless paving are the herringbone bond and the running bond. Walls, curbs, or planters surrounding the paved area also help to restrain movement, so any of the designs shown can be used with careful planning.

Laying a border of clay brick at the edges also helps prevent the shifting of mortarless paving or flooring. It also serves as a guide for elevation and slopes needed on the paving to drain surface water. The edging can be a soldier course of clay brick set in concrete or mortar, or some type of design that is in a different position than the paving bond. The edging should be laid before the paving clay brick is installed.

2.1.2 Clay Brick Paver Patterns

Joint patterns play an important part in the appearance of any clay brickwork. Because of gravity, clay brickwork on vertical surfaces is usually limited to patterns that have vertical and horizontal mortar joints. This is not a limitation in clay brick paving, where there is more opportunity for imaginative patterns.

Some bond patterns are directional. That is, they direct the eye along the line of unbroken joints. Examples of directional patterns are running bond and herringbone laid diagonally. Other patterns are static. This type of pattern does not give the impression of movement. Examples of static patterns are basket weave and stack bonds. When creating horizontal surfaces, try to use the movement in the pattern to direct the eye to a central point.

Whatever the paving situation, there is a color of clay brick and a paving pattern to suit it. Regardless of whether the structure is for commercial or residential use, or the area to be paved is large or small, the clay brick must be chosen and laid with an eye to durability and design.

Before deciding which paving pattern to use, consider the following questions:

- Is the clay brick size compatible with the scale of pattern desired, or should a second pattern more compatible with the size of the clay brick be used?
- Will the pattern involve special units or an excessive amount of cutting? Edge-cutting becomes less important as areas increase in size.
- Is a directional pattern required to emphasize a particular route, such as through a garden, or to emphasize a statue or other special feature?
- Is a good interlocking pattern required to keep the paving from shifting as a result of heavy loads?

Once these questions have been answered, a pattern can be selected. Some of the more popular bonds are running, stack, herringbone, basket weave, and circular.

The running bond is one of the simplest patterns to lay. It is a directional pattern. It can be used to emphasize a garden feature such as a waterfall, or to direct the eye toward doorways. It could, for instance, be used with sliding glass doors to visually unite the inside and the outside. It is also a good pattern to use for narrow paths where there is not enough room to make a large pattern feasible. The running bond resists movement in mortarless paving.

Running bond can also be laid nondirectionally for larger areas, such as patios and terraces within defined areas. It does not require any major clay brick cutting, and can be attractively edged with stack bond. If you are using a basket-weave or herringbone pattern over a large area, use a simpler pattern to break up the main pattern into smaller areas. This will often create a more interesting scale for a large area.

Although the clay brick may be laid flat or on edge, laying them flat is more economical. Running bond may also be used to aid drainage from a patio or a terrace. The water will flow in the direction of the joints.

Another very simple pattern that requires little cutting is the stack bond. It can be laid with standard mortar joints, or with sand in the joints and the clay brick closely butted together. It is, however, difficult to maintain the regularity of the pattern through a large area. Because of this, large areas are often broken into smaller areas using different patterns.

The herringbone bond also resists movement in mortarless paving. Over many centuries, herringbone has been one of the most popular paving patterns. It is a completely natural way to lay small units. It gives a secure, interlocking pattern that is static, but has a certain movement in it as well. The reason for the popularity of herringbone is that it is never tedious; from whichever direction you look at it, a different regularity of pattern results—either strong rows or zigzags (see *Figure 12*).

The herringbone pattern is appropriate for any size area. It can be used on a narrow path or to pave a large terrace. This pattern has many irregular edges, which will require more cutting. Because of this, the paved area is often edged with a row of running bond or a header course.

The basket-weave pattern is an adaptable bond that can be varied in a number of ways. This bond gives a good interlocking static pattern and works well with a terrace or a patio, but it is not particularly effective on a narrow path or walkway. It can be used in small patches, broken up with concrete or with bands of running or stack bond.

A 4-inch × 8-inch actual clay-brick size should be used for the basket-weave pattern to allow for proper joint alignment in a mortarless setting. The use of other clay brick sizes will require an excessive amount of cutting. The basket-weave pattern is one of the more difficult patterns to lay in both mortarless and mortar settings. Because of this, regardless of the type of setting, you may want to begin by laying the clay brick dry in order to establish the pattern. The clay brick in this pattern can be laid flat, on edge, or as a combination of both.

With its restful radiating lines and ever-increasing circles, a circular paving pattern can be one of the most satisfying to the eye. This pattern is best used in small areas, surrounding a tree, statue, birdbath, or some other favorite garden object. The pattern creates an instant focal point. If a circular pattern is laid on a sand bed with sand in the joints around a young tree, the pattern presents no problem when the tree begins to grow and thicken. The inner ring may be removed with no damage to the roots of the tree.

Laying a circular pattern is not as difficult as many people think. But the visual appeal of the pattern will be affected by the radius of the circle in relation to the length of the clay brick. With a small radius, the outer ends of wedge-shaped joints can look unsightly if they become very wide. It is a good idea to start the circles with half clay brick and then change to full clay brick as the radius increases, as shown in *Figure 13*. Alternatively, specialty pavers are available that are precut for circular patterns (*Figure 14*).

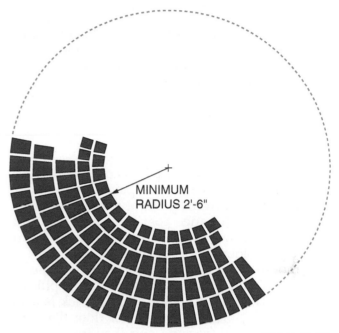

MINIMUM
RADIUS 2'-6"

28202-14_F13.EPS

Figure 13 Laying a circular pattern with standard clay brick.

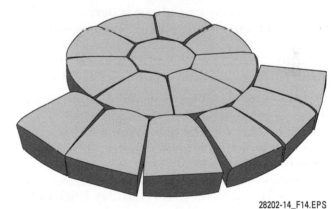

28202-14_F14.EPS

Figure 14 Precut circular pavers.

28202-14_F12.EPS

Figure 12 Herringbone pattern in a pedestrian area.

Clay Brick Roads

In the United States and some countries in northern Europe, clay brick paving has been used for roads bearing heavy traffic. Clay brick roads are more flexible than macadam and are still used in certain areas for steep hills. Lombard Street in San Francisco is probably the most famous clay brick-paved road in the world.

Today, because of the labor cost and craftsmanship involved in laying clay brick, and the development of high-speed automatic road-building equipment, clay brick has been replaced by concrete and asphalt for roadways.

These patterns are not the only ones that can be used. You should be able to produce many unusual patterns by combining or varying the standard patterns. Specialty pavers can be used to create additional patterns. In this respect, clay brick-paving craftsmanship has developed into a completely different trade relative to vertical clay brickwork.

2.2.0 Installing Clay Brick Pavers

The popularity of clay brick pavers can be attributed to the flexibility of installation in addition to their appearance and durability. Clay brick pavers can be installed as an interior or exterior surfacing; with or without mortar joints; over a sand or concrete base, or over a wood subfloor.

2.2.1 Interior Floors

Interior clay brick flooring is recommended for any area in a house. Clay brick pavers offer a floor surface that is extremely durable and therefore has a low maintenance cost. The patterns and color can be varied to enhance any interior. It can be used in residential or commercial structures.

A smooth, hard clay brick has a good appearance and does not absorb moisture from spills or mopping. The natural color of clay brick is also attractive. Clay brick in an interior floor does not have to withstand freezing and thawing. The decision about size is important; if the floor is laid mortarless, then the clay brick should be actual 4 inch × 8 inch. A clay brick floor is permanent; this must be kept in mind when choosing size and color.

The following are the principal elements to consider when installing clay brick pavers:

- *Base or subgrade* – The principal support. It may be a concrete slab or well-tamped earth.
- *Cushion* – An intermediate layer, usually a layer of sand, which aids leveling and placing of the clay brick.
- *Clay brick* – The wearing surface.
- *Joints* – The space between the clay brick or the material that will fill the space.

The best base for floors in an interior setting is a concrete slab. The slab should include any reinforcing or insulation as required by the local soil and climatic conditions.

There are two ways of installing clay brick flooring: laying with mortar (mortared) or without mortar (mortarless). Mortared clay brick paving is the traditional method and has been the most popular over the years. Mortarless paving, however, is becoming more popular. Both methods have advantages and disadvantages. There are good reasons for the increased demand for mortarless paving: it offers ease of construction, and there are no mortar joints to crack or break up. Mortarless clay brick can also be laid faster and with less cost.

Interior floors are often placed in a bed of mortar and have mortar joints. The base is usually a concrete slab. The cushion is a bed of mortar about ⅜- to ½-inch thick.

Both the cushion and the joints are created as each individual unit is leveled and placed. Cleaning may be unnecessary if excess surface mortar is brushed or wiped away with burlap rags as it begins to stiffen. At a minimum, this will make cleaning easier. The face mortar joints are tooled with a flat jointer to ensure maximum hardness and density.

Thin clay brick pavers can be used over any type of rigid base, such as a concrete slab. These pavers are normally installed using a thin-set method or a thin mortar bed. With the thin-set installation, an epoxy adhesive is troweled into the base and the paver is placed, allowing for the proper joint spacing. Protect the surface with a release agent. Grout designed for this purpose is then floated over the pavers, filling in all the mortar joints. The surface must be carefully washed to remove all grout from the face of the pavers.

With any installation using the thin-set method, the subgrade must be very accurately leveled. Any imperfections in the subgrade will be seen on the finished surface.

Mortarless clay brick floors are faster, easier, and cheaper to place than floors laid in mortar. One obvious disadvantage is that mortarless floors tend to shift. A pattern, such as a herringbone or running bond, should be selected for mortarless floors to limit any shifting of the clay brick.

The bedding is most often mortar with a concrete base. Be sure to use clay brick with actual dimensions of 4 inches × 8 inches. After the pattern is laid, fine sand should be swept over the surface to fill any cracks. The floor is then sealed and waxed if desired.

2.2.2 Exterior Surfaces

Clay brick is an excellent choice for exterior surfaces because of its durability and low maintenance (*Figure 15*). It is important that the clay brick selected for exterior pavers have low-absorption qualities and be extra hard and well-burned, since these areas are exposed to moisture and weathering. They should meet the requirements in *ASTM C902* and *C1272*.

Figure 15 Exterior clay brick floor.

Generally, exterior paving surfaces follow the same construction as interior floors, requiring a base, a bedding, and the final clay brick layer. There are other things to consider that influence design and construction. These include the following:

- The presence of moisture
- The possibility of freezing and thawing
- The chance of movement, particularly at the edges

Good drainage must be provided because exterior clay brick paving is subject to changing weather and moisture conditions. When exposed to excessive moisture, clay brick is susceptible to efflorescence, stains, and the growth of fungus or mold. Regular freezing and thawing may cause clay brick to disintegrate. In cold climates, water collecting between clay brick can freeze and push them out of place. Fungi can grow in areas of high humidity and/or with a high water table.

Walks, patios, and other surfaces should be sloped to prevent moisture damage. They may be sloped either to one side or to both sides by making a crown in the approximate center of the paved area. Patio floors should be sloped away from the adjoining structure or any other place that can trap water. Otherwise a gutter must be provided. The slope must be adequate to drain the entire area—usually ⅛ to ¼ inch per foot. However, large areas may need a greater slope. Where necessary, gutters can be placed at the edge of clay brick pavements to prevent adjacent areas from draining onto the paving.

In areas having a high water table, the main moisture problem may be the movement of water upward through the soil rather than downward drainage. Water from the soil is often heavily laden with soluble salts and may contribute to efflorescence or cause staining.

Some restraint is necessary at the perimeter of clay brick pavings because they tend to spread or shift at the edges, even under normal traffic and weather conditions. The clay brick edging is laid after the concrete slab is placed. It bonds better if laid in a soldier or rowlock course (*Figure 16*) and bonded to the edge with mortar. This edging can also be used as a guide for elevation and slopes.

Type S mortar should be used where there is a concrete base. Masonry cement, with the proper

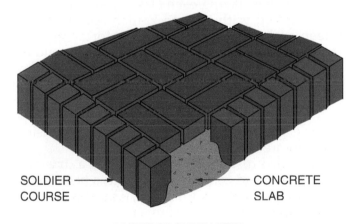

SOLDIER COURSE   CONCRETE SLAB

UNDERGROUND VIEW

Figure 16 Edging.

Dampness Breeds Mold

Molds or fungi can develop in damp areas. They are unsightly. They may also cause a dangerously slippery surface when wet. Where conditions are conducive to the growth of molds or fungi, the mason should use a capillary break, such as a layer of gravel, to prevent the upward flow of moisture. Clay brick pavers should not be placed directly on the existing soil in these conditions.

amount of lime already added, may be substituted. There are three basic methods of installing clay brick paving with mortar joints. The first is the conventional way, spreading mortar on the base and applying the mortar to the unit using the trowel.

You should butter the clay brick with mortar, being sure to form a solid head and cross joint. Then, lay the clay brick in the mortar bed and level it. Slope the base to achieve proper water drainage.

The most accurate method of leveling clay brick in the mortar bed is by using a long, wooden straightedge. Level the clay brick from one side of the edging to the other. Tap the clay brick down or bedding up until they are all touching, with no high or low spots. A rubber tip on the handle of the trowel is often used when tapping on the clay brick. A standard crutch tip can be used. This prevents denting and splitting of the trowel handle. A rubber mallet may also be used to tap the clay brick into position (*Figure 17*).

On a large or long floor, a nylon line can be stretched tightly before the clay brick is laid to serve as a guide for the work. Trigs can be used to keep the line stable. Care must be taken not to lay to a sagging line, or the finished surface will have a low spot where water will collect.

The second method of laying mortared clay brick involves laying each clay brick on a mortar leveling bed with a ⅜- to ½-inch space left between each unit for head joints. Mortar grout is then placed between the spaces. Care should be taken not to stain the surface of the grouted clay brick because it will be difficult to clean. This is especially true if the clay brick has a high absorption rate. It is recommended that the exposed face of the clay brick be lightly coated with a releasing agent to prevent grout stains.

The third installation method uses a dry mixture of portland cement and sand as a base, in the same proportions as for grout. The clay brick is laid on a damp cushion of this mixture. Then, the joints between the units are filled with the same mixture. Clean any excess mixture from the paving surface with a brush. Dampen the surface with a fine mist of potable water until the joints are thoroughly moist. Keep paving damp for two or three days to allow for proper curing.

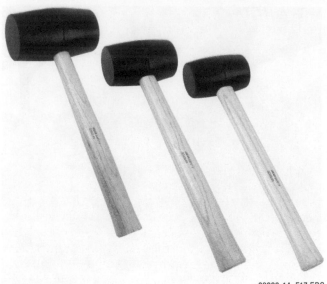

28202-14_F17.EPS

Figure 17 Rubber mallets.

Only mortarless paving should be placed over a gravel or sand base. The base for paving laid in mortar should be a concrete slab.

The key to mortarless paving is the proper preparation and compaction of the subgrade and base before laying any clay brick. Ample drainage and slope must be provided to drain water from the surface. Constructing the proper edging for mortarless clay brick is even more important than for mortared clay brick surfaces. The edging should be set in mortar or concrete, because mortarless clay brick tends to shift underfoot and the edging keeps the main floor from moving.

Except where there is a special subsurface drainage problem, sand can be used as a base. A leveling cushion of 1 to 1 ½ inches is spread out on the tamped earth and screeded off with a wooden straightedge, much like concrete is screeded. Low spots are filled after the screeding is done.

A layer of geotextile fabric is often placed over the base before the sand cushion is spread. There are several advantages to using geotextile fabric, for example:

- It helps prevent grass and weeds from growing up between the joints.
- It confines the sand so that it does not erode over time.

It is worth the extra time and money to install a layer of this material when paving without mortar.

Easy Does It

Spread only enough mortar for the bed joints to settle the clay brick in position firmly without too much tapping. Although the trowel handle may be protected, tapping too hard or too much may crack the clay brick.

The paving units are placed on the leveling bed and tightly fitted together while maintaining proper elevation and slope. Sand is then swept into the joints. After the sand is swept into place, it should be dampened and set with a fine spray of potable water. After the sand settles, more sand may need to be added.

2.2.3 Cleaning and Finishing Clay Brick Pavers

When a mortared clay brick floor or walk is laid, it is almost impossible to prevent mortar from getting on the face of the clay brick. To keep this to a minimum, you should work with care. Most of the mortar can be removed by wiping the clay brick with a trowel and brush. However, not all of the mortar and mortar stains can be removed by simple brushing before the mortar is completely set. The clay brick must be cleaned after the laying is complete.

Clay brick pavers typically need cleaning after an installation is completed. This can be done by washing with potable water and a sponge or by brushing with a proprietary cleaner. Usually, washing with water using a sponge or brush will be sufficient to remove fresh mortar splashes and spilled concrete.

> **WARNING!**
>
> Proprietary cleaners are harmful to human skin and especially dangerous to the eyes. They are capable of damaging clothing. Always wear appropriate personal protective equipment when using proprietary cleaners. Refer to the safety data sheet (SDS) for the proprietary cleaner for information on hazards and prevention.

When cleaning interior floors, extra precautions are necessary. At times, sufficient ventilation must be provided. Before using any proprietary cleaning solution, read the label and safety data sheet (SDS). Rubber gloves, goggles, and respirators may be required. Before applying any solution, thoroughly wet the masonry by spraying it with a fine mist of potable water. This layer of water reduces the absorption of proprietary cleaner into the clay brick—and into the mortar joints in particular—to reduce the possibility of discoloration and staining. Solutions should be applied with a stiff fiber brush. Scrub small areas at a time. Limit your area to about 100 square feet maximum. Rinse thoroughly with potable water.

Clay brick should be allowed to set for a few days to make sure the mortar has properly cured before finishing. Exterior clay brick is generally only cleaned. No other finishing is required because most finishes will not stand up to exterior exposure. Interior pavers are typically finished after cleaning. The finish provides absorption protection, easier cleanup, and a finished appearance.

Interior surfaces can be finished with a sealer or a wax. A sealer protects the clay brick from absorption. It is easily placed by mopping or spraying. Silicone sealer, for example, has a long life, prevents absorption of water, and is available in a solvent-based or water-based solution. Solvent-based sealers dry faster than water-based ones. Waxing provides the higher-gloss surface required in some installations. Waxed surfaces are the easiest to clean and require periodic refinishing in order to maintain the gloss finish.

2.3.0 Identifying Concrete Pavers

Concrete pavers are manufactured to lock into a tight pattern that will not shift under heavy foot or vehicular traffic. Concrete pavers are available in a variety of sizes and shapes, such as cobblestone, chamfered, multiweave, and grass pavers. Most concrete pavers used for foot traffic and limited vehicle traffic (such as driveways) are $2\frac{3}{8}$ inches thick, about the same thickness as a standard clay brick. Concrete pavers used on streets and for industrial applications should be at least $3\frac{1}{8}$ inches thick. The pavers shown in *Figure 18* are suitable for high-traffic pathways.

Concrete pavers are highly resistant to freeze-thaw cycles and are not damaged by salt-based deicing materials or petroleum-based products. They are also highly resistant to abrasion, skids, and cracking. Concrete pavers do not require curing once installed and can handle traffic immediately. Their design permits easy removal

Allow for Expansion

Some provision must be made for the movement of paving or flooring when laid in mortar. Expansion joints will allow for such movement. The expansion joint for large areas should be located parallel to curbs and edgings, at 90-degree (i.e., right-angle) turns, and around any type of interruption or break. Fillers for expansion joints must be compressible and made from materials that do not rot. Generally, pavement joint fillers are either made solid or preformed of materials such as butyl rubber, neoprene, or other elastic compounds.

Figure 18 Interlocking concrete pavers used on a street.

Concrete pavers are installed so that each unit is interlocked with its neighbors for structural stability and the ability to transfer loads. Proper installation requires three types of interlock: vertical, horizontal, and rotational (see *Figure 19*). A properly installed concrete paver, therefore, will not rise above or sink below other pavers, slide from side to side, or rotate. Some pavers have small tabs on one side and one end to ensure consistent joint spacing. Some manufacturers make special edging pieces to fill the voids and create a straight edge. Otherwise you would have to cut the pavers or fill the voids with mortar to make straight edges.

Permeable interlocking concrete pavement (often abbreviated PICP) systems use concrete pavers along with permeable joint materials such as crushed stone. PICP is often used in sustainable development projects because of its ability to permit infiltration of storm water below the road or sidewalk surface, thereby reducing storm-water runoff, and its ability to reflect radiant heat from sunlight.

for access to underground utilities, and the pavers that have been removed for access to utilities can be used again once the underground work is complete. They are also available in many colors, textures, and finishes.

ASTM C936, *Standard Specification for Solid Interlocking Concrete Paving Units*, is the standard that applies to concrete pavers in the United States. The specification requires a minimum average compressive strength of 8,000 psi, a maximum average absorption of 5 percent, the ability to withstand a minimum of 50 freeze-thaw cycles without losing an average of more than 1 percent of its material, and ability to pass a battery of abrasion-resistance tests.

2.4.0 Installing Concrete Pavers

The installation process for concrete pavers can be broken down into a series of tasks that must be completed in order. The steps may vary depending on the job, but in general they are as follows:

- Planning and layout
- Soil-subgrade excavation and compaction
- Geotextile application
- Subbase and base aggregate spreading and compaction
- Edge-restraint construction
- Bedding-sand placement and screeding
- Concrete paver placement

Don't Let Sand Block Drainage

A layer of sand is often used directly under clay brick to aid leveling and placement. If the sand is placed over gravel, it will sift down into the larger material, causing uneven settling and blocking drainage. To prevent this, add a layer of small-sized stone screenings, gravel, or roofing felt between the gravel and the sand cushion.

Ensure Proper Drainage

In areas where groundwater is present or where poor drainage exists, drain tiles and gravel may be placed under the base to carry the water away.

After the gravel is placed, a leveling bed of stone screenings or gravel about 1 inch thick is laid on top. The drain tile is installed at the point in the base where the worst drainage problem is located. The drain tile is first covered with a strip of felt roofing paper to keep out dirt, and then it is covered with gravel. The drain tile must lead to a drain area away from the finished area to carry the water away. If it does not, the tile becomes a trap for water.

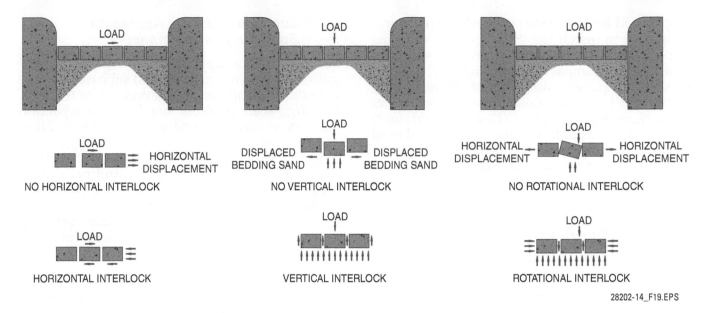

NO HORIZONTAL INTERLOCK   NO VERTICAL INTERLOCK   NO ROTATIONAL INTERLOCK

HORIZONTAL INTERLOCK   VERTICAL INTERLOCK   ROTATIONAL INTERLOCK

28202-14_F19.EPS

Figure 19 The three types of interlock for a concrete paver.

Larger jobs may require the use of mechanical equipment to place pavers. This involves motorized or hydraulic clamps that can pick up a layer of interlocked pavers and place them in the desired location (see *Figure 20*). The following sections will focus on the techniques required for the manual installation of concrete pavers.

2.4.1 Planning and Layout

Always begin by identifying underground and overhead utilities to ensure that the installation process does not interfere with them. Mark the excavation area with stakes, taking care to set them back slightly from the area to be excavated, to ensure that they are not accidentally removed by the excavator. Mark the grade using marks on the stakes or by running a line. For roads, the minimum slope is 1.5 degrees.

2.4.2 Soil-Subgrade Excavation and Compaction

During the excavation process, inspect the soil and remove any organic materials such as roots, rocks, and debris. Replace organic material with clean and compacted backfill material. Stabilize low and wet areas with a layer of crushed stone or cement. Wrap drainage pipe in geotextile fabric

28202-14_F20.EPS

Figure 20 Hydraulic clamp used for picking up a layer of pavers.

Test First

Proprietary cleaners can stain masonry. Test a small patch before cleaning the entire area with a proprietary cleaner. Choose an area that is hidden or easily replaced. Allow the test area to dry to ensure that there is no discoloration.

as specified by the manufacturer's instructions, being sure to install the drainage pipe with the proper slope. Then ensure that the soil subgrade is sufficiently compacted to minimize the settlement of the pavers. *ASTM D698, Standard Test Methods for Laboratory Compaction Characteristics of Soil Using Standard Effort (12 400 ft-lbf/ft³ ([600 kN-m/m³])*, is the applicable standard for soil used as fill for concrete pavers. During the excavation and compaction process, monitor soil humidity and compaction to ensure that it complies with this specification.

2.4.3 Geotextile Application

The project specifications may call for the installation of a layer of woven or nonwoven geotextile fabric to serve as a separation layer to protect concrete pavers from saturated soil, freeze/thaw cycles, or rutting from heavy loads. Geotextile fabric is often used to extend the loadbearing capacity of the base. *Table 1* lists the separation requirements for geotextile fabrics as specified by the Interlocking Concrete Pavement Institute (ICPI). The values in *Table 1* are expressed in terms of the minimum average roll value, or MARV. This is a statistical calculation of the fabric's physical, mechanical, hydraulic, and strength properties.

When installing geotextile fabric, turn up the fabric along the sides of the opening so as to allow the sides of the base layer to be covered when the base layer is placed. Ensure that the fabric is free of wrinkles before placing the subbase and base layer. Do not allow vehicles to drive on the fabric after it has been laid and smoothed, to avoid creating ruts and wrinkles.

2.4.4 Subbase and Base Aggregate Spreading and Compaction

Refer to the local applicable code for grading requirements in your area. Spread subbase and base material and compact it in layers of 4 to 6 inches. On large-scale projects, use heavy compaction equipment such as vibratory rollers (*Figure 21*) to compact the aggregate material.

ICPI recommends a minimum 4-inch base thickness for sidewalks, patios, and pedestrian areas after compaction if the concrete pavers are being installed over well-drained soils. Bases for driveways should be at least 6 inches thick after compaction, and bases for streets should be at least 8 inches thick after compaction. Add between 2 and 4 inches to the base thickness if the pavers are being installed in a cold climate or in areas with soil that is constantly wet or weak.

Ensure that the base, including any subbase, has been compacted according to the appropriate standard throughout the excavation, but pay special attention to the level of compaction near edge restraints, catch basins, and utility structures. It may be necessary to apply geotextile fabric over these features to prevent bedding sand from migrating.

Aggregate should be at the required optimum moisture level when compacted, to avoid additional settling or expansion after compaction. The finished surface of the compacted aggregate in the subbase, if any, and base should be able to prevent the migration of bedding sand into it. A final course of fine material or a coat of bitumen may be required to achieve this.

Table 1 Geotextile Fabric Separation Requirements

Geotextile Class		Class I[a]		Class II[a]		Class III[a]	
Elongation	ASTM Standard	< 50%	> 50%	< 50%	> 50%	< 50%	> 50%
Grab Strength[b]	ASTM D4632	315 lb [1400 N]	202 lb [900 N]	247 lb [1100 N]	157 lb [700 N]	180 lb [800 N]	112 lb [500 N]
Sewn Seam Strength[b,c]	ASTM D4632	283 lb [1260 N]	182 lb [810 N]	223 lb [990 N]	142 lb [630 N]	162 lb [720 N]	101 lb [450 N]
Tear Strength[b]	ASTM D4533	112 lb [500 N]	79 lb [350 N]	90 lb [400 N][d]	56 lb [250 N]	67 lb [300 N]	40 lb [180 N]
Puncture Strength[b]	ASTM D6241	618 lb [2750 N]	433 lb [1925 N]	495 lb [2200 N]	309 lb [1375 N]	371 lb [1650 N]	223 lb [990 N]
Permittivity[b,e]	ASTM D4491	0.02 sec⁻¹					
Apparent Opening Size	ASTM D4751	0.024 in [0.60 mm] maximum average roll value					
Ultraviolet Stability	ASTM D4355	> 50% after 500-hr exposure					

[a] The severity of the installation conditions generally dictates the required geotextile class. Class I is the most severe and Class III is the least severe.
[b] All numeric values represent minimum average roll value (MARV) in the weaker principal direction.
[c] When sewn seams are required.
[d] The required tear strength for woven monofilament geotextiles if 250 N.
[e] Default Value. Permittivity of the geotextile should be greater than the soil.

Figure 21 Vibratory roller used to compact subbase and base aggregate.

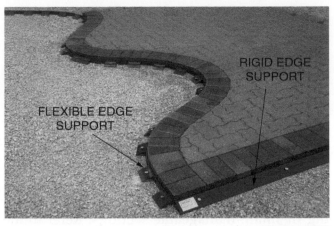

Figure 22 Rigid and flexible edge restraints for concrete pavers.

2.4.5 Edge-Restraint Construction

Once the subbase and base aggregate has been spread and compacted, but before the placement and screeding of bedding sand, install edge restraints around the edges of the pavement area. Edge restraints provide resistance to lateral loads transmitted by the concrete pavers. They can be made from a variety of materials, including concrete, plastic, steel, and aluminum. They are available in rigid and flexible styles (*Figure 22*).

Ensure that edge restraints are placed at the correct level. This is particularly important if the restraint is to be used to mark the height of the bedding sand for screeding. Always verify the height of restraints prior to laying the bedding sand.

2.4.6 Bedding-Sand Placement and Screeding

Spread bedding sand and screed it to a nominal thickness of 1 inch. Particles of bedding sand should be symmetrical and sharp, and the sand should be washed and free of foreign material. Begin by placing screed pipes or rails on the bed and fill the pavement area with sand to the level of the screed pipes or rails. Then use a level board to screed the bedding sand. When this is done, remove the screed pipes or rails and fill in the resulting gaps with bedding sand. Bedding sand should not be disturbed once it has been screeded.

Never use sand that is frozen or saturated. Do not use sand intended for use in masonry mortar, or other materials such as crushed limestone or stone dust. Do not use bedding sand to compensate for unevenness in the subbase or base.

2.4.7 Concrete Paver Placement

Finally, after the other preparations have been made, the concrete pavers can be placed on the bedding sand. If the pavers come with spacing ridges, the pavers can be butted up against each other. Otherwise, maintain a consistent joint width between pavers. Joints should be between $\frac{1}{16}$ inch and $\frac{3}{16}$ inch. Use cut pavers to fill gaps along the edge of the pavement area.

Once a section of pavers has been placed, compact the pavers to set them into the bedding sand and to force the bedding sand into the joints along the bottom of the pavers. Then sweep dry joint sand into the joints and compact the pavers again. Joint sand is often finer than bedding sand, to ensure that it fills any gaps left by the bedding sand. Remove excess bedding sand following the compaction. The final elevation of the pavement surface should not vary by more than $\frac{3}{8}$ inch every 10 feet. Follow the local applicable code for the height of pavers near catch basins, utility covers, drain channels, and other ground features or installations.

The ICPI recommends using a 45- or 90-degree herringbone pattern on streets and other areas with vehicular traffic. This pattern provides the maximum possible loadbearing support.

2.4.8 Cleaning and Finishing Concrete Pavers

When the pavers have been installed and compacted, use liquid or dry joint-sand stabilizer to secure the joint sand in the joints between pavers. Joint-sand stabilizer also helps make pavers easier to clean and can prevent staining. Liquid stabilizers are applied with a low-pressure sprayer, while dry stabilizers are swept into joints and then wetted to activate them. Water- and solvent-based surface sealers are liquids that provide additional protection once the joint-sand stabilizer has been added. Apply sealers to the surfaces of concrete pavers according to the manufacturer's instructions. Most sealers darken and enhance the colors of concrete pavers (*Figure 23*), and they also provide additional stabilization to joint sand.

Over time, concrete pavers can require cleaning due to stains, efflorescence, dirt, and wear from normal use. Use proprietary cleaners designed for concrete pavers to remove stains. Remove stains from the lowest area of the pavement area and work up to the highest, proceeding in small, manageable sections. Refer to the manufacturer's specifications for the proprietary cleaner and for the concrete paver to identify how to clean certain types of stains, and to learn what precautions to take to ensure the paver is

28202-14_F23.EPS

Figure 23 Concrete pavers before and after the application of a liquid surface sealer.

not damaged or discolored. Always wear appropriate personal protective equipment when cleaning concrete pavers.

Additional Resources

ASTM D698, Standard Test Methods for Laboratory Compaction Characteristics of Soil Using Standard Effort (12 400 ft-lbf/ft³ ([600 kN-m/m³]), Latest Edition. West Conshohocken, PA: ASTM International.

ASTM C902, Standard Specification for Pedestrian and Light Traffic Paving Brick, Latest Edition. West Conshohocken, PA: ASTM International.

ASTM C936, Standard Specification for Solid Interlocking Concrete Paving Units, Latest Edition. West Conshohocken, PA: ASTM International.

ASTM C1272, Standard Specification for Heavy Vehicular Paving Brick, Latest Edition. West Conshohocken, PA: ASTM International.

Tech Spec Number 2, *Construction of Interlocking Concrete Pavements.* 2011. Herndon, VA: Interlocking Concrete Pavement Institute. **www.icpi.org**

Technical Note TN14, *Paving Systems Using Clay Pavers.* 2007. Reston, VA: The Brick Industry Association. **www.gobrick.com**

Vocational Skills Training for Segmental Paver Installation, First Edition. 2003. Stephen Jones. Prior Lake, MN: Pave Tech, Inc.

2.0.0 Section Review

1. The mortar thickness that manufacturers plan for between nominal sizes of clay brick for paving and floors is _____.

   a. ¼ inch
   b. ⅜ inch
   c. ½ inch
   d. ¾ inch

2. When installing clay brick on exterior surfaces over a concrete base, the mortar that should be used is _____.

   a. Type M
   b. Type N
   c. Type O
   d. Type S

3. PICP stands for _____.

   a. permeable interlocking concrete pavement
   b. permanently installed concrete pavement
   c. permeable interchangeable cement pavement
   d. porous interstitial concrete platform

4. ICPI recommends that when installing concrete pavers on streets, the thickness of the base after compaction should be at least _____.

   a. 10 inches
   b. 8 inches
   c. 6 inches
   d. 4 inches

SECTION THREE

3.0.0 MASONRY STEPS, PATIOS, AND DECKS

Objective

Lay out and build steps, patios, and decks made from masonry units.

   a. Describe the various types of steps.
   b. Explain how to recognize patterns and tread designs.
   c. Explain how to build a concrete base.
   d. Explain how to set clay brick in steps.
   e. Explain how patios are constructed.
   f. Explain how decks are constructed.

Performance Tasks 1 and 2

Lay out and construct a set of steps with three risers.

Lay out and construct a 5-foot by 7-foot clay brick patio section.

Steps are the most common method of moving from one level to another. If a clay brick walk traverses a sharp change in grade, a set of steps may be needed. You could build a set of cast-in-place concrete steps, or if the dimensions are right, use precast concrete. However, this may detract from the beauty of the surrounding clay brickwork.

Steps for private homes are typically a minimum of 48 inches wide. Some codes allow 30- and 36-inch widths. However, steps should be at least as wide as the door or walkway they serve. A landing is desirable to divide flights of more than 5 feet. The landing should be no shorter than 3 feet in the direction of travel. The top landing should be no more than 7½ inches below the door threshold. For maximum stability and economy, clay brick steps should be installed in a mortar bed over a concrete base, even if the walk is dry-laid paving on a sand base.

To build clay brick steps, you need to figure the total height the steps must climb, called the rise, and the total length the steps traverse, called the run. The rise and run of the steps must then be broken down into the number and height of the vertical risers and the depth of the horizontal treads. Certain rules of thumb dictate the relationship between riser height and tread depth based on what is safe and comfortable for walking. *Table 2* gives typical measurements for risers and treads.

Table 2 Typical Measurements for Stairs

When Riser Is:	Tread Should Be:
4" to 4½"	18" to 19"
5" to 5½"	16" to 17"
6" to 6½"	14" to 15"
7" to 7½"	10" to 11"

Paving sidewalks, patios, and driveways with masonry units are among the easiest masonry work (*Figure 24*). The variety of materials available that were described in the previous section and the combinations possible are almost unlimited. Since most on-ground construction using pavers is similar, this section will cover techniques using the patio as a typical structure. Decks require some special considerations.

3.1.0 Identifying Masonry Steps

Steps can be designed to fit any change of grade. The final design will depend on the total height and the run available to achieve the change in elevation. *Figure 25* shows a clay brick staircase. The risers and treads are arranged in an attractive pattern that fits with the overall clay brickwork.

There are many ways to create clay brick risers and treads, such as using different thicknesses of clay brick, laying them flat, setting them on edge, or varying the mortar joint thickness. The treads can be clay brick, as shown in *Figure 26*, or topped with stone, concrete slab, or other material. These options provide some flexibility in achieving the

28202-14_F24.EPS

Figure 24 Sidewalk and patio.

---

required dimension so that the risers add up to the correct overall height. The same options can be used in achieving the desired length of the tread.

For flights of steps less than 30 inches high, make each riser 7½ inches or less and each tread at least 11 inches. For flights of steps higher than 30 inches, risers should be 6 inches or less and treads at least 12 inches. The sides of the steps can be left as exposed concrete or covered with clay brick. If you want to cover the sides with clay brick, simply make the concrete base thinner by the width of two clay brick plus two mortar joints.

These guidelines are a general approach to the design of clay brick steps. With a garden step you have some leeway in the step design, as there are often no hard boundaries defining the top and bottom of the steps.

3.2.0 Identifying Patterns and Tread Designs

Steps should be at least as wide as the sidewalk leading up to them. A width that is a multiple of 8 inches will allow the use of whole clay brick. *Figure 27* shows two tread designs that are divisible by eight, making use of the modular standard.

Two 3⅜-inch-wide clay brick laid flat plus two ⅜-inch mortar joints equals 8 inches. Three 2¼-inch-thick clay brick laid on edge plus three ⅜-inch mortar joints is also nominally 8 inches. If you draw a plan of the treads, you will be able to figure the number of clay brick you will need.

3.3.0 Building a Concrete Base

There are two ways to build a concrete base for clay brick steps. A 4-inch-thick slab reinforced with wire mesh can be used to support tiers of clay brick stacked and mortared to form the steps. If you need only two or three steps, use a flat slab, but this method takes a lot of clay brick.

For a more economical installation, a 4-inch-thick stepped base reinforced with wire mesh is used to support a single layer of clay brick pavers. The concrete base is built in the same way as a concrete step, but you must account for the di-

28202-14_F25.EPS

Figure 25 Clay brick staircase.

28202-14_F26.EPS

Figure 26 Simple clay brick steps .

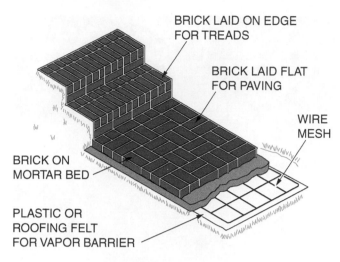

BRICK LAID ON EDGE
FOR TREADS

BRICK LAID FLAT
FOR PAVING

WIRE
MESH

BRICK ON
MORTAR BED

PLASTIC OR
ROOFING FELT
FOR VAPOR BARRIER

BASKET WEAVE PATTERN

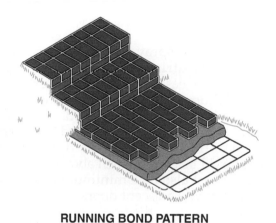

RUNNING BOND PATTERN

28202-14_F27.EPS

Figure 27 Common stair patterns.

mension of the clay brick when forming the base. The bottom and top treads of the concrete base are flush with the top of the 2-inch sand base for the walkway at either end.

3.4.0 Setting Clay Brick in Steps

Once the concrete base is built, the remaining work requires laying the clay brick in the correct pattern and keeping the joints a consistent size. Carry out the following steps to lay the clay brick:

Step 1 To check the size of the base and the spacing of the pavers, lay out the units without mortar on at least two steps. Adjust the width of the mortar joints as necessary to get the best fit.

Step 2 Clean off the concrete surface and apply a mortar bed using a Type S mortar. This setting bed will hold the pavers in place

and help to compensate for minor irregularities in the concrete surface. Start with the bottom step and mortar one tread at a time.

> **WARNING!**
>
> Mortar is caustic. Avoid getting mortar on your skin.

Step 3 Butter the sides of the pavers with a trowel, and place them on the mortar setting bed, forming joints that correspond to the dry bond. Cut off excess mortar with the edge of the trowel. Check the surface with a level, and tap the units gently with the trowel handle, if necessary, to bed them in the mortar.

Step 4 The joints are ready for tooling when the mortar is thumbprint hard. Using a concave jointer, tool the short joints first, and then the long joints. Then finish the remaining steps. Mortar the next tread and lay the clay brick as previously described until you finish the top step.

Step 5 Brush the surface with a stiff brush to remove mortar drips and dust. Use a plastic or wooden scraper, then a brush to remove large mortar splatters. If necessary, clean the complete project with a suitable proprietary cleaner. A ratio of 1:10 is sufficient.

> **WARNING!**
>
> Proprietary cleaners can cause chemical burns to skin. Wear appropriate personal protective equipment to protect your skin, eyes, and lungs.

3.5.0 Constructing Patios

The construction of a patio requires several design considerations, including slopes, widths, and drainage. These factors may also determine what type of paver you use for the project. Patio pavers can be installed with and without mortar. Either method can be used to create a durable, long-lasting surface. Regardless of the method used, the final job will only be as stable as the subgrade underneath the sand or mortar. Subgrade work for patios should be done as carefully as work for footings and other substructure support.

3.5.1 Materials and Patterns

Dry-laid pavers do not require mixing mortar or building a concrete foundation. The required materials include a suitable base material, a bedding material, and the pavers. Gravel is often used as the base material. Sand also makes a good bedding material.

If the pavers are to be laid on a concrete slab, mortar will be needed to bond the pavers to the slab. You will need to build the slab first, using cement, aggregate, and sand according to the local codes and site requirements. Construction of the slab will require forming.

The pattern design and environmental requirements will determine the type of paving material to use. A variety of pavers is available. Most are easy to use and can be handled with normal masonry tools and techniques. Clay brick and concrete masonry units are addressed in this module. Other materials commonly used are flagstone, slate, rock, and wooden block.

3.5.2 Excavation

Drainage is the most important factor to consider in preparing the subgrade. Water must be carried away from the paved area. The subgrade should be uniform, hard, free from foreign matter, and well-drained. The best masonry paving installations are made by removing all organic matter such as grass, sod, and roots before preparing the subgrade (*Figure 28*). Loosen hard spots. Tamp the area to provide the same uniform support. Dig out mucky spots and fill them with solid material similar to the rest of the subgrade. Granular material can also be used as fill. Either must be compacted thoroughly. All fill material should be uniform and free of vegetative matter, large lumps, large stone, and frozen soil.

Excavation of the area should not start until it has been properly laid out. This involves determining the proper dimensions of the patio and its finished elevation and slope. These requirements are usually defined in the residential plans.

Measure and lay out the size and shape of the patio using wooden or metal stakes and string. Set the stakes adjacent to any existing construction to mark the edge of the paving. Set stakes for the outside corners a little beyond the paving edge. Remember to account for the thickness of any edging material. Corners should be checked for square using the 3-4-5 method, which was explained in the *Masonry Level One* module *Measurements, Drawings, and Specifications*.

Once the layout has been completed and checked, you can transfer the alignment to the

28202-14_F28.EPS

Figure 28 Excavation of area for sidewalk.

ground by using spray paint or by sprinkling sand under the string lines. Then, remove the string lines for easier digging. The excavation method will depend on the size of the area, the type of soil, and the depth required.

Several factors must be considered to determine the depth of the excavation. The finished paved surface should be about 1 inch above the adjacent ground. It must also be a minimum of 1 inch below the sill of any adjacent doors. The patio must be sloped away from the house or other adjacent structures to provide proper drainage. This slope should be 1 inch for every 4 feet of length. Plan your excavation depth to these parameters, considering the gravel subbase, the sand bed, and the paver thickness below the ground.

Excavate along the paint or sand marking lines. Remove all grass, sod, roots, and large rocks to the depth necessary to accommodate the thickness of the base, sand bed, and pavers.

Although compacted soil can be an adequate subbase, the patio will last much longer and require less maintenance when built on a gravel subbase. A gravel subbase should consist of 2 inches of gravel ranging in size from fine sand to approximately ¾ inches in diameter. A poorly drained site will make the use of gravel mandatory.

Only compactible gravel can be used to form a well-drained, firm subbase. Crushed limestone of ¾ inch or less is ideal. You should avoid river run, as it will not compact sufficiently. Figure on using about 1 cubic foot of material for every 50 square feet of patio area.

Spread and compact the gravel in layers for maximum density (*Figure 29*). The general method

---

NCCER – *Masonry Level Two* 28202-14

is to lay half the gravel over the entire area. Tamp with a hand tamper or compactor. Then add the rest of the material and tamp again.

3.5.3 Edging

Patios must have retaining edges to prevent horizontal movement of mortarless clay brick or other paving units. Several types of edging can be constructed using concrete, clay brick, wood, or plastic.

Clay brick edging is set in soil. It is made by standing the clay brick on end with faces toward the pavement. Clay brick can also be placed at an angle with the face against the pavement. *Figure 30* shows two types of edges that can be constructed with clay brick.

Wooden edges can be made of 2-inch × 6-inch redwood, cedar, or pressure-treated wood. These should be held in place by stakes placed approximately every 4 feet. Stakes should be driven down 2 inches below the top edge of the frame and beveled sharply down from that edge so they can be covered with soil.

Preformed plastic edging is designed to secure mortarless pavers after they are installed (*Figure*

31). Metal spikes are driven into precut holes and secure the edging into the gravel base.

3.5.4 Base Material

Base preparation will vary depending on the type of subgrade and the method of construction. If the soil is well-drained and not subject to frost, a base of 1 to 1½ inches of sand should be used. The sand used for the base should be clean, naturally occurring material with angularly shaped particles and a maximum size of ³⁄₁₆ inch. Concrete sand conforming to *ASTM C33, Standard Specification for Concrete Aggregates*, is preferred. When ordering sand for bedding and joint filling, allow for

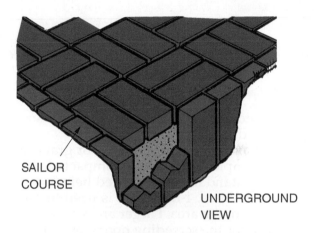

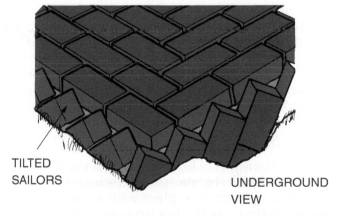

Figure 30 Using clay brick for edging.

Figure 29 Tamp gravel in layers.

Use of the 3-4-5 Rule

The 3-4-5 rule is a based on the Pythagorean theorem, which has been used in construction for centuries. Its use was covered in detail in *Masonry Level One*.

Explain how, armed only with a tape measure, you would use the 3-4-5 rule to lay out and/or check 90-degree angles (right angles), such as those needed to lay out square lines and corners for porches.

28202-14_F31.EPS

Figure 31 Plastic edging.

28202-14_F32.EPS

Figure 32 Screeding.

approximately 1 cubic yard for every 150 square feet of paved area.

Before pavers are laid, the sand or gravel base must be completely flat and compacted. *Figure 32* shows a standard 2 × 4 screed being worked over a gravel base. 1-inch pipe is used to set the boundaries for the area. Larger areas may require two people for the screeding operation—one person to handle each end—working the screed in a back-and-forth motion.

For mortared construction, paving units must be placed on a new or existing concrete slab. Constructing the slab is similar to the construction of footings for a foundation. Follow the procedure described in the section *Understanding Spread Foundations*.

3.5.5 Setting Clay Brick and Pavers

Begin placing clay brick pavers on the sand with joint widths between ¹⁄₁₆ and ³⁄₁₆ inch (*Figure 33*). Then tamp them into place with a rubber mallet. Lay the first few paving units carefully because they determine the line and grade for placing the remainder of the units. As you proceed with each row, use a string line as a guide to keep the units straight. You should begin laying the units from one corner and place them from only one direction. If you have a helper, you should both be working together at one place and in the same direction.

When working with interlocking concrete pavers, check to see how they fit together. They may have to be laid in a certain sequence to stay even. Mortar should not be used with the interlocking pavers. They are designed to fit hand-tight without mortar.

After setting several square feet of pavers, lay a 32-inch length of 1 × 6 on the surface of the pavers. Tap the wood over the entire surface with a hammer or mallet to uniformly bed the pavers into the sand. If individual units are slightly tilted or too high, tap them gently into place with the rubber mallet. If a unit is too low, lift it out, place sand underneath, and reseat the paver. Periodically check the surface of the area with a 4-foot level to ensure that the pavers are level or are sloping uniformly in the correct direction. After all of the pavers have been laid, double-check the slope across the entire area. Adjust any individual pavers.

The final step for mortarless pavers is to fill voids between pavers with sand. Concrete-quality sand should be used. Spread dry sand over the

Keep Off the Sand

Once the sand or gravel base has been screeded, stay off it. If the prepared bed does get walked on, rework and screed the area again. Footprints and other irregularities in the surface will make the pavers unstable and diminish the quality of the work.

28202-14_F33.EPS

Figure 33 Placing pavers.

entire surface. If the sand is not completely dry, allow it to sit until it has dried. Sweep it across the surface into the cracks between the pavers. For large areas, use a vibrating tamper to compact the sand bed and settle the pavers into place. After tamping, fill remaining cracks with sand and sweep clean.

When laying clay brick pavers that will be mortared, maintain a ⅜- to ½-inch gap between all units. Before filling the joints, check all units with a level. Even them up by tapping with the trowel handle.

There are two ways to mortar clay brick paving units once they are laid. The first method is a dry method where a 1:4 mix of cement and dry sand is mixed, placed on the pavement, and brushed into the joints. When using this method, be sure to use a kneeling board, and carefully sweep all mortar from the upper surface of the clay brick. Next, take a ½-inch-thick piece of wood and tamp all the mortar in the joints. Add more dry mix if needed to completely fill the joints.

Again, sweep any mortar on the units into the joints. Finally, wet the surface using a very fine spray of potable water. Be careful not to splash any mortar out of the joints, and do not allow any water to pond on the surface. Keep the mortar damp for approximately 3 hours. When the mortar is firm, tool the joints with a jointer as you would when finishing a masonry wall. After several more hours, scrub off any mortar on the surface with burlap.

An alternative method is to mix mortar in the standard fashion and trowel the mix into the joints. A grout bag may be used as well. When using this method, work carefully to minimize spilling. Clean all excess mortar off the surface of the pavers as soon as possible. Tool the joints in the normal fashion.

3.6.0 Constructing Decks

The use of clay brick pavers on decks and roof plazas is popular in some parts of the United States. The deck is supported by suspended wood joists, a steel frame, or a reinforced concrete base. The structural design must take into consideration the weight of the masonry and the maximum allowable deflection for the finished surface. The deadweight for mortared or mortarless clay brick pavers is approximately 10 pounds per square foot per inch of thickness. For example, for 1⅝-inch-thick units, the weight would be 16.25 pounds per square foot. For 2¼-inch units it would be 22.50 pounds per square foot.

Too much give in the subfloor can cause cracks in the mortar joints. As a rule of thumb, deflection should be limited to ¹⁄₆₀₀ of the span. This means that for an 8-foot span the allowable deflection of the subfloor would be approximately ³⁄₁₆ of an inch. Mortarless clay brick paving may be installed over bases designed for deflections of ¹⁄₃₆₀ of the span. Materials and patterns for this type of structure would be basically the same as discussed in the section *Interior Floors*, except for the substructure and the subfloor, which can be made out of wood, metal, or concrete.

A metal subfloor is supported by joists that are spaced to support the structure within minimum deflection requirements. The metal subfloor is laid on the joists and fastened according to the specifications. Once the supporting surface is in place, the preparation of the base and cushion for the clay brick pavers can now progress in the same manner as any other clay brick flooring installation. As in other outdoor applications, adequate drainage must be provided. This will prevent damage from alternating freezing and thawing of trapped water.

Additional Resources

ASTM C33, Standard Specification for Concrete Aggregates, Latest Edition. West Conshohocken, PA: ASTM International.

Patios & Walkways. 2010. Peter Jeswald. Newtown, CT: Taunton Press.

3.0.0 Section Review

1. For flights of steps less than 30 inches high, each tread should be at least _____.

   a. 7 inches
   b. 9 inches
   c. 11 inches
   d. 13 inches

2. To allow the use of whole clay brick when installing steps, make the width of the steps _____.

   a. exactly 8 inches
   b. a multiple of 32 inches
   c. a multiple of 8 inches
   d. the same as the door or entryway

3. To support clay brick when building a concrete base for clay brick steps, use _____.

   a. wire mesh
   b. edge restraints
   c. crushed stone or other aggregate
   d. a stepped footing

4. When applying a mortar setting bed for clay brick steps, start _____.

   a. from the outside edges and work inward
   b. with the bottom step
   c. from the inside edges and work outward
   d. with the top step

5. When excavating for a patio, the most important factor to consider in preparing the subgrade is _____.

   a. proper edging
   b. pattern design
   c. drainage
   d. environmental requirements

6. The rule of thumb when installing mortared subfloors in decks is that deflection should be limited to _____.

   a. approximately 10 pounds per square foot per inch of thickness
   b. $3/16$ of an inch
   c. $1/360$ of the span
   d. $1/600$ of the span

4.0.0 FIREPLACES AND CHIMNEYS

Objective

Explain how to lay out and build fireplaces and chimneys.

   a. Explain the basic theory of the fireplace.
   b. Describe the parts of a fireplace.
   c. Explain the key points of workmanship.
   d. Explain how to lay out chimneys and fireplaces.
   e. Explain how to begin the fireplace.
   f. Explain how to finish the fireplace.
   g. Describe a multi-opening fireplace.

Trade Terms

Corbel: The process of laying masonry units to form a shelf or ledge.

Cove: A concave area between two perpendicular planes; the area at the intersection of a wall and floor, or wall and ceiling.

Damper: A metal device used for regulating the draft in the flue of a chimney, usually made of cast iron.

Downdraft: A current of air that moves with force down an opening in a chimney.

Slushing: The process of using a trowel to fill collar joints and other wide openings with mortar.

28202-14_F34.EPS

Figure 34 Outdoor patio fireplace.

Fireplaces have gained renewed popularity as a source of heat. In addition, they enhance the atmosphere and appearance of a room or a backyard patio, as shown in *Figure 34*. They can increase the value of a home or building.

A fireplace must be built to be strong, stable, durable, and fireproof. If it is to work efficiently, each part must be designed and built properly. A mason must understand the functions of the parts of a fireplace in order to build them correctly. Unlike foundations, the fireplace is often the central focus of the primary living room in a home. It should reflect the highest-quality craftsmanship.

4.1.0 Understanding the Basic Theory of the Fireplace

A fireplace is a fuel-burning combustion chamber. It heats a room by warming the air in the room. When a fire is burning in the fireplace, the cool air within the room is drawn across the burning fuel. The air is heated and expands. It becomes lighter than the cooler air nearby.

The lighter, heated air rises up the chimney. This creates a draft that removes smoke and combustion gases. At the same time, the draft draws fresh air to the fire, supplying it with oxygen that is necessary for continued combustion. Some of the heat generated by the fire is reflected back into the room by the shape of the firebox, the principal heating element in the fireplace. A **damper** regulates the draft so that smoke will be drawn up the chimney and out of the house.

4.1.1 Importance of the Draft

The draft is the current of air that passes through the firebox. This air current is created by the difference in weight between the hot gases in the flue and the cooler air outside the chimney. The movement and intensity of the draft depend on the amount of the temperature difference and on the height of the chimney. If not designed properly, strong winds will cause downdrafts, sending smoke into the house.

A fireplace and chimney in the center of a house will normally provide a better draft than a chimney on an exterior wall. This is because the central chimney remains warmer. This creates a greater temperature difference between the cool air in the room and the heated air in the chimney.

Three common mistakes are often encountered in chimney design: the flue is sized improperly, the flue is not high enough to produce a good draft, and the vertical ascent of the flue is altered to the left or right too quickly. Any of these mistakes can result in a chimney that does not draw properly. Without an adequate draw, smoke will exhaust into the house.

4.1.2 Draft Operation

The actual operation of draft in the fireplace occurs as cool air is drawn down one side of the flue while warm air, smoke, and gases return up the other side. This process is shown in *Figure 35*. When the air coming into the chimney reaches the smoke shelf behind and slightly below the bottom of the damper, the air is deflected up the chimney on the opposite side of the flue. Smoke and gases are taken up the chimney, causing air to be drawn from the room into the firebox. As the air enters the firebox, it quickly becomes heated. It rises through the throat of the damper, where it meets the colder air bouncing off the smoke shelf.

The air drawn from the room can be replaced by outside air that enters the room through cracks around the windows, doors, or vents. To eliminate this type of draft, an opening can be made in the sidewalls of the fireplace that connects to an outside vent specially designed to provide the needed air flow.

As the fire burns, the damper is adjusted so the fire burns freely but does not smoke. The maximum amount of heat is then radiated into the room. If the damper is left completely open when

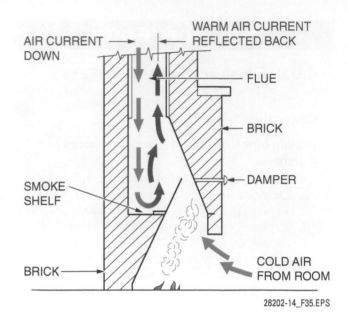

Figure 35 Fireplace draft.

the fireplace is in use, a great amount of heat escapes up the chimney. After the fire is entirely out, the damper should be closed to prevent entry of insects, animals, and rain, and to prevent heat from escaping. An open damper will draw out a large amount of heat when not in use, increasing the cost of heating the home.

4.1.3 Heat Radiation

A fire in the fireplace generates radiant heat. Heat radiation, like light, travels in straight lines. Unless a person is in the direct line of radiation, little heat will be felt. An ordinary fireplace will produce almost no heating effect from moving air currents such as those produced by a warm-air duct in a room.

4.2.0 Identifying the Parts of a Fireplace

You must know the major parts of a masonry fireplace. You must also understand the purpose

Fireplaces

Fire pits have been used for heating and cooking for thousands of years. As building construction developed, fire pits were replaced with fireplaces. These included a chimney or smoke chamber in addition to the burning pit. The first fireplaces in the United States can be traced back to the early 1600s. Before the development of the damper, these early structures were not very efficient. Without a damper, there was no control over the burning process or draft. Much of the heat escaped up the chimney.

Fireplaces are not as efficient as stoves or other heating devices. However, modern improvements have made them an attractive alternative to oil- or gas-burning appliances. Some improvements include dampers to control the draft, fireproof flue linings, highly improved portland cement mortars, and special firebrick for lining the firebox. In design, the outer hearth on many fireplaces is raised off the floor to provide a bench seat in the room and a higher level of heat radiation.

of each part in order to properly interpret plans and specifications and build the fireplace. A typical fireplace is shown in *Figure 36*. All fireplaces may not have all of the parts shown. Some may be combined into one unit. However, as you become familiar with these parts, you will develop a basic understanding of most types of fireplaces.

This sectional view shows that the fireplace is composed of inner and outer shells bonded together with mortar or grout. The inner shell surrounds the fire and draft. It includes the firebox, smoke chamber, and flue lining. Fireproof materials are normally used to form this inner shell, because control of fire, heat, and smoke are the principal functions of this area.

The outer shell, made from standard masonry units, supports and surrounds the inner shell. The foundation, ash pit, hearth, and chimney walls are parts of the outer shell that provide structural support and stability for the fireplace.

- *Foundation* – A fireplace should be supported by and built on its own foundation or footing. The fireplace foundation should be constructed so that it is separate from the foundation of the building, with the proper dimensions and reinforcement.
- *Ash pit* – In an elevated fireplace, ashes from the fire can be swept into the ash pit and removed at a later time. The ashes are swept into the ash pit through an ash dump in the inner hearth and removed through a cleanout door. Build the ash pit so that the ashes can be deposited and removed. It must also be highly resistant to moisture penetration.
- *Hearth* – The inner hearth is the surface directly under the fire, forming the floor of the firebox. The outer hearth is placed directly in front of the firebox opening. It protects the floor from sparks that may fall or fly out of the firebox. The inner and outer hearth work together as a fire-protecting surface. To be completely fireproof, the hearth must be free of any cracks or voids.
- *Fireplace butt wall* – The butt walls, also called breastwork, are the walls that support and enclose the firebox. They extend from the foundations to the bottom of the flue where the chimney starts. Standard masonry units are used for the butt wall and the chimney.
- *Firebox* – The openings and surfaces directly adjacent to the fire form the firebox. The opening into the room (fireplace opening) and the opening into the chimney (flue opening) must be accurately dimensioned for efficient heating and smoke removal. The side and back walls of

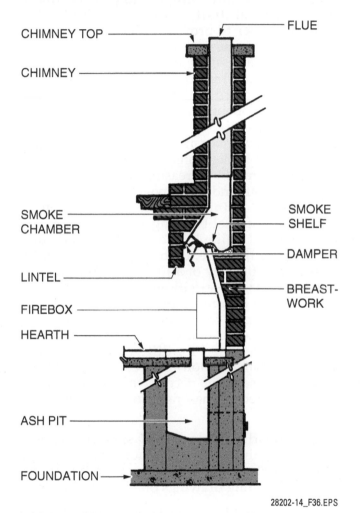

CHIMNEY TOP
FLUE
CHIMNEY
SMOKE CHAMBER
SMOKE SHELF
DAMPER
LINTEL
BREAST-WORK
FIREBOX
HEARTH
ASH PIT
FOUNDATION

28202-14_F36.EPS

Figure 36 Parts of a fireplace.

Rumford Fireplaces

Count Rumford was born in 1753 in Massachusetts. He was a loyalist and left with the British in 1776. He worked for the Bavarian government most of his life and received his title from them.

Rumford was primarily known for the work he did on the nature of heat. He applied his knowledge to the improvement of the fireplace. The Rumford fireplace was taller and shallower with widely angled covings to reflect more heat. He streamlined the throat to eliminate turbulence to better carry away smoke.

In the 1790s the Rumford fireplace was state of the art worldwide. Jefferson had them installed in Monticello. They were common from 1796 to 1850. They are still known for their elegance and heating efficiency.

the firebox must be properly dimensioned and sloped to reflect heat into the room. Firebrick is used for most fireboxes.

- *Smoke shelf* – Downdrafts are reflected back up the flue by the smoke shelf. The smoke shelf projects out from the rear butt wall and has a curving upper surface.
- *Damper* – The damper is a mechanical device that is opened or closed to regulate the draft. Install the damper so that it can be opened and closed freely, and so that heat and smoke do not escape around the sides.
- *Lintel* – The lintel is a unit or device that is placed across the span of the fireplace opening. It supports the masonry units placed over the opening. Install the lintel with the proper bearing surface and insulation to prevent failure due to imposed loads or heat expansion.
- *Smoke chamber* – The smoke chamber extends from the throat to the bottom of the flue. It collects and discharges smoke and combustion gases and reflects downdrafts.
- *Flue* – The flue is a vertical opening through the chimney that extends from the smoke chamber to the top of the chimney. The principal functions of the flue are to create the draft and discharge smoke. Round flues are more efficient than rectangular flues. The specified size of the flue opening must be maintained for the height of the flue. The rule of thumb for optional draw is for the square-inch measurement of the flue to be $\frac{1}{10}$ that of the firebox opening. All joints within the flue must be completely filled to prevent leaks.
- *Chimney* – The chimney includes the masonry laid around the flue liner and other objects such as anchors, reinforcement, and flashing. The masonry chimney walls and sections of the flue liner are laid up as part of one operation when constructing the chimney.
- *Chimney top* – The top of the chimney must be a minimum of 3 feet above the roof and can include such items as a chimney cap or hood for moisture protection and to reduce downdrafts. Other items, such as spark arresters or screens, may also be added.

4.3.0 Identifying the Key Points of Workmanship for a Fireplace

A fireplace must be built to the highest workmanship standards. It is usually the focal point of a room. The fireplace must not only work correctly, it must also show that neat, careful craftsmanship was used in laying the masonry units. Any inferior work will be obvious and will reflect poorly on the mason's craft.

The dimensions of the various parts of a fireplace are carefully designed to provide the required natural draft and to reflect heat into a room. In order to provide efficient heating and draft, the fireplace opening and the size of the firebox vary according to the size of the room.

You must build each part of the fireplace according to its specified dimension. For example, both the fireplace opening and the flue opening must be a specified size to create the required draft. The smoke chamber must also be constructed to its specified dimensions. This chamber directly affects the draft by increasing the speed, or velocity, of the draft and the removal of smoke.

There are definite relationships between the principles of fuel-burning heating chambers, the design of a fireplace, and the methods used to build the fireplace. Each of these factors influences the efficiency of a fireplace. You must follow the plans and specifications carefully.

There are a number of design and building techniques used for fireplace construction. You should be aware that different building techniques are required for different design styles. Fireplace construction that results in efficient, economical, safe, and decorative heating units can be constructed by masons who are able to do the following:

Step 1 Read the plans and specifications to obtain proper instructions on how to perform the work. This includes understanding drawings of fireplace details.

Step 2 Lay out the fireplace at its specified location, maintaining the required clearances.

Trees Affect Chimney Draft

High trees or a hill near the chimney can influence the draft and should be considered when building a chimney. Trees can cause downdrafts, which result in puffs of smoke coming from the firebox into the room. These obstructions may also prevent the fireplace from drawing properly.

If possible, build the chimney higher than any nearby obstructions. The chimney should be built a minimum of 3 feet above the top of the ridge of the roof or the point where the chimney penetrates the roof. This is high enough to allow for adequate air movement and protect against a fire caused by combustion in the chimney flue.

Step 3 Lay up the various parts of the fireplace in strict accordance with dimensions specified for each part and the overall dimensions of the fireplace.

Step 4 Build a foundation for the fireplace that is independent of the building and capable of supporting the fireplace.

Step 5 Lay the foundation walls, forming a usable ash pit in the process.

Step 6 Lay up or install the specified inner and outer hearth, using acceptable trade practices for the materials and methods specified.

Step 7 Build the firebox with the proper opening, wall thickness, and reflecting walls while maintaining tight joints.

Step 8 Bond with standard or fireclay mortar, using the proper procedures for placing and aligning units with the different types of mortar.

Step 9 Install the damper and lintels according to the manufacturer's instructions and accepted trade practices.

Step 10 Corbel masonry to form the smoke shelf, throat, smoke chamber, and liner support.

Step 11 Build the chimney, using the proper trade practices to install the liner, maintain clearance, place fire-stops, set anchors, terminate the chimney, and cap the chimney.

Step 12 Apply the specified fireplace face or finish so that it adds to the attractive appearance and efficient performance of the fireplace.

Step 13 Conduct a smoke test after construction and before the fireplace is put into use.

In order to build a fireplace, you must be able to perform these tasks. In addition, you must understand the relationships between the different parts. Some masons do not attempt fireplace construction due to the difficulty. However, it can be very rewarding. A well-trained mason should be capable of building a sound fireplace.

4.4.0 Laying Out Chimneys and Fireplaces

The fireplace is usually the center attraction in a room; therefore, serious consideration must be given to its location. When deciding the location of the fireplace, remember that it is a permanent structure. Fireplaces built near doorways should be designed in such a way that their draft pattern takes into account disturbances caused by the door. Traditionally, fireplaces have been built as part of an outside wall, but this is not always the case. A fireplace can also be placed on an inside wall, in the center of a room, or to provide structural support.

The size of the fireplace depends on the size of the room. A fireplace that is too small will not supply enough heat for the room. A fireplace that is too large will provide too much heat.

A fireplace with an opening 30 to 36 inches wide will accommodate a room with 300 square feet of floor area. The dimensions of the fireplace opening are increased for larger rooms. The height and width must be in correct proportion for the fireplace to function properly. *Table 3* contains standard fireplace dimensions based on the width of the front opening. *Figure 37* shows the location of each dimension.

The front of the fireplace can be built either flush with the wall or offset into the room. Many are built flush with the wall because it speeds up construction and thereby saves money and time; however, a fireplace that extends into the room adds a dimension of depth. The major problem to consider when offsetting the fireplace into the room is that the flue lining must be laid plumb against the back wall. You can position the flue to the left or right if necessary.

4.4.1 Locating and Laying Out the Fireplace

Deciding where to put the fireplace is not usually the mason's task. However, an experienced mason can provide valuable insight if consulted early in the process.

Usually, the mason's duty is reading the plans and specifications to locate and lay out the fireplace. This varies from job to job, depending on what tasks are performed by other trades. For example, at some sites the masons building the fireplace may also lay the foundation. On other jobs, the mason may have to build the fireplace around a foundation and frame already in place. If this is the case, you must carefully check the location of the fireplace foundation. If there is any discrepancy between this location and the location shown on the drawings, notify the job supervisor immediately.

When the frame is in place, different procedures are used to locate and lay out exterior and interior fireplaces.

An exterior fireplace and chimney are located outside the building's exterior wall. Generally, an

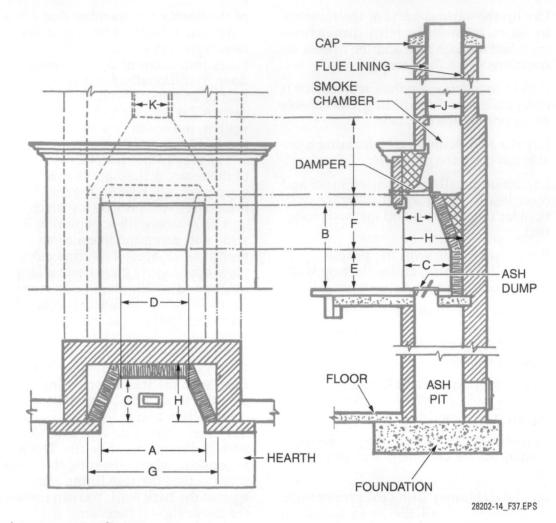

Figure 37 Fireplace construction diagram.

28202-14_F37.EPS

opening will be left in the building for the fireplace to head in or intersect the building. The opening is braced with double studs on each side and double headers above. This ensures that the frame has adequate structural strength. The opening is usually as wide as the fireplace, and at a minimum, extends vertically from the hearth to the smoke chamber.

You must square the butt of the fireplace (the lower and wider section of the fireplace) at the opening, and lay out the face of the fireplace

Table 3 Standard Fireplace Dimensions

Finished Fireplace Opening						Rough Dimensions			Flue Lining Size		Damper Depth
A	B	C	D	E	F	G	H	I	J	K	L
24	24	16	11	14	15	32	20	19	8 × 12		10¼
30	29	16	17	14	20	38	20	24	12 × 12		10¼
33	29	16	20	14	20	41	20	24	12 × 16		10¼
36	29	16	23	14	21	44	20	27	12 × 16		10¼
42	32	16	29	14	23	50	20	32	16 × 16		10¼
48	32	18	34	14	23	56	22	37	16 × 16		10¼
54	36	20	37	16	26	62	24	44	16 × 20		10¼
60	38	22	42	16	28	68	26	45	16 × 20		11¾
72	40	22	54	16	29	80	26	54	20 × 24		11¾

Refer to Figure 37 for dimension location.

along the interior face of the frame wall. The following six steps explain how to locate and lay out the fireplace in relation to the existing wall and wall opening, as shown in *Figure 38*.

Step 1 Mark a line (A-B) at a right angle to the frame.

Place a framing or steel square against the frame wall with the tongue on the fireplace wall line. Hold the framing square in a position that marks a 90-degree corner, or lay it flat in this position. The blade of the square should be flush against the frame wall, but make sure that the wall is true. Allow for clearance space at the sides.

With the square in the proper position, plumb down with a level at two different points along the square, and make two marks. Use a straightedge or a level to extend this line for the length of the fireplace.

Step 2 Mark a parallel line (C-D) for the opposite side.

Use the level and a 6-foot rule or tape to mark the width of the fireplace. Use the level to keep the measurement straight and the 6-foot rule to measure the specified width. Connecting the two marks made in this manner results in a line that is parallel to the first line. Extend the line for the length of the fireplace.

Step 3 Mark the wall opening line (A-C) of the fireplace.

Read the plans to determine where the face of the fireplace opening must be located. This dimension is stated frequently in terms of a given distance from the frame wall. Measure out the specified distance from the frame wall at two or more points, and then draw a straight line connecting the points. If the frame wall is not true, place the line at a distance that adjusts for variations in the frame-wall dimensions.

Step 4 Mark the rear wall line (B-D) of the fireplace.

Measure the length of the fireplace. Start the measurement at the front line, and make separate measurements on both sides of the fireplace. After the two length marks are in position, draw a connecting line. The four wall lines and corners of the fireplace are now in place.

Step 5 Check the corners for square.

Use the level to check each corner for squareness. If necessary, make a slight adjustment (¼-inch) in the lines to square the corners. In most cases, large adjustments in the location of the lines should not be made. This could indicate that an error has been made in marking one of the lines. If

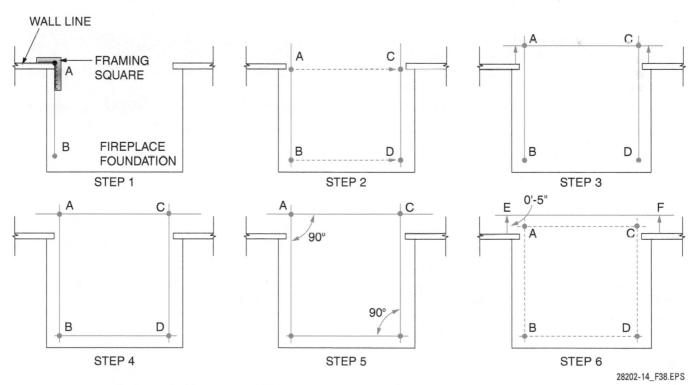

Figure 38 Layout of exterior fireplace.

that is the case, establish the lines again, beginning with Step 1. Each corner should now form a 90-degree angle.

Step 6 Mark the wall line (E-F) for the face of the fireplace.

Measure out from the wall (or front line) to accommodate the width of the face units plus 1 inch, approximately 5 inches at two points, and make marks. Using the marks, snap a chalkline to mark the face line. The extra inch allows for working space and variation in the frame wall. If the face brick is available, do a dry layout. This will determine where the front end of the firebox opening will be.

Locating and laying out an interior fireplace and chimney requires a slightly different procedure. Workers in related crafts prepare openings in the floor, ceiling, and roof. Masons must locate and lay out the fireplace so that it is centered in each of these openings. The mason begins work after the craft responsible for constructing the openings has cut and braced the joists, headers, plates, and rafters.

There are different ways of locating and laying out a fireplace that is partially or entirely within a structure. Experienced masons work off the first-floor opening (and other openings) to measure and sight from the opening down to the fireplace foundation. Measurements made at the opening to center the fireplace are transferred to the foundation so that the fireplace will be centered properly when it is constructed.

Use a plumb bob, laser, or straightedge to transfer measurements from the opening to the foundation. When the first-floor opening is 6 feet or more above the fireplace foundation, these devices will accurately locate and lay out the fireplace. The level can be used for distances less than 4 feet. To use the plumb bob procedure, drive a nail into the first-floor joists and headers at the point of the fireplace corners. Tie a line to the nail and lower a plumb bob to the foundation. The point of a steady plumb bob marks the corner of the fireplace on the foundation, as shown in *Figure 39*.

> **NOTE**
>
> Before marking the final dimensions on the foundation, you should always check the plans against the local code requirements concerning the required amount of space between the exterior surface of the fireplace wall and any combustible material.

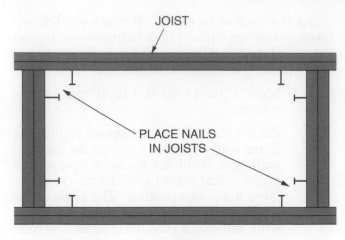

PLAN VIEW

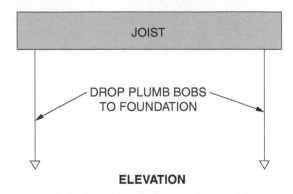

ELEVATION

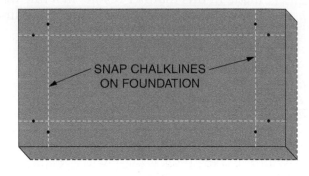

TOP VIEW

28202-14_F39.EPS

Figure 39 Marking the fireplace on the foundation.

Steps in the procedure are as follows:

Step 1 Locate the fireplace foundation and the fireplace opening in the first floor.

Step 2 Check dimensions of the opening and foundation to ensure that there is space for the fireplace and required clearances. Usually, the minimum clearance will be determined by local building codes.

Step 3 Mark the lateral dimensions of the fireplace on the openings. Place the marks on the wood headers and joists so that there will be sufficient clearance between the fireplace and the wood framing.

Step 4 Drive nails into the wood at the marks. The head of the nail must project out far enough for a line to be tied to the nail.

Step 5 Lower the plumb bob from each nail point. Suspend the plumb bob to within ½ inch above the fireplace foundation, and mark the point directly under the point of the plumb bob after the plumb bob is steady.

Step 6 Mark the wall lines.

Step 7 Using the marks made from the plumb bob, snap chalklines on the fireplace foundation to outline the exterior face of the walls. Square the corners, and check the overall dimensions.

4.4.2 Chimneys

After you have located and laid out the fireplace, there are two ways to proceed. One way is to complete the fireplace, including the firebox and all facing work, as the chimney is built. The second method is to build the chimney completely to the top, leaving a rough opening for installing the fireplace later.

Only masonry materials should be used in constructing a fireplace chimney. If clay brick is used as the facing material, concrete block can be used inside as the fill materials because they will not be subjected to direct heat. You should always use a hard-fired clay brick for the firebox and the smoke chamber. Flue liners should also be surrounded by at least 4 inches of hard-fired clay brick for fire protection.

Masonry units should be laid using a Type N mortar. In some locations, Type S may be specified due to wind or seismic conditions. All of the masonry units should be bonded together into the main outside walls either by using metal wall ties or by interlocking the units with each other.

The footings must be of a high-grade concrete. For a proper load distribution, the footing should be wider on all sides than the actual chimney. Steel reinforcement can be added to the concrete for extra strength.

There must be a separate flue for every fireplace or heat source. The most important points to remember for flue installation are to adequately support the flue, keep the joints as tight as possible, and completely fill all joints. Use of some type of long-lasting material, such as high-strength mortar or fine clay, is recommended. The lowest section of the flue lining should be supported on at least three sides with corbeled clay brick projecting from the inside wall of the chimney and flush with the inside wall of the flue lining. Set the flue lining in a bed of mortar and cut it off flush with a trowel. Wipe the joint with a cloth to make sure it is smooth and full of mortar. Corbeling support eliminates the chance of the flue slipping down into the chimney as the construction proceeds.

Set any additional flues in a bed of mortar with as tight a joint as possible. Finish in the same manner with a smooth joint. If possible, set the flues slightly ahead of the clay brickwork, level and plumb. This process results in a correctly placed flue lining with an excellent mortar joint.

No wood should be in contact with the chimney at any time. Leave a minimum 2-inch space between the outside face of the chimney and all woodwork. This rule also applies at the point where the chimney passes through the roof. Always observe the requirements of local codes.

The mortar bed joints should be raked out to receive a weatherproof flashing slightly above the roof line, or the flashing can be built into the masonry work.

The chimney is capped off with a portland cement mortar wash coat on an angle, to form a wind cove. Use Type M mortar because of its strength. To prevent the mortar from drying out too fast, cover it with wet burlap and dampen periodically until it has cured. This process prevents cracking that sometimes takes place due to sun and wind exposure.

The flue linings should extend at least 4 inches above the top of the finished cove. If more than one flue is in the chimney, the top section of each flue is leveled. If the chimney is rather large and only two flues are being used, sometimes a

Fill Voids Properly

Do not build up the outside walls of the chimney and then fill the space between the chimney and the outside wall with discarded material from the job site. Using leftover mortar particles, scraps of nonmasonry materials, or sand invariably results in an inferior chimney. Remember, only masonry materials should be used to fill in around the chimney.

dummy flue is added between the two flues to balance out the top of the chimney. This addition is made by cutting off a section of flue and building it into the top of the chimney. Fill a dummy flue to the top with mortar to prevent water from accumulating.

The termination height of the chimney will depend on the type of chimney and its intended use. For fireplaces, the *International Building Code®* specifies that the chimney outlet must be located a minimum of 3 feet above the highest point that the chimney penetrates the roof. Also, outlets must be a minimum of 2 feet higher than any portion of the building within 10 feet.

4.5.0 Beginning the Fireplace

The foundation distributes the weight of the fireplace and provides it with support and stability. A fireplace should have a foundation that is independent of the foundation for the structure. This is necessary because of the weight of the fireplace and because of the excessive thermal movement of the fireplace structure. Fireplace foundations, consisting of the footing and foundation wall, must be designed and constructed to support the fireplace and reduce settling, shifting, tipping, and cracking.

The footing will normally be placed by the general contractor or a subcontractor specializing in concrete work. Once the footing is in place, the masonry contractor can begin work on the foundation wall.

A summary for building a fireplace foundation includes these necessary steps:

Step 1 Mark the location of the foundation on the ground.

Step 2 Excavate the earth to the required dimensions and depth (based on local codes).

Step 3 Spread and compact a 6-inch layer of gravel or cinders under the footing.

Step 4 Place steel reinforcement.

Step 5 Pour, screed, and rough-finish the concrete.

Step 6 Allow the concrete to cure before proceeding with the work.

Step 7 Grade the ground so that it slopes away from the foundation on each side.

The first step in laying out the foundation wall is to study the plans and mark a chalkline on the footing to correspond with the size indicated on the plans. The masonry should be dry-bonded so that full units can be used without any cutting. If the base is to be concrete block, be sure to use block and clay brick as the layout material, to make sure the bond will work when clay brick is started at the finish grade line.

The units are then pushed aside from the layout line, and mortar is spread on the base. The chimney should be designed to allow the first course layout to use full units. If it is not, a slight adjustment in the chimney size may be required in order to use full-size units. (Always make the chimney slightly larger if adjustment is needed.) Lay the foundation masonry, using standard masonry practices, to the height of the cleanout door (usually four or five courses). Install the cleanout door in the masonry work, if required. Then build the foundation wall of the chimney to the first-floor height.

4.5.1 Foundation

The section of the fireplace between the footing and the hearth is the foundation wall. It supports upper sections of the fireplace and encloses a void space. In some fireplaces this void space is used as an ash pit. Ashes from the fire are stored in the ash pit until they can be disposed of.

The foundation walls are laid up according to accepted procedures for laying walls. For example, the units are laid out dry to check bond pattern and spacing. Corner leads and lines are

Separate Flues Properly

In some cases two or more flues may need to be installed in one chimney. Separate flues should be divided by a minimum 4-inch partition wall to prevent air in one flue from leaking over and affecting the draft in the flue next to it. Mortar should not, however, be slushed tightly down between the flue and clay brick. A small gap must be left to allow for expansion due to heat. Because of this, full mortar joints are even more important when building chimneys or fireplaces.

When more than one flue is required, it is often necessary to offset one of the flues for proper fit in the chimney so as not to restrict draw. These offsets or bends should not vary more than 30 degrees. This restriction eliminates the chance that the flue might be choked off. The flue lining should be cut neatly on a miter to fit the angle of the slope. This cut can best be done by first filling the flue with sand, tamped tightly. A sharp chisel and light hammer are used to make the cut along the marked line. After working all the way around the flue, the tools break the flue cleanly.

used on wide fireplaces. The first and following courses are properly aligned and bonded.

Normal dimensioning practices are used on fireplace foundation walls. Each wall must have the specified height, length, and thickness. For example, the walls will be 4 inches, 6 inches, 8 inches, or 12 inches thick, depending on material and load requirements. Some local codes require a cross wall in wide fireplaces. The cross walls are spaced at a required distance, such as eight times the thickness of the wall. Using this requirement, the cross walls in a 4-inch wall would be spaced at intervals of 32 inches or less; cross walls for 6-inch walls would be spaced at 48 inches or less. This principle is shown in *Figure 40*.

The empty space between the foundation walls and the cross walls can be used as an ash pit, left empty as a crawl space, or filled with rubble. If the area is to be filled, mortar and broken masonry units can be used as rubble. If rubble is used, it should be placed loosely into the space as the work progresses.

Ashes from fuel burned in the firebox can be brushed into the ash pit through a hole in the floor of the firebox. Ash pits have three main parts: the ash dump, the storage space, and a cleanout door. Foundation walls that contain ash pits must be properly tooled to provide maximum resistance to moisture penetration; wet ashes can cause a disagreeable odor to permeate the house or building.

The cleanout door is an opening in the foundation wall that permits the removal of ashes. Usually, a tight-fitting 10-inch × 12-inch door and frame is placed over the opening. This size of door is large enough to allow a shovel to be used to remove the ashes. The sill of the door is placed approximately 10 inches above grade or above the basement floor. The door is attached to the wall by bonding the frame in the rough masonry opening with mortar. The door must be properly aligned and bonded so that it fits tightly.

The storage space for the ash pit extends from the foundation slab to the hearth. The top and bottom of the storage space is shaped so that the ashes can fall freely and be removed easily. At the bottom, the ash pit is sloped from the back toward the cleanout door. Mortar is spread along the back, sides, and floor of the ash pit, providing smooth surfaces that make it easier to remove the ashes with a shovel. Sharp corners at the top of the ash pit may hold ashes and can be reduced by corbeling toward the ash dump. Corbeling is not, however, specified for all ash pits.

The ash dump is an opening through the floor of the firebox or hearth. The opening is covered by a metal frame and plate that is hinged or pivoted so that it can be opened and closed. The ash dump is located over the center of the ash pit. Usually, this placement means that it is near the rear center of the firebox. The opening for the ash dump is constructed while laying up the inner hearth. The ash-dump frame and plate can be placed after the fireplace is completed or during progress of the work.

4.5.2 Hearth

Placed on top of the foundation walls, the hearth provides a bearing surface for upper sections of the fireplace. The part of the hearth under the firebox, called the inner hearth, forms the floor of the firebox and holds the burning fuel. The outer hearth extends from the fireplace opening into the room to protect adjacent flooring from sparks and combustion.

The seven basic dimensions shown in *Figure 41* are used to make measurements for the hearth. Dimensions used to build the inner and outer hearth are:

- *Depth of fireplace (A)* – Measured from the face of the fireplace opening to the exterior face of the rear wall.
- *Width of fireplace (B)* – Measured from the exterior face of one sidewall to the exterior face of the opposite wall.
- *Depth of firebox (C)* – Measured from the face of the opening to the interior face of the rear wall of the firebox.
- *Width of firebox (D, D1)* – Measured from the interior face of one sidewall to the corresponding point on the opposite wall. Separate measurements are required at the opening and at the rear of the firebox.

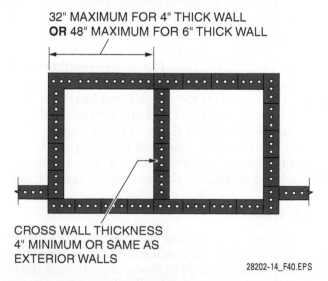

32" MAXIMUM FOR 4" THICK WALL
OR 48" MAXIMUM FOR 6" THICK WALL

CROSS WALL THICKNESS
4" MINIMUM OR SAME AS
EXTERIOR WALLS

28202-14_F40.EPS

Figure 40 Layup of foundation walls.

- *Front projection of outer hearth (E)* – Measured from the face of the opening to the front edge of the outer hearth. Depending on the number of square feet in the firebox and size of the opening, this dimension may be from 16 inches to 20 inches.
- *Side projection of outer hearth (F)* – Measured from the jamb of the opening to the side edge of the outer hearth, usually a distance of 8 inches to 12 inches.
- *Thickness of hearth (G)* – Measured vertically from the bottom to the top of the hearth.

A concrete rough hearth is constructed by using plywood or lumber formwork. This is installed under the hearth area by placing it on wood strips nailed to the headers that surround the hearth area in the floor. Steel reinforcing rods are placed over the forms in a grid pattern and tied together with tie wire. Metal panels may also be used to support the concrete and the rough hearth.

Concrete is placed in the forms and vibrated so that it settles around and under the steel rods. The

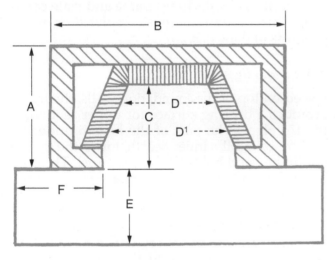

PLAN

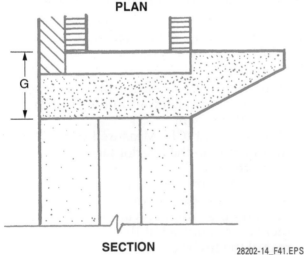

SECTION

28202-14_F41.EPS

Figure 41 Hearth dimensions.

wood forms can be left in place after the concrete hardens, or they may be removed from under the part of the hearth that projects out from the main chimney. The builder usually leaves the forms in place. Formwork must also be used to build the ash-dump opening. Short pieces of wood or sheathing that can be knocked out easily after the concrete has hardened are recommended.

The finished hearth can be constructed of quarry tile, clay brick, stone, or other similar materials. It may be built flush with the floor or raised. In many modern fireplaces the hearth is raised in order to elevate the heat source. The raised hearth also provides a bench seat close to the fire.

Before the hearth can be built, the rough opening of the fireplace must be determined in order to make sure that there will be enough room to install the finished hearth and firebox. *Table 3* lists finished sizes for different fireplaces. When building a rough opening for a fireplace that is to be installed later, allow 8 inches more than the finished width shown on the table. Allow 13 inches more than the opening shown on the table for the damper height. These allowances provide enough working room to build the finished firebox and set the damper.

For example, a standard fireplace having finished dimensions of 36 inches × 29 inches should have a rough opening constructed 44 inches wide and 42 inches high. If a raised hearth is built, the additional inches must, of course, be added to the rough-opening height.

The hearth is usually built of firebrick, which is made of special clay that is highly resistant to heat. A standard firebrick measures 4½ inches × 9 inches × 2½ inches. Firebrick is buff or off-white in color, and for best results should be laid in either Type S portland cement mortar or fireclay mortar.

The hearth is always laid out using the center of the fireplace as the reference point. After the width of the opening has been determined from the plans, the overall distance of the inner hearth can be marked off with a pencil. Then, the outside wythe of firebrick is laid. The course should be longer than the measured distance, since a masonry jamb will be laid on top of the inner hearth, hiding the rough edges. Small head and bed joints should be struck with either a flat or concave jointer.

The ash dump is installed in the opening left when the rough hearth was built. The ash-dump frame has a projecting flange that holds it in place.

4.5.3 Walls

Fireplace walls surround the firebox and support the chimney. The walls should never be less than 8 inches thick. If made of stone, they should never be less than 12 inches thick. Check your local codes to make sure of the required thickness. These restrictions prevent the inside walls of the fireplace from getting too hot, which would result in cracking. Thickness is especially important for the walls of a fireplace and chimney built in the center of a building, where flammable building materials may be in contact with the walls.

The four walls of the fireplace are constructed in stages. The section known as the butt extends from the hearth (or foundation) to the chimney. The butt is usually constructed before the firebox is built; it must be in place before the smoke chamber and chimney are constructed. The butt is laid to the proper height as the firebox is built so that the damper can be set properly.

After the hearth is ready, lay out and lay up the rear and side walls of the butt. These exterior walls are called rough walls when they are not tooled. Butts may be built directly on a foundation slab when there are no foundation walls.

Lay out the butt and firebox walls at the same time to ensure that each wall is dimensioned properly. Place, align, and bond units and courses of the butt using accepted trade practices for the clay brick or block material. Standard masonry units are normally specified for this work. Lay the butt to a predetermined height, such as the lintel height, the start of the smoke chamber, or the start of the flue. Form and tool the joints to make them full, solid, and compact. All joints exposed to view, weather, or combustible gases should be tooled and finished. Place wall ties for the face of the fireplace in the specified locations during the progress of the work.

The entire butt wall need not be constructed before constructing the firebox. However, you should decide in advance how much of the butt is to be constructed so that the walls will be in place when needed and will not restrict the work area or interfere with the work that must be done on other parts of the fireplace.

4.6.0 Finishing the Fireplace

The firebox is constructed with materials that are different from those used to lay up the surrounding butt walls. Firebrick and other approved materials that are resistant to extremely high heat

are used for fireboxes. In contrast, standard clay brick, block, stone, and other approved materials are used to build the surrounding backup and facing for fireplace walls.

4.6.1 Firebox Dimensions

Seven basic dimensions may be used to provide instructions for constructing the firebox. You must be able to read these dimensions from the construction drawings and use them to make the necessary on-site measurements. These dimensions are illustrated in *Figure 42*.

- *Depth (A)* – Measured along the center line of the fireplace opening from the opening to the interior face of the rear wall.
- *Width at opening (B)* – Measured from one jamb to the opposite jamb at the face of the fireplace opening.
- *Width at the rear (log width) (C)* – Measured from one sidewall to the opposite wall in a straight line along the rear wall.
- *Height of opening (D)* – Measured from the hearth vertically; this perpendicular measurement marks the location of the lintel.
- *Height of vertical back wall (E)* – Measured vertically from the hearth to the point where the back wall begins inclining.
- *Height of inclined back wall (F)* – Measured along the sloped or inclined back wall from the top of the vertical back wall to the finished height of the back wall.
- *Firebox wall thickness (G)* – Measured as the distance between the interior and exterior faces of the wall at right angles.

4.6.2 Firebox Walls

The sidewalls of the firebox should be angled (though not more than 3 or 4 inches per foot of depth), and the back wall should be sloped. The purpose of the slope and angled sides is to reflect heat from the fire out into the room. The slope of the back wall also helps to move the smoke and gases upward.

The firebox walls should be marked on the inner hearth and the masonry units dry-bonded and marked before mortar is applied. The follow-

Hearths

In the past, when the rough hearth was constructed, a clay brick arch was laid on a wooden form beneath the hearth. This arch was called a trimmer arch. The concrete or clay brick rough hearth was then placed over the arch, and the arch carried the load or weight of the finished hearth.

In present-day construction, clay brick trimmer arches are seldom used because they require a great deal of time to install.

ing procedure explains how to lay out the walls as shown in *Figure 43*:

Step 1 Establish the center line for the fireplace opening (Line O-C).

Step 2 Square and center the opening width (Line A-B).

Step 3 Measure off the specified depth along the center line (Line O-C).

Step 4 Square and center the rear opening width at the specified depth (Line D-E).

Step 5 Connect the proper points on the front and rear opening lines by using a level or 6-foot rule (Points A-D-E-B).

Step 6 Mark off wall thickness points for the firebox (Lines A^1D^1 and B^1E^1).

The first course of the firebox is laid in mortar, completely around the firebox. This establishes the bond. Then, allowing approximately ½ inch for expansion, the mason backs this course with rough clay brick as a filler.

The firebox walls are continued up to the height of the vertical back wall (normally 14–16 inches). It is a good idea to sprinkle the hearth with sand to prevent mortar from sticking to the firebrick while the walls are laid. The head and bed joints should be laid as tightly as possible because the heat of a fire has a tendency to burn out large joints.

At the four-course height, the back wall begins a slope or roll that brings the firebrick up under the rear flange of the damper. Because of this slope, the back wall presents special problems. To lay out and lay up the rear wall properly, you must determine (a) the proper finished height and depth for the rear

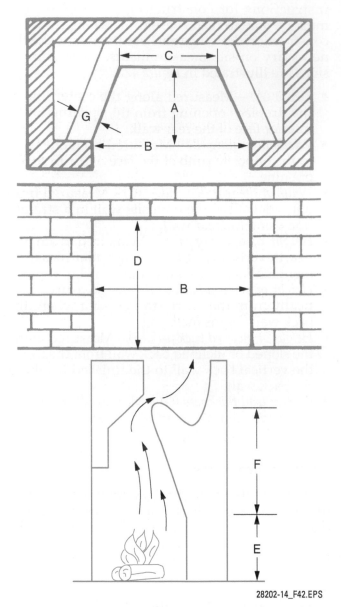

28202-14_F42.EPS

Figure 42 Firebox dimensions.

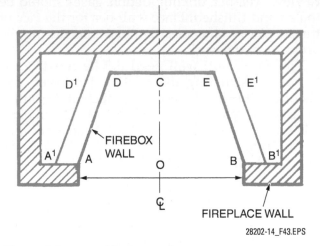

28202-14_F43.EPS

Figure 43 Layout of firebox walls.

Fill Voids Properly

A void exists between the butt wall and firebox wall. The void will vary in size according to the dimensions of the fireplace. Fill the void with noncombustible materials, such as chipped masonry units, sand, or dried mortar. In seismic areas, reinforcing rods and grout are placed in the void.

When two or more combustion chambers connect to the same chimney, a flue or cleanout duct for one chamber can pass through the void. Cover the void space with a noncombustible material such as a concrete cap or masonry course. The void cover must be strong enough to support the loads imposed by the chimney.

wall, (b) the angle for sloping the upper courses, and (c) the course that starts the sloping section.

The rear firebox wall must meet at least two requirements. It must support the damper 6 inches or more above the fireplace opening. The interior face of the rear wall must be in line with the interior edge of the damper. If the rear wall is finished too low or too close to the opening, it may cause smoke to enter the room and reduce the draft.

To determine the slope of the inclined wall, mark the location of and set the damper at the wall height now constructed. The horizontal distance between the rear wall and the location of the damper is the distance the rear wall must slope. The slope is determined by dividing the vertical height of the inclined wall by the horizontal distance measured.

The rear wall must fit flush with the sidewalls. This is accomplished by cutting the masonry units at the end of the sidewalls on the slope determined to fit corresponding courses of the rear wall. The angle to cut these masonry units can be measured by dry bonding the sidewalls. The following directions will explain this process. Refer to *Figure 44* as you review each step.

Step 1 Dry-bond the remaining courses of the sidewalls. Place the units to the specified depth, height, and thickness, and in the established pattern bond. Use a stick or wedge to maintain the ³⁄₁₆-inch mortar joint.

Step 2 Mark the sloping section. Place a mark at the course that starts the sloping section of the rear wall (point A).

Step 3 Measure off damper dimensions. Mark what will be the final location of the damper by measuring the damper depth or actually placing the damper in position. Place the front of the damper in its proper location, 6 inches above the opening and at the face. Level the damper, and mark the leading edge of the section that will sit on the rear wall (point B).

Step 4 Use a level or straightedge to draw a line connecting the points A and B. Use a heavy-duty pencil to mark this line on the sidewall masonry units. This is the slope for the rear wall and line of cut.

Step 5 Label the firebrick. The masonry units must be picked up and bonded in the same position that they are dry bonded. Label each unit with a number and letter combination. For example, use letters for courses and numbers for position in the course.

Step 6 Cut the end units. Follow the marked line on the end units to make the cuts. When both sidewalls are identical, a duplicate cut can be made for the other sidewall.

Step 7 Lay up the sidewalls. Place, align, and bond the sidewalls to the finished height. Check the overall alignment and dimensions. Place the damper in position on the sidewalls, and check the closure at the rear wall. Use the level to check course alignment and the slope. The corresponding slopes on both sidewalls permit the rear wall to be laid flush.

Step 8 Lay the rear wall. After checking for rear-wall alignment, lay the rear wall tight against the sidewalls and straight across the width of the firebox. To bring the rear wall flush against the sidewalls and up against the damper, bevel the mortar joints or cut the firebrick. Align the courses and level the supporting surface for the damper during the progress of the work.

The procedures for laying up and bonding firebrick are slightly different from procedures used to lay up and bond standard units. For example, the nominal course height of firebrick is 2¾ inches or 4¾ inches. It is necessary to work out separate course plans and course heights for the firebox and

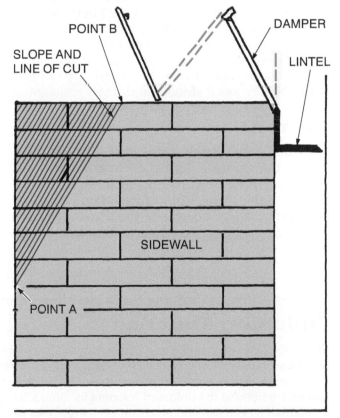

28202-14 F44.EPS

Figure 44 Layup of sidewalls.

the surrounding fireplace walls because of differences in the dimensions of the masonry units. A spacing rule can be used to control course height for the firebox as well as the fireplace.

Generally, fireclay mortar is applied by dipping the bonding surface into the mortar or by using a trowel to butter the surface. Spread mortar over the entire bonding surface of the bed and head joints. Remove excess mortar with a trowel, and tool the joints before the mortar sets. Use a piece of burlap to clean the face of the firebrick. Rubbing with burlap fills small holes in joints and removes mortar from the face of the units when using fireclay. Firebrick is normally laid flat (on the 4-inch side) with ³⁄₁₆-inch- or ¼-inch-thick joints as specified. This is a tight joint, and it must be solid, compact, and full.

The smoke shelf is formed in the area behind the inclined rear wall of the firebox at the rear of the fireplace. The smoke shelf is located behind and slightly below the damper. The smoke shelf is one of the more important parts of the fireplace. If not properly installed, the fireplace will not draw adequately. The smoke shelf turns the air coming down the flue back up the chimney, carrying the smoke and gases from the fireplace. If the smoke shelf is too high or too low, the smoke swirls around the damper area, resulting in a fireplace that functions incorrectly.

The smoke shelf must be concave; mortar is placed and shaped in the void area between the two rear walls, creating the ability for the smoke to curl and rise up the flue. Any holes or voids in the smoke shelf should be filled with mortar. Troweling a coat of mortar over the entire area of the smoke shelf makes it smooth and water resistant.

4.6.3 Throat

Smoke and combustible gases pass from the firebox into the smoke chamber through the throat. The throat must be constructed in strict accordance with the plans, to control draft and remove smoke efficiently. The area of the throat opening must be no less than the effective area of the flue. *Figure 45* shows this layout.

The dimensions used to build the throat are as follows:

- *Height* – Locate the bottom of the throat 6 to 8 inches above the lintel. The walls of the firebox and fireplace must be finished to accurately place the throat and damper.
- *Width* – The width of the throat is the same as the width of the damper. If the damper is 36 inches wide, the throat should be the same. The width measurement is taken parallel to the span of the firebox opening.
- *Depth* – Make the depth measurement at a right angle to the span of the opening. Measure from the throat opening to the interior surface of the rear firebox wall at a point 6 inches above the lintel. Usually, the depth of the throat and damper are the same.

Some damper assemblies have a built-in throat that is dimensioned accurately when the damper is placed. When the throat must be dimensioned, it is a good idea to mark the dimensions on the hearth while laying out the firebox and butt walls. Build the firebox up to the throat, and check the opening before proceeding with the work.

4.6.4 Damper

The damper regulates drafts in the fireplace by opening and closing the throat. When a fire is burning, the damper is fully or partially open. This regulates the heat and the smoke removal. When there is no fire, the damper can be closed to prevent drafts in the room, reduce the loss of heat from the central heating system, and keep rodents, birds, insects, and downdrafts from entering. Since the damper is exposed to direct heat from the fire, it must be made of a material that withstands high temperatures over a long period of time, such as cast iron. For heating and draft efficiency, the area of the damper opening must be equal to at least 90 percent of the effective flue area.

There are two main types of dampers: the poker and the rotary control. The poker type is operated by inserting a poker into a ratchet arm located un-

Finishing Touches

The firebox can be finished in different patterns to create distinctive appeal. The herringbone pattern adorned fireplaces in palaces and castles throughout Europe. Firebrick can be laid in the herringbone pattern, or the firebox walls can be purchased as constructed panels. Several manufacturers offer prepanelized patterned fireboxes. The panels are bonded to reinforced concrete forming a 5-inch-thick panel.

Alternately, you can create a distinctive firebox by using firebrick in contrasting colors. Firebrick is available in red, dark red, ivory, and gold.

der the damper and pushing upward. This action opens the damper to the desired width.

The rotary control is operated by turning a projecting handle built into the face of the finished fireplace. This motion causes the same opening operation as the poker type, but a person does not have to work in the hot area of the fireplace to open the damper.

You must assemble the damper before installing it. It should be put together according to the manufacturer's instructions, and all moving parts should work freely before being installed in the fireplace. On a rotary damper, check the worm gear that operates the lid. Sometimes, rough metal edges are left from the casting process; these should be filed smooth to permit the smooth operation of the damper. Once the damper is walled in, mistakes are difficult to correct.

Installation procedures vary with the type of damper and design of the fireplace, but in general the steps are as follows:

Figure 45 Throat dimensions.

Step 1 Dry-set the damper in its proper position to check dimensions before bonding it into place. The damper should be at least one course of firebrick higher than the lintel.

Step 2 Spread conventional mortar on the top course of the firebox.

Step 3 Place the damper so that the plate is positioned as required and the flange or frame can be sealed in the mortar bed. Tap the flange of the damper gently with the handle of a clay-brick hammer to form a tight bed joint. Cast-iron dampers can be cracked, so tap carefully. Some dampers require that insulation, such as glass wool, be wrapped around the flange before it is bonded in mortar; others can be loosely bonded in mortar without the insulation. Always follow the manufacturer's instructions.

Step 4 Cut excess mortar from the face of the damper bed joint.

Step 5 Open the blade of the damper, and cover it, if necessary, to prevent mortar damage when working above the damper. Place an empty cement bag in the open damper to protect working parts.

Step 6 Lay up several courses above the damper to secure it in position.

4.6.5 *Smoke Chamber*

The space above the smoke shelf and damper to the bottom of the flue lining is called the smoke chamber. It has a funnel shape that compresses the heated air as it rises. This increases the velocity of the draft.

High-quality, hard clay brick should be used to build the smoke chamber. The back wall is built straight and parged with a coat of mortar. After slushing all flatwork at the damper height with mortar, the clay brickwork is corbeled in from the front until it narrows at the flue. Leave a ½-inch space between the damper and the corbeled clay brick to allow for expansion. The corbelling, of course, must not interfere with opening the damper.

The clay brickwork of the smoke chamber should be tied into the main chimney with metal wall ties. The slope of the sides and front should be approximately 30 degrees. Corbeling uniformly to the center of the chimney prevents the draft from drawing on only one side and allows the fire to burn evenly in the firebox.

Corbeling should not exceed ¾ inch per course. After all the corbeling is finished and the flue lining is ready to set, the underside of the corbeling should be smooth so the smoke has an even flow to the flue, as shown in *Figure 46*. This can be done by either parging or knocking off the ends of the clay brick.

The following procedure is recommended for determining the correct slope:

Step 1 Locate the center of the chimney on the back inside wall.

Step 2 Measure the distance from the top of the smoke shelf to the point where the flue lining is to be set.

Step 3 Mark the width of the flue on the wall.

Step 4 Draw a line from the flue mark to where the corbeling will start at the smoke-shelf level. The line can be marked clearly on the parged wall. The corbeling is then laid to match this line.

4.6.6 Front Facing

The front facing gives the fireplace its decorative and architectural appearance. Clay brick, stone, and similar materials are anchored to the rough fireplace walls to form the face. Since it is the part of the fireplace that will be seen, the work should be of the highest quality.

The first course of masonry is laid to a line across the front of the fireplace to ensure that it is perfectly straight. If clay brick is used, it should be dry-bonded to make sure that whole clay brick will be used at the lintel. If a half clay brick were to appear in the center of the lintel because of improper bonding, the work would need to be torn down and rebuilt.

If stone is used for the facing, dry bonding is not necessary. The different sizes of the stone enable the masonry work to be varied without coursing problems. However, care must be used to achieve the pattern symmetry or randomness that is desired. Wall ties projecting from the rough masonry are used to tie the stonework to the backup work.

A lintel is placed across the opening of the fireplace to support the masonry that will be laid above. It should be set one course of firebrick lower than the damper and have a minimum bearing surface of 6 inches on each side of the opening. The lintel should be at least ⅜ inch thick, 3½ inches high, and 3½ inches wide. If the opening of the fireplace is 48 inches or more, a heavier angle iron must be used to prevent sagging.

Before building the first fire, let the fireplace cure at least 7 to 10 days. Start the fire slowly and keep it back to the rear of the firebox. The air in the flue must heat up before the fireplace can draw smoothly. Never start a fireplace off with a big roaring fire, because it can crack the exterior face of the fireplace or damage the firebox. Build the fire up gradually.

4.7.0 Understanding Multiple-Opening Fireplaces

The conventional fireplace, open on only one side, is the most popular type because it is the simplest to build. When fireplaces are open on more than one side, they are called multiple-opening fireplaces. *Figure 47* shows different designs for a multiple-opening fireplace.

Although the multiple-opening or multiple-face fireplace may appear to be a modern design, it is actually old in origin. The corner fireplace, which has two adjacent sides open, has been popular in Scandinavian countries for centuries. The fireplace with opposite faces open, a necessary feature when a fireplace is to serve two rooms with one chimney, has been used in this country for some time. When a fireplace is located in the center of a room, it is known as an island fireplace.

Although attractive, multiple-opening fireplaces present many more problems concerning draft and operation than do conventional fireplaces. Cross drafts blowing from face to face can cause smoke to be blown into the room. Oversized flues and glass fire screens on one or more of the face openings may be necessary to control these problems.

Regular dampers can be used for multiple-opening fireplaces, but special dampers are also available. These dampers are more efficient than the standard throat damper. A regular damper has a long, narrow throat designed to make the fire burn evenly along the back wall of the fireplace with only one side open. A multiple-opening fireplace has a larger area of opening that requires not only a larger flue but also a larger damper throat. The larger throat creates a stronger draft. This helps prevent problems

---

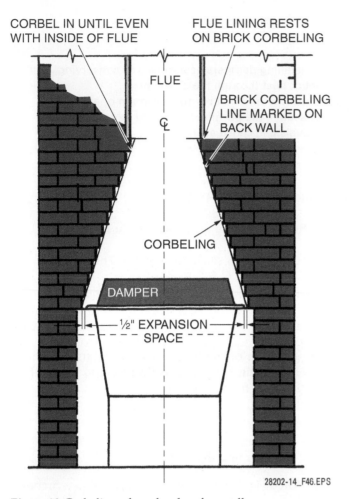

Figure 46 Corbeling of smoke chamber walls.

from crosscurrents. Special dampers also have a high funnel shape to promote even burning of the fire.

The most important advantage of the multiple-opening damper is that it permits a choice of flue locations. When a multiple-opening fireplace is built, often the flue cannot be located directly above the fireplace. Therefore, the flue lining must be slanted or angled. The smoke shelf can be built on any of four sides when using the special multiple-opening damper. Which side to use depends on the best location for the flue.

Dampers for multiple-opening fireplaces are available with smooth, high-funnel throat shapes that offer as little obstruction to draft as possible. Some have removable valve plates with adjustable tension devices to permit draft adjustments. Once the adjustments are made, it is rarely necessary to change them. The multiple-opening damper is also useful for single-sided fireplaces with very large openings. These allow more control of the draft than standard single-opening dampers.

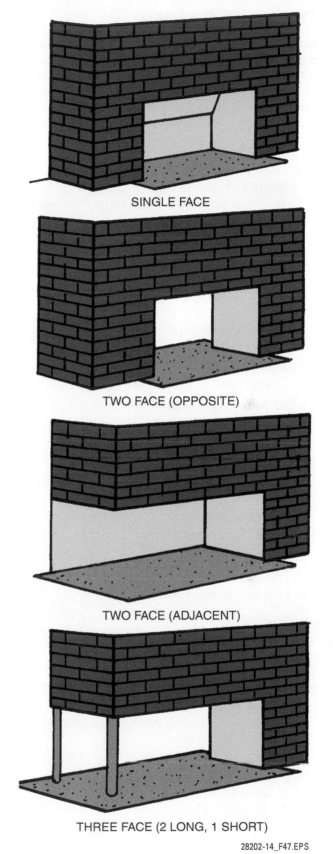

Figure 47 Multiple-opening fireplace design.

Damper Safety

Most industry standards and local building codes recommend installing dampers as a means of conserving energy. A fireplace without a damper loses a significant amount of heat through the chimney.

As a safety precaution against asphyxiation, most standards recommend that dampers remain open at least 10 percent when in a closed position. If a damper is completely sealed, the burning fire would release carbon monoxide into the room. The air in the room could reach toxic levels and cause asphyxiation.

Additional Resources

Technical Note TN19, *Residential Fireplace Design*. 1993. Reston, VA: The Brick Industry Association. **www.gobrick.com**

Technical Note TN19A, *Residential Fireplaces, Details and Construction*. 2000. Reston, VA: The Brick Industry Association. **www.gobrick.com**

Technical Note TN19C, *Contemporary Brick Masonry Fireplaces*. 2001. Reston, VA: The Brick Industry Association. **www.gobrick.com**

4.0.0 Section Review

1. A fireplace and chimney will normally provide a better draft if it is placed _____.

   a. as far as possible from windows and doors
   b. along an exterior wall
   c. in the center of the house
   d. along an interior wall

2. The walls that support and enclose the firebox are called _____.

   a. butt walls
   b. smoke shelves
   c. lintels
   d. ash pits

3. Fireplace foundations should be _____.

   a. mortared on top of the building foundation
   b. twice the thickness of the building foundation
   c. independent of the building foundation
   d. integral with the building foundation

4. To accommodate a room with 300 square feet of floor area, a fireplace should have an opening of _____.

   a. 24 to 30 inches
   b. 26 to 32 inches
   c. 28 to 34 inches
   d. 30 to 36 inches

5. The part of the hearth under the firebox that forms the floor of the firebox and holds the burning fuel is called the _____.

   a. inner hearth
   b. outer hearth
   c. upper hearth
   d. lower hearth

6. The height of the vertical back wall in a firebox is measured _____.

   a. vertically from the hearth to the base of the smoke chamber
   b. vertically from the hearth to the point where the back wall begins inclining
   c. along the sloped or inclined back wall from the top of the vertical back wall to the finished height of the back wall
   d. vertically from the face of the opening to the front edge of the outer hearth

7. In a multiple-opening fireplace, the primary advantage of a multiple-opening damper is that it _____.

   a. encourages crosscurrents
   b. permits a choice of flue locations
   c. is more efficient than standard throat dampers
   d. allows the use of a smaller damper

SUMMARY

Residential masonry involves not only the construction of the walls and partitions of the structure, but also foundations, patios, walks, steps, decks, fireplaces, and chimneys. These components of a residence require special expertise to build effectively and efficiently.

The foundation is the component of a structure that supports its weight and resists settling. Footings at the bottom of the foundation transmit the weight from the foundation walls to the soil beneath. The two general types are spread foundations and pile foundations. Footings and foundation walls may be constructed of reinforced concrete or masonry units. Masonry walls for foundations can be reinforced with steel bars and grout, or pilasters. Exterior drainage should be provided to keep hydrostatic pressure on the walls to a minimum and prevent water from getting inside.

Paving materials include clay brick, concrete paver units, and interlocking concrete pavers. There are many sizes and shapes to choose from. You must carefully select the type of unit to use for a particular project, based on such factors as environment, traffic, loading, and soil conditions. The two basic methods of laying clay brick pavers are mortarless construction and the conventional use of mortar. Mortarless construction uses a base of sand or gravel to lay the pavers on. The spaces between the pavers are filled with sand or other fine material to keep shifting to a minimum.

Masonry units can be used to build steps and decks. For maximum stability and economy, clay brick steps should be installed in a mortar bed over a concrete base. You must figure out the rise and run of the steps in order to build a uniform series of steps that will be safe and comfortable for walking.

Fireplace construction involves many different masonry techniques. You must be able to read and understand plans and drawings, know how heat radiation and drafting work, and be able to lay out and build clay brick walls to tight specifications. Fireplace construction is complicated by the fact that there are a number of dimensions that need to be satisfied in relation to each other. If this is not done, the structure may be out of proportion to the surrounding room and may not function as intended. Fireplace construction uses firebrick; the application of mortar is different, and joints in firebrick are much smaller because the heat of a fire has a tendency to burn out large joints.

1. The purpose of the footing is to resist _____.
   a. groundwater infiltration
   b. settling of the building
   c. termite infestation
   d. stress on floor joists

2. When a pile foundation is used, the foundation walls are placed _____.
   a. directly upon the pilings
   b. on a concrete footing
   c. on a beam or other footing that rests upon the pilings
   d. on a concrete slab

3. Unformed footings are produced by pouring concrete _____.
   a. in a trench cut into the earth
   b. in a metal mold, then delivering it to the site
   c. along the edges of the foundation walls
   d. in a thick layer at each foundation corner

4. Footings that support a foundation wall on a sloping grade are called _____.
   a. intermittent
   b. sloped
   c. isolated
   d. stepped

5. For a raft foundation, the walls and slab floor are cast _____.
   a. separately on site, then assembled
   b. in large metal forms
   c. as a single unit
   d. in a factory, then assembled on site

6. The horizontal stiffness of a foundation wall can be increased by _____.
   a. adding pilasters every 16 inches
   b. incorporating bond beams into the design
   c. grouting the hollow cells
   d. using a thicker slab

7. Salvaged clay brick should not be reused in outdoor structures because _____.
   a. it contains asbestos
   b. its sizes are irregular
   c. it is too soft and will quickly start flaking
   d. it will not bond properly with new mortar

8. Clay brick sizes described as actual are based on the _____.
   a. 4-inch modular system
   b. 32-mm system
   c. 6-inch modular system
   d. 3-4-5 ratio

9. An example of a static bond pattern is the _____.
   a. Flemish bond
   b. herringbone
   c. running bond
   d. stack bond

10. When laying a basket-weave pattern, using a 4-inch × 8-inch actual brick size will _____.
    a. require frequent alignment adjustment
    b. minimize the need for cutting
    c. simplify leveling of the surface
    d. require excessive cutting

11. When clay brick pavers are used for interior flooring, face mortar joints are tooled with a _____.
    a. trowel
    b. concave jointer
    c. groover
    d. flat jointer

12. In a thin-set installation, thin clay pavers are bedded in _____.
    a. epoxy adhesive
    b. fine sand
    c. portland cement mortar
    d. grout

13. To provide proper drainage, walks and patios should be sloped _____.
    a. ¹⁄₁₆ to ⅛ inch per foot
    b. ⅛ to ¼ inch per foot
    c. ¼ to ½ inch per foot
    d. ⅜ to ⅝ inch per foot

14. To discourage weed and grass growth between the joints of a mortarless patio or walkway, the installation often includes a layer of _____.

    a. sand
    b. roofing paper
    c. geotextile fabric
    d. gravel

15. To level clay brick in a mortar bed, use a rubber mallet or _____.

    a. a weighted roller
    b. a trowel handle fitted with a rubber tip
    c. a mason's hammer
    d. vibrating tamper

16. Properly installed interlocking concrete pavers must provide _____.

    a. only horizontal interlock
    b. both horizontal and vertical interlock
    c. only rotational interlock
    d. horizontal, vertical, and rotational interlock

17. When installing concrete pavers, the base material should be compacted in layers that each have a thickness of _____.

    a. 2 to 3 inches
    b. 3 to 5 inches
    c. 4 to 6 inches
    d. 5 to 7 inches

18. For maximum loadbearing support, ICPI recommends laying street pavers in a _____.

    a. double basket-weave pattern
    b. herringbone pattern
    c. running-bond pattern
    d. single basket-weave pattern

19. A flight of steps should include a landing if its length is greater than _____.

    a. 5 feet
    b. 6 feet
    c. 7 feet
    d. 8 feet

20. The width of a flight of masonry steps should be _____.

    a. one-half its length
    b. 36 inches
    c. at least as wide as the sidewalk leading to it
    d. four times the riser height

21. For proper drainage, a patio should slope away from the house 1 inch for every _____.

    a. 1 foot of length
    b. 4 feet of length
    c. 6 feet of length
    d. 8 feet of length

22. Concrete pavers used for a patio with mortared construction must be placed on a base consisting of .

    a. screeded sand
    b. ¾" gravel
    c. compacted stone chips
    d. a concrete slab

23. Pavers may be mortared by brushing into the joints a dry mix of cement and sand with a cement/sand ratio of _____.

    a. 1:3
    b. 1:4
    c. 1:5
    d. 1:6

24. In a fireplace, smoke may flow into the room instead of up the chimney if the fireplace has an inadequate _____.

    a. draft
    b. hearth
    c. damper
    d. smoke shelf

25. The vertical opening that extends from the smoke chamber to the top of the chimney is called the _____.

    a. smoke chute
    b. vent
    c. flue
    d. cleanout

26. To ensure that a fireplace provides efficient heating and draft, the firebox size and fireplace opening will vary according to _____.

    a. the location of the building
    b. the size of the room
    c. the materials used for the firebox
    d. the fuel being burned

27. Use a plumb bob, laser, or straightedge to transfer measurements from the opening to the foundation when the height of the first-floor opening above the fireplace foundation is at least _____.

a. 6 feet
b. 5 feet
c. 4 feet
d. 3 feet

28. For fire protection, flue liners should be surrounded by hard-fired clay brick to a thickness of at least _____.

a. 10 inches
b. 8 inches
c. 6 inches
d. 4 inches

29. An ash-pit cleanout door is usually installed in the fireplace foundation wall when the wall height reaches _____.

a. 2 or 3 courses
b. 24 inches
c. 4 or 5 courses
d. 18 inches below the firebox location

30. The smoke shelf must be located slightly below _____.

a. the fireplace opening
b. the bottom of the damper
c. the center of the throat opening
d. the top of the damper

Trade Terms Quiz

Fill in the blank with the correct term that you learned from your study of this module.

1. A wall portion projecting from a wall face and serving as a vertical column and/or beam is called a(n) _____.

2. _____ is the placement of paving brick or block on a horizontal surface in some pattern to form a smooth, flat surface without the use of mortar.

3. To lay masonry units to form a shelf or ledge is to _____.

4. _____ is the drainage of storm water and other runoff into the ground beneath a paved surface.

5. Solid rock that cannot be easily dislodged or removed from the soil is called _____.

6. _____ are brick, solid concrete block, or patterned concrete block that are used to build smooth, horizontal surfaces.

7. The process of using a trowel to fill collar joints and other wide openings with mortar is called _____.

8. A(n) _____ is a current of air that moves with force down an opening in a chimney.

9. An enlargement at the bottom of a wall that distributes the weight of the superstructure over a greater area to prevent settling is called a(n) _____.

10. The pressure at any point in a liquid at rest, equal to the depth of the liquid multiplied by its density is called _____.

11. _____ is the load on a bearing surface divided by its area, expressed in pounds per square inch.

12. Any structure that supports any vertical load in addition to its own weight is said to be _____.

13. _____ stone block or brick has beveled edges that do not go all the way across the edge or end of the block.

14. The process of applying a thin coat of plaster on a masonry surface is called _____.

15. A(n) _____ is a metal device used for regulating the draft in the flue of a chimney, usually made of cast iron.

16. The concave area between two perpendicular planes, such as the area at the intersection of a wall and floor or of wall and ceiling is called a(n) _____.

Trade Terms

Bearing pressure
Bedrock
Chamfered
Corbel
Cove
Damper

Downdraft
Footing
Hydrostatic pressure
Infiltration
Loadbearing
Mortarless paving

Parge
Pavers
Pilaster
Slushing

Trade Terms Introduced in This Module

Bearing pressure: The load on a bearing surface divided by its area, expressed in pounds per square inch.

Bedrock: Solid rock that cannot be easily dislodged or removed from the soil. Typically, the rock that forms the outer crust of the earth.

Chamfered: Stone block or brick with beveled edges that do not go all the way across the edge or end of the block.

Corbel: The process of laying masonry units to form a shelf or ledge.

Cove: A concave area between two perpendicular planes; the area at the intersection of a wall and floor, or wall and ceiling.

Damper: A metal device used for regulating the draft in the flue of a chimney, usually made of cast iron.

Downdraft: A current of air that moves with force down an opening in a chimney.

Footing: An enlargement at the bottom of a wall that distributes the weight of the superstructure over a greater area to prevent settling.

Hydrostatic pressure: The pressure at any point in a liquid at rest, equal to the depth of the liquid multiplied by its density.

Infiltration: The drainage of storm water and other runoff into the ground beneath a paved surface.

Loadbearing: Any structure that supports any vertical load in addition to its own weight.

Mortarless paving: The placement of paving brick or block on a horizontal surface in some pattern to form a smooth, flat surface. No mortar is used to bond the units to the surface underneath or to each other.

Parge: To apply a thin coat of mortar or grout on the outside surface of a masonry surface to prepare it for the attachment of veneer or tile, or to waterproof it.

Pavers: Brick, solid concrete block, or patterned concrete block that are used to build smooth, horizontal surfaces. Pavers are manufactured in many different thicknesses and shapes.

Pilaster: A wall portion projecting from a wall face and serving as a vertical column and/or beam.

Slushing: The process of using a trowel to fill collar joints and other wide openings with mortar.

Additional Resources

This module presents thorough resources for task training. The following resource material is suggested for further study.

ASTM C33, Standard Specification for Concrete Aggregates, Latest Edition. West Conshohocken, PA: ASTM International.

ASTM D698, Standard Test Methods for Laboratory Compaction Characteristics of Soil Using Standard Effort (12 400 ft-lbf/ft³ (600 kN-m/m³)), Latest Edition. West Conshohocken, PA: ASTM International.

ASTM C902, Standard Specification for Pedestrian and Light Traffic Paving Brick, Latest Edition. West Conshohocken, PA: ASTM International.

ASTM C936, Standard Specification for Solid Interlocking Concrete Paving Units, Latest Edition. West Conshohocken, PA: ASTM International.

ASTM C1272, Standard Specification for Heavy Vehicular Paving Brick, Latest Edition. West Conshohocken, PA: ASTM International.

Bricklaying: Brick and Block Masonry. 1988. Brick Industry Association. Orlando, FL: Harcourt Brace & Company.

Concrete Masonry Handbook, Fifth Edition. W. C. Panerese, S. K. Kosmatka, and F. A. Randall, Jr. Skokie, IL: Portland Cement Association.

Patios & Walkways. 2010. Peter Jeswald. Newtown, CT: Taunton Press.

Tech Spec Number 2, *Construction of Interlocking Concrete Pavements*. 2011. Herndon, VA: Interlocking Concrete Pavement Institute. **www.icpi.org**

Technical Note TN14, *Paving Systems Using Clay Pavers*. 2007. Reston, VA: The Brick Industry Association. **www.gobrick.com**

Technical Note TN19, *Residential Fireplace Design*. 1993. Reston, VA: The Brick Industry Association. **www.gobrick.com**

Technical Note TN19A, *Residential Fireplaces, Details and Construction*. 2000. Reston, VA: The Brick Industry Association. **www.gobrick.com**

Technical Note TN19C, *Contemporary Brick Masonry Fireplaces*. 2001. Reston, VA: The Brick Industry Association. **www.gobrick.com**

Vocational Skills Training for Segmental Paver Installation, First Edition. 2003. Stephen Jones. Prior Lake, MN: Pave Tech, Inc.

Figure Credits

Section Review Answers

Answer	Section Reference	Objective
Section One		
1. d	1.1.1	1a
2. a	1.2.0	1b
3. b	1.3.0	1c
Section Two		
1. c	2.1.1	2a
2. d	2.2.2	2b
3. a	2.3.0	2c
4. b	2.4.4	2d
Section Three		
1. c	3.1.0	3a
2. c	3.2.0	3b
3. a	3.3.0	3c
4. b	3.4.0	3d
5. c	3.5.2	3e
6. d	3.6.0	3f
Section Four		
1. c	4.1.1	4a
2. a	4.2.0	4b
3. c	4.3.0	4c
4. d	4.4.0	4d
5. a	4.5.2	4e
6. b	4.6.1	4f
7. b	4.7.0	4g

NCCER CURRICULA — USER UPDATE

NCCER makes every effort to keep its textbooks up-to-date and free of technical errors. We appreciate your help in this process. If you find an error, a typographical mistake, or an inaccuracy in NCCER's curricula, please fill out this form (or a photocopy), or complete the online form at **www.nccer.org/olf**. Be sure to include the exact module ID number, page number, a detailed description, and your recommended correction. Your input will be brought to the attention of the Authoring Team. Thank you for your assistance.

Instructors – If you have an idea for improving this textbook, or have found that additional materials were necessary to teach this module effectively, please let us know so that we may present your suggestions to the Authoring Team.

NCCER Product Development and Revision
13614 Progress Blvd., Alachua, FL 32615

Email: curriculum@nccer.org
Online: www.nccer.org/olf

❏ Trainee Guide ❏ Lesson Plans ❏ Exam ❏ PowerPoints Other _____

Craft / Level: _____ Copyright Date: _____

Module ID Number / Title: _____

Section Number(s): _____

Description: _____

_____

_____

_____

_____

Recommended Correction: _____

_____

_____

_____

_____

Your Name: _____

Address: _____

_____

Email: _____ Phone: _____

28203-14

Reinforced Masonry

Reinforced masonry is masonry that has been reinforced with grout or with a combination of grout and rebar. Grout is a fluid cementitious mixture used to fill masonry cores or cavities to increase structural performance. Grout is used to bond wythes together, transfer lateral stress from masonry to rebar reinforcement, and help support vertical loads. The grout bonds the steel and masonry together to carry loads as a single unit. This module presents information on the manufacture and use of grout in grouted and reinforced walls and reinforced masonry elements such as bond beams, lintels, piers, pilasters, and columns.

Module Three

Objectives

When you have completed this module, you will be able to do the following:

1. Name and describe the primary ingredients in grout and how it is prepared.
   a. Explain the characteristics of coarse and fine aggregates.
   b. Explain the characteristics of admixtures.
   c. Explain the role of water content in grout.
   d. Explain why compressive strength is important.
   e. Explain what mix specifications are and why they are important.
   f. Explain the procedures for mixing grout.
2. Describe how grout is placed.
   a. Explain what low-lift grouting is and how to place grout using this technique.
   b. Explain what high-lift grouting is and how to place grout using this technique.
   c. Explain why mortaring of joints for grouted masonry is important.
   d. Explain how to use mechanical vibrators with grout.
3. Describe how to construct reinforced walls and masonry elements.
   a. Explain how to cut and bend rebar.
   b. Explain how to place rebar in reinforced walls.
   c. Explain how to install bond beams and bond-beam lintels.
   d. Explain how to install precast lintels.
   e. Explain how to install piers, pilasters, and columns.

Performance Tasks

Under the supervision of your instructor, you should be able to do the following:

1. Place grout in a hollow block wall and properly consolidate it.
2. Construct shoring for a masonry lintel.
3. Build a masonry lintel out of CMU.
4. Build a pier or pilaster.

Trade Terms

Blowout	Key	Rebar
Bond beam	Lift	Reinforced masonry element
Bridging	Lintel	Reinforced wall
Column	Pier	Rodding
Grouted wall	Pilaster	Vibrating

Industry-Recognized Credentials

If you're training through an NCCER-accredited sponsor, you may be eligible for credentials from NCCER's Registry. The ID number for this module is 28203-14. Note that this module may have been used in other NCCER curricula and may apply to other level completions. Contact NCCER's Registry at 888.622.3720 or go to **www.nccer.org** for more information.

Code Note

Codes vary among jurisdictions. Because of the variations in code, consult the applicable code whenever regulations are in question. Referring to an incorrect set of codes can cause as much trouble as failing to reference codes altogether. Obtain, review, and familiarize yourself with your local adopted code.

Contents

Topics to be presented in this module include:

Figures and Tables

SECTION ONE

1.0.0 GROUT INGREDIENTS AND PREPARATION

Objective

Name and describe the primary ingredients in grout and how it is prepared.

   a. Explain the characteristics of coarse and fine aggregates.
   b. Explain the characteristics of admixtures.
   c. Explain the role of water content in grout.
   d. Explain why compressive strength is important.
   e. Explain what mix specifications are and why they are important.
   f. Explain the procedures for mixing grout.

Trade Terms

Grouted wall: A hollow masonry wall in which the voids are filled with grout but not reinforcing bar.

Lift: One continuous placement of grout or cement without interruption, equivalent to one layer.

Pilaster: A vertical reinforced masonry element consisting of a thickened section of a wall to which it is structurally bonded.

Reinforced masonry element: A hollow masonry structure other than a wall in which the voids in the masonry units are filled with grout and reinforcing bar.

Reinforced wall: A hollow masonry wall in which the voids in the masonry units are filled with grout and reinforcing bar.

G rout is a fluid mix of cementitious material, water, admixtures, and aggregate, that is used to reinforce loadbearing walls or enclosed spaces that contain steel reinforcement (*Figure 1*). Grout is used in grouted walls, reinforced walls, and reinforced masonry elements. A grouted wall is a hollow masonry wall in which the voids in the masonry units are filled with grout only. If reinforcing bar is added to the inside of the wall along with grout, the wall is classified as a reinforced wall. Other masonry structures that are filled with grout and reinforcing bar are called reinforced masonry elements.

Grout must be loose enough that it does not stick to a trowel. It must have enough water to

28203-14_F01.EPS

Figure 1 Reinforced foundation wall.

allow it to be poured or pumped, without segregating, to fill hollow masonry cores, or cavities. *Figure 2* shows the relative consistencies of concrete, mortar, and grout. To test the material consistency, a cone is filled and inverted. The stiffer cement and mortar will keep the shape of the cone better than the more fluid grout. An 8- to 11-inch fall, or slump, is usually specified for grout. The more fluid in the grout, the looser the consistency, and the greater the slump. In some situations, grout can be made using Type 3 portland cement to make a mortar-like mixture.

The American Society for Testing and Materials (ASTM), American Concrete Institute (ACI), the American Society of Civil Engineers (ASCE), and The Masonry Society (TMS) have published industry-wide specifications for various types of grout:

- *ASTM C476, Standard Specification for Grout for Masonry*

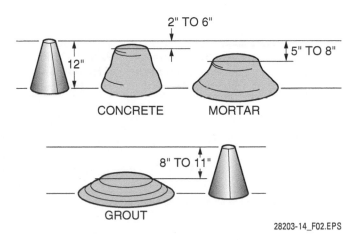

28203-14_F02.EPS

Figure 2 Slump test for concrete, mortar, and grout.

- *ACI 530/ASCE 5/TMS 402, Building Code Requirements for Masonry Structures*
- *ACI 530.1/ASCE 6/TMS 602, Specifications for Masonry Structures*

Grout must flow readily and fill masonry cores completely. The flow is affected by the choice of ingredients and the way they are mixed. It is important to select the right grout mixture and mix the ingredients properly. Failure to do so may cause voids, segregation, or poor bonding. Grout testing is covered in more detail in the *Masonry Level Two* module *Construction Inspection and Quality Control*.

1.1.0 Understanding Coarse and Fine Aggregates

Grout used in concrete masonry walls should comply with *ASTM C476, Standard Specification for Grout for Masonry*. The major difference between fine and coarse grout is not the consistency, but the aggregate used. The aggregate size in fine grout is 3/8 inch. Aggregates used in grout must comply with *ASTM C404, Standard Specification for Aggregates for Masonry Grout*. The specification also includes requirements for grading and cleanliness.

Aggregate size is important in grout because grout must fit into very tight spaces. Aggregate size does not affect overall compressive strength. *Table 1* shows mixing formulas by volume for fine and coarse grout. Note that both types of grout take the same proportions of cement to lime.

1.2.0 Understanding Grout Admixtures

Grout admixtures perform the same functions as mortar admixtures. They change the properties of the grout to meet site conditions.

The most common admixtures for grout are shrinkage compensators. Grout usually shrinks 5 to 10 percent as it cures and the surrounding masonry units absorb water. Air-entraining admixtures are not commonly used for grout because it is not usually exposed to moisture under freeze-thaw conditions.

There are five admixtures commonly used in grout:

- *Shrinkage compensators* – Cause a controlled expansion of the grout intended to offset shrinkage due to initial water loss. This improves the bond between the grout and masonry units and is especially useful for high-lift grouting.
- *Accelerators* – Increase the heat of hydration of grout. They are used in cold weather to decrease setting time and increase the rate of strength gain.
- *Retarders* – Keep the grout workable for longer time periods. They are used in hot weather or when grout must be transported from off site and will not be placed right away.
- *Fly ash* – Provides a greater slump with less water. It is used as a pumping aid or to replace some of the portland cement.
- *Superplasticizers* – Reduce the water content while maintaining high flow consistency. They are not often used in grout, as they limit the time the grout can be placed.

The use of any admixture must be approved by the project engineer. All grouts with admixtures should be tested in trial batches to ensure they meet construction specifications.

> **NOTE**
> Antifreeze compounds, used to lower the freezing point of grout, are prohibited by *ASTM C476*.

> **NOTE**
> Chloride-based admixtures cannot be used in grout. They will corrode the steel reinforcement. This can cause efflorescence in the wall.

Table 1 Grout Mix Formula

Type	Parts by Volume of Portland Cement or Blended Cement	Parts by Volume of Hydrated Lime or Lime Putty	Aggregate, Measured in a Damp, Loose Condition	
			Fine	Coarse
Fine grout*	1	0 to 1/10	2¼ to 3 times the sum of the volumes of the cementitious materials	___
Coarse grout	1	0 to 1/10	2¼ to 3 times the sum of the volumes of the cementitious materials	1 to 2 times the sum of the volumes of the cementitious materials

* Only sand can used as an aggregate in fine grout.

1.3.0 Understanding the Role of Water in Grout

The amount of water in grout determines its consistency and relates to its compressive strength. Water affects grout in several key ways:

- It affects compressive strength.
- It ensures that grout will pour or pump easily without segregating.
- It allows grout to flow around reinforcement.
- It allows grout to fill corners and recesses without voids.

Grout water content and consistency will depend partly on external factors such as the air, the masonry unit, and the cavity size. Air and masonry units absorb the moisture from grout, thus reducing its water-to-cement ratio and increasing its strength. If the masonry is very dry and highly absorptive, the grout needs more water and a 10-inch slump. If the masonry is not absorptive, the grout needs less water and an 8-inch slump. If the air is hot and dry, the grout will need more water than if the air is damp and cool.

If the cavity has a small surface area relative to its volume, the grout will need less water than if the cavity has a large surface area relative to its volume. This is because the water is more readily absorbed by larger surface areas.

1.4.0 Understanding Compressive Strength

The mix proportions in *Table 1* produce grouts with compressive strengths ranging from 1,000 to 2,500 psi (pounds per square inch) after 28 days of curing, as tested in the laboratory according to ASTM specifications. The variation depends on the amount of mixing water used.

On the job, the in-place compressive strength of grout is usually over 2,500 psi because of the absorption of water by the masonry units. The water held in the masonry keeps the grout moist so it can hydrate slowly. Slow hydration allows the grout to develop strength that increases beyond the specifications over time.

1.5.0 Defining Mix Specifications

The particular grout mix will be part of the project specifications. Unless otherwise specified, grout mix proportions must conform to the ASTM proportions given in *Table 1*. The fineness or coarseness of grout is chosen on the basis of the size of the spaces to be grouted and the height of the pour. These spaces are typically collar joints, cavities in cavity walls, concrete masonry unit (CMU, commonly known as block) cells, hollow **pilasters**, and other enclosed recesses. *Table 2* shows the recommended cavity spacing for fine or coarse grout.

The height of the cavity to be filled is known as the grout pour height. The size of the cavity limits the grout pour height. A larger cavity is needed if it is to be poured from a greater height. Fine grout is typically used to fill cavities that are smaller than 2 inches. Coarse grout is typically used to fill cavities that are 2 inches or larger.

1.6.0 Mixing Grout

Grout can either be mixed at the job site or at a central batching plant that is designed to blend the materials in predetermined proportions ac-

Table 2 Allowable Grout Pour Heights

Grout Type[1]	Max. Grout Pour Height, ft	Min. Width of Grout Space[2,3], in	Min. Grout Space Dimensions for Grouting Cells of Hollow Units[3,4] in × in
Fine	1	¾	1½ × 2
Fine	5	2	2 × 3
Fine	12	2½	2½ × 3
Fine	24	3	3 × 3
Coarse	1	1½	1½ × 3
Coarse	5	2	2½ × 3
Coarse	12	2½	3 × 3
Coarse	24	3	3 × 4

[1] Fine and coarse grouts are defined in *ASTM C476* (ref. 2).
[2] For grouting between masonry wythes.
[3] Grout space dimension is the clear dimension between any masonry protrusion and shall be increased by the diameters of the horizontal bars within the cross section of the grout space.
[4] Area of vertical reinforcement shall not exceed 6 percent of the area of the grout space.

Aggregate Size

The smaller the grout space, the smaller the maximum aggregate size that will fit. Although ASTM specifications limit maximum aggregate size to ½ inch, some job specifications call for aggregates up to ¾ inch. This larger aggregate is used in large cavities such as columns and pilasters. Grout with larger aggregate takes up more space, shrinks less, and requires less cement to meet strength standards.

cording to a specification. When grout is batched at a plant off the job site, it requires continual agitation to keep it from segregating. This is the main reason that grout is usually delivered in a truck-mounted mixer that agitates it continually. Grout is mixed on the job site under unusual circumstances or only in very small amounts that can be used quickly.

For large jobs, it is recommended that grout be mixed and delivered in conformance with *ASTM C94, Specifications for Ready-Mixed Concrete*. Typically, premixed grout will be transferred from the truck mixer into a pump hopper. It is pumped from the hopper to its deposit site. *Figure 3* shows a typical grout pump and hopper. Grout should be placed within 90 minutes of the initial mixing at the plant.

Grout can be mixed at the job site in a two-speed cement or mortar mixer, using the same general procedures as for mixing mortar. The two-speed mixer will have a high mixing speed and a low agitating speed to keep the grout from segregating without overmixing it.

Before using any power mixer, review the equipment's operating manual, as well as any general safety procedures for working with power equipment. When mixing grout, use the following procedure:

Step 1 Check the formula.

Step 2 Premeasure the ingredients.

Step 3 Add one-third of the amount of water.

Step 4 Add one-half of the sand.

Step 5 Start the mixer at high speed. On some mixers this is marked as mix speed.

Step 6 Add the cement and lime.

Step 7 Add the remaining sand.

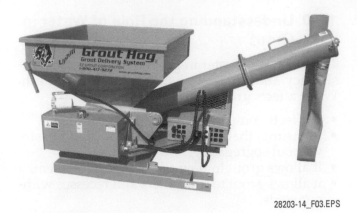

28203-14_F03.EPS

Figure 3 Grout pump and hopper.

Step 8 Add one-third of the remaining water.

Step 9 Blend for one to two minutes.

Step 10 Add the remaining water.

Step 11 Blend for an additional three to four minutes; for a total of five minutes, minimum.

Step 12 Empty the grout into a bucket or other container with a lip to make it easier to pour.

Step 13 Switch the mixer to low (agitate) speed if there is any grout remaining in the mixer.

Step 14 Discard all grout not placed within 90 minutes after the water is added.

Step 15 Clean the mixer thoroughly after emptying the grout.

> **WARNING!**
>
> Grout is caustic. You should always wear appropriate personal protective equipment when mixing or placing grout.

Check Local Codes

ASTM C476 and *ASTM C404* specify grout and aggregate gradations, but local codes sometimes differ on specific items. Items such as maximum aggregate size allowed per opening size may vary locally, so it is important to check codes as well as specifications.

Additional Resources

ACI 530/ASCE 5/TMS 402, Building Code Requirements for Masonry Structures, Latest Edition. Reston, VA: American Society of Civil Engineers.

ACI 530.1/ASCE 6/TMS 602, Specifications for Masonry Structures, Latest Edition. Reston, VA: American Society of Civil Engineers.

ASTM C94, Specifications for Ready-Mixed Concrete, Latest Edition. West Conshohocken, PA: ASTM International.

ASTM C404, Standard Specification for Aggregates for Masonry Grout, Latest Edition. West Conshohocken, PA: ASTM International.

ASTM C476, Standard Specification for Grout for Masonry, Latest Edition. West Conshohocken, PA: ASTM International.

1.0.0 Section Review

1. The size of aggregate in fine grout must not exceed _____.

   a. ⅓ inch
   b. ⅜ inch
   c. ½ inch
   d. ⅝ inch

2. Admixtures that keep the grout workable for longer time periods are called _____.

   a. accelerators
   b. shrinkage compensators
   c. superplasticizers
   d. retarders

3. If the surface area of a cavity is small relative to its volume, the amount of water required by the grout compared to a cavity with a large surface area relative to its volume will be _____.

   a. proportional
   b. the same
   c. more
   d. less

4. Because of the absorption of water by the masonry units, the in-place compressive strength of grout usually exceeds _____.

   a. 2,500 psi
   b. 4,000 psi
   c. 5,500 psi
   d. 7,000 psi

5. The grout pour height is limited by the _____.

   a. thickness of the grout
   b. size of the aggregate
   c. mix specification
   d. size of the cavity

6. Once initially mixed at the plant, grout should be placed within _____.

   a. 30 minutes
   b. 60 minutes
   c. 90 minutes
   d. 120 minutes

2.0.0 GROUT PLACEMENT

Objective

Describe how grout is placed.
   a. Explain what low-lift grouting is and how to place grout using this technique.
   b. Explain what high-lift grouting is and how to place grout using this technique.
   c. Explain why mortaring of joints for grouted masonry is important.
   d. Explain how to use mechanical vibrators with grout.

Performance Task

Place grout in a hollow block wall and properly consolidate it.

Trade Terms

Blowout: The swelling or rupture of a cavity wall from too much pressure caused by pouring liquid grout into the cavity.

Bond beam: A course of masonry units with steel rebar inserted and held in place by a solid fill of grout or mortar; used as a lintel or reinforcement beam to distribute stress.

Bridging: The mounding of grout or cement over an obstruction, creating a void under the obstruction.

Key: A recess or groove in one placement of grout or concrete that is later filled with a new placement of grout or concrete so that the two lock together in a tongue-and-groove configuration.

Rebar: Reinforcing bar embedded in concrete, mortar, or grout in such a manner that it acts together with the other components to resist loads.

Rodding: Poking the grout with a rod in order to consolidate it.

Vibrating: Consolidating grout with the use of a mechanical vibrator.

It is important to place grout slowly. Because of its viscosity and aggregates, grout can trap air bubbles as it fills a space. You must consolidate grout by vibrating or rodding to ensure that it fills all voids and adheres to all reinforcement without air gaps. Vibrating involves the use of a mechanical vibrator to consolidate grout. Rodding involves poking the grout with a rod to consolidate it.

The amount of grout placed continuously without interruption is called a lift. Grout is applied in either a low lift or a high lift. Low-lift grout is poured to fill a space less than 5 feet high, while high-lift grouting fills a space over 5 feet high. A lift of 12 inches or less can be consolidated by rodding, while a lift over 12 inches must be consolidated by mechanical vibration.

Before grouting, the mortar between the masonry units must be strong enough to contain the wet grout. The mortar should be allowed to harden before high-lift grouting. The time can vary depending on factors such as the number of wythes and the weather conditions.

If the mortar is not firm enough, the act of grouting may cause a blowout. A blowout occurs when the structure buckles from the weight of the grout. A movement of even one-quarter of an inch means that the wall has to be taken down and rebuilt. This is why, even if the mortar has firmed, it is important to proceed carefully and slowly. Blowouts may also be caused by excessive vibration and the excessive use of cut masonry units.

Protrusions of mortar into the cavity will prevent the grout from bonding properly. Conversely, grout on surfaces to be mortared will keep the mortar from bonding properly. It is important to handle grout carefully. Keep it off any surface that is to be mortared and keep it off the face of the masonry units. Apply grout carefully. Grout on masonry surfaces is very hard to clean.

WARNING!

Working with dry cement or wet grout can be hazardous to your health. Dry cement dust can enter open wounds and cause blood poisoning. The cement dust, when it comes in contact with body fluids, can cause chemical burns to the membranes of the eyes, nose, mouth, throat, or lungs. It can also cause a fatal lung disease known as silicosis. Wet cement or concrete can also cause chemical burns to the eyes or skin. Repeated contact with cement or wet grout can also cause an allergic condition known as cement dermatitis. Make sure to wear appropriate personal protective equipment when working with dry cement or wet grout. If wet grout enters waterproof boots from the top, remove the boots and rinse your legs, feet, boots, and clothing with clear water as soon as possible.

2.1.0 Using Low-Lift Grouting Techniques

Single-wythe block walls to be grouted with a low lift are first built to a height of no more than 5 feet. The hardening of the mortar can take up to 24 hours, and grouting begins afterward. If the cores are to be reinforced, the steel reinforcement is positioned and secured. The reinforcing steel bars, called rebars, overlap and can be bound to the bars in the foundation. They protrude above the grout level to provide the specified lap into the following masonry lift, as shown in *Figure 4*. Place wire mesh or other grout stop device under bond beams to contain grout, or use a solid-bottom unit. Cells containing steel are filled solidly with grout. Vertical cells should provide a clean, continuous cavity.

The grout should not be poured to the very top of the block, but to within 1 inch of the top. This will form a key or keyway that permits the next lift of grout to bond with the block and the grout below it. The key allows the two lifts to lock together in a tongue-and-groove joint.

As the grout is placed, it should be directed against the inner face of the block to avoid getting grout splatter on the outside face of the wall (*Figure 5*). As one person places the grout, another should rod or vibrate the grout to make sure that no voids remain.

The next section of the wall is then built, and reinforcements are installed. The mortar should be allowed to harden for a predetermined amount of time before the next lift of grout is poured. This is established by the specifications. If this is to be the final section of wall, the grout will be poured all the way to the top of the last

28203-14_F05.EPS

Figure 5 Grout directed to inner face of block.

block. If there is to be another lift, the grout will be poured to within 1 inch of the top of the last block.

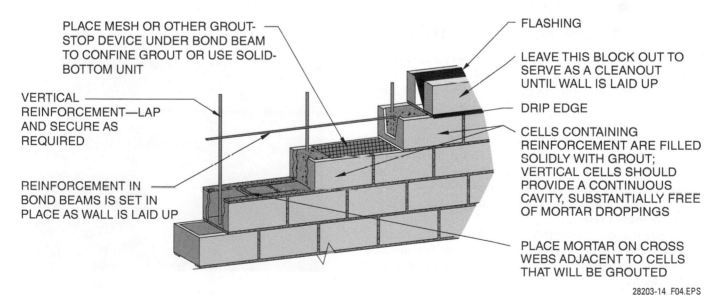

PLACE MESH OR OTHER GROUT-STOP DEVICE UNDER BOND BEAM TO CONFINE GROUT OR USE SOLID-BOTTOM UNIT

VERTICAL REINFORCEMENT—LAP AND SECURE AS REQUIRED

REINFORCEMENT IN BOND BEAMS IS SET IN PLACE AS WALL IS LAID UP

FLASHING

LEAVE THIS BLOCK OUT TO SERVE AS A CLEANOUT UNTIL WALL IS LAID UP

DRIP EDGE

CELLS CONTAINING REINFORCEMENT ARE FILLED SOLIDLY WITH GROUT; VERTICAL CELLS SHOULD PROVIDE A CONTINUOUS CAVITY, SUBSTANTIALLY FREE OF MORTAR DROPPINGS

PLACE MORTAR ON CROSS WEBS ADJACENT TO CELLS THAT WILL BE GROUTED

28203-14_F04.EPS

Figure 4 Typical reinforced concrete masonry construction using low-lift grouting.

2.2.0 Using High-Lift Grouting Techniques

High-lift grouting is complicated because of the way the wall must be built. If there is nothing in the wall to prevent the grout from flowing top to bottom without interruption, the full height of the wall can be built. Cleanout openings must be made at the bottom of the wall or at the bottom of every core containing reinforcement. *Figure 6* shows spaced cleanout openings. *Figure 7* shows a continuous cleanout opening for walls to be grouted in every vertical section of block cores.

Cleanout openings allow the masons and inspectors to be sure that all inside surfaces to be grouted are sufficiently clean and free of excessive mortar protruding inside the cavity. The openings also allow for checking the rebar ties.

Seal the cleanout openings once the cavity has passed inspection. A continuous cleanout opening can be sealed with dimension lumber secured across it. Spaced openings can be sealed with dimension lumber or with cut faces of matching masonry units. The cut faces can be secured with dimension lumber.

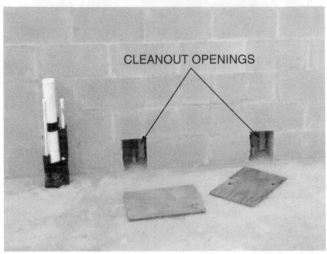

28203-14_F06.EPS

Figure 6 Spaced cleanout openings.

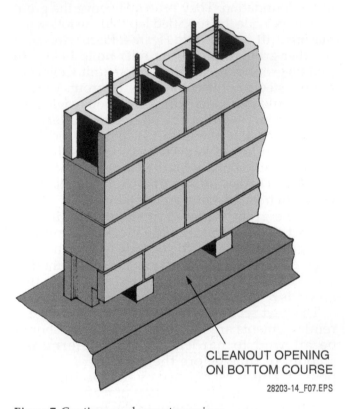

CLEANOUT OPENING ON BOTTOM COURSE

28203-14_F07.EPS

Figure 7 Continuous cleanout openings.

Innovators Make a Difference

Grout is often pumped using hoses. Heavy hoses filled with grout must be moved around the wall as cavities are filled. Damian Lang, an innovative masonry contractor based in Ohio, developed a tool that makes the mason's job easier and faster.

The Grout Hog® is designed to hold and pump grout from above the mason. A crane or a forklift lifts the pump and the hose is suspended from above. The flow is controlled by the mason and the hose can be moved easily as grouting progresses. This reduces the physical demands on the mason and increases safety.

Every mason can contribute to safety, quality, and production on the job site. Some, like Lang, will develop ideas to increase productivity for his own company and for all masons.

28203-14_SA01.EPS

2.2.1 High-Lift Grouting for Cavity Walls

High-lift grouting for cavity walls requires building the wall to the specified height (*Figure 8*). The cavity is inspected, and the cleanout openings are sealed. The grout is then poured into the cavity in lifts using a cement bucket or a pump. The grout is then vibrated and allowed to stiffen for about an hour before the remaining lift is poured. The average lift should be around 4 feet high. If the lift is higher, it will be difficult to work out the voids.

The cavity should not be filled until the mortar has hardened enough to hold the weight of the grout. Project specifications or local building codes set hardening time, which can vary significantly depending on the work and locale.

High-lift grouting requires that walls be temporarily braced until the mortar and grout have fully set. Partially completed walls should also be braced during construction against lateral loads from winds or other forces. Dimension lumber can be used for bracing. All Occupational Safety and Health Administration (OSHA) requirements for strength, spacing, and height of bracing must be met in addition to any local building codes.

2.2.2 Grouting around Reinforcement Steel

Take extra care in pouring grout when there is hardware in the cavity. Do not allow the grout stream to separate over rods, spacers, rebars, or other embedded items. This leads to a condition called bridging, where the grout mounds over the reinforcement and creates a void underneath. Further placing and consolidating may not fill that void.

To avoid bridging, fill the space underneath the obstruction first, then fill on top. Sometimes you must rod or vibrate the grout before finishing the lift to make sure there is no bridging. If bridging does occur, using a vibrator can correct it.

2.3.0 Using Mortar Joints for Grouted Masonry

Mortaring of joints becomes critical when the wall is to be grouted. Poor mortaring allows grout to

28203-14_F08.EPS

Figure 8 Wall prepared for grouting.

leak, compromising the integrity and strength of the wall. To make sure head joints are completely filled with mortar, use a shoved joint. This technique consists of buttering the ends of each unit and shoving them forcefully into place to produce a solid joint with no voids in the mortar.

Mortaring block for walls to be grouted also calls for precision. If alternate cells are to be grouted, mortar must be spread for full bed joints on shells and webs. This will keep the grout from oozing out of the cells.

Angle the inside of the mortar joint with the edge of the trowel. This bevel helps prevent mortar from excessive protrusion beyond the masonry unit. Carefully cut off mortar that protrudes into the cavity. Excessively protruding mortar will

Stronger Bonds

If at any time there will be at least an hour's delay in the grouting, the grout should be poured to 1 inch below the top of the block. Leaving that space forms a keyway that allows the new grout to bond with the block and the old grout below it.

The key method is also useful for filling cavity walls with no horizontal obstructions in the cavity. If the specifications call for horizontal reinforcement every 3 or 4 feet, the section of wall can be built that high. Then the grout can be poured and the reinforcement laid after the grout is poured.

prevent the grout from filling the space beneath the protrusion. This will result in an incomplete bond.

When working with brick or block cavity walls, always keep the cavity clear of mortar. Mortar protruding into the cavity or collected along the bottom of the cavity will keep the grout from properly bonding to the masonry and to any reinforcing steel. As you are building a cavity wall, use a trowel to clear excess mortar off the internal faces of the masonry units.

2.4.0 Using Mechanical Vibrators for Grout

Mechanical vibration consolidates grout and concrete. Vibrating forces the air out of the grout and compacts it within the cavity and around the reinforcement. Air voids rise to the surface during vibration so the cavity will be solidly filled. Mechanical vibrators (also called pencil vibrators), as shown in *Figure 9*, come in a range of sizes. The smaller machines can be carried as a backpack.

The process of using a mechanical vibrator is straightforward. As with other power equipment, read the manufacturer's instructions and review safety procedures before operating any equipment. Here are some helpful tips:

- Turn the vibrator on before you insert it and turn it off as you remove it from the grout, to avoid splatter.
- Lower the head and shaft of the vibrator vertically into the full depth of the fresh layer of grout. Keep the shaft as vertical as possible.
- If there is a previous lift that has not set up, insert the head of the vibrator into the previous lift.

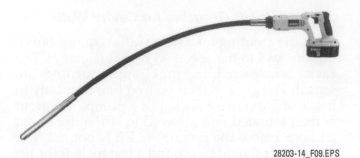

28203-14_F09.EPS

Figure 9 Mechanical vibrator.

- Manipulate the shaft up and down for 5 to 15 seconds. Do not run the vibrator any longer than 15 seconds in one location as this may cause segregation.
- Withdraw the vibrator slowly with a series of up-and-down motions.
- Keep the vibrator off any reinforcing bars.
- In a cavity, reinsert the vibrator so that it overlaps the area affected by the previous insertion. Look carefully to be sure of an overlap.

Vibration is complete when air bubbles rise to the surface and the grout settles into place.

WARNING!

Vibrators can be hazardous. To avoid accidents, remember the following rules:

- Keep hoses and electrical cords from snagging on masonry or reinforcements.
- Keep electrical wires, connectors, and equipment out of water, oil, or concrete.
- Clean the equipment thoroughly when you are finished.

The Grout Grunt™

Veteran mason Giovanni Agazzi invented the Grout Grunt™ as a way for masons to save time, money, and effort when performing grouting operations. His invention is lighter and more ergonomic than other grout delivery methods such as pumps, transit trucks, or buckets. The Grout Grunt™ features an injection-molded plastic scoop with a trapezoidal lower half and rectangular upper half. The scoop is designed to provide masons with the same amount of grout every time while also allowing faster pours, reduced lifting strain, and easier cleaning.

The original metal Grout Grunt™ weighed 10 pounds. The current models weigh just 2 and 3 pounds each, while carrying more grout per scoop. It's this type of continual innovation that has helped Grout Grunt™ become a popular and essential tool for masons in the field.

The Grout Grunt™ motto says it all: "Designed BY a mason FOR a mason."

28203-14_SA02.EPS

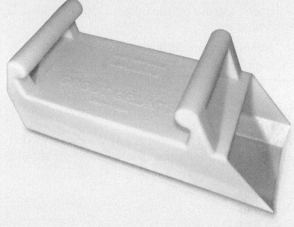

28203-14_SA03.EPS

Additional Resources

Bricklaying: Brick and Block Masonry. 1988. Brick Industry Association. Orlando, FL: Harcourt Brace & Company.

Concrete Masonry Handbook for Architects, Engineers, Builders, Fifth edition. 1991. W. C. Panerese, S. K. Kosmatka, and F. A. Randall, Jr. Skokie, IL: Portland Cement Association.

Technical Note 11E, *Guide Specifications for Brick Masonry, Part 5, Mortar and Grout.* 1991. Reston, VA: The Brick Industry Association. **www.gobrick.com**

2.0.0 Section Review

1. When placing grout using a low lift, to avoid splattering grout on the outside face of the wall, direct the grout _____.

   a. against the inner face of the block
   b. into the center of the block
   c. in a slow undulating pattern from side to side
   d. from the top of the cavity down

2. When grouting a cavity wall using high-lift technique, the height of the average lift should be around _____.

   a. 5 feet
   b. 4 feet
   c. 3 feet
   d. 2 feet

3. When using mortar joints on grouted masonry, to help ensure the grout fills the space, the inside of the mortar joint should be _____.

   a. angled with a trowel edge
   b. flush with the face of the masonry unit
   c. allowed to protrude slightly for better bonding
   d. built up with a trowel

4. Segregation of the grout may occur if a mechanical vibrator is allowed to run in one location for longer than _____.

   a. one minute
   b. 30 seconds
   c. 15 seconds
   d. five seconds

SECTION THREE

3.0.0 REINFORCED WALLS AND MASONRY ELEMENTS

Objective

Describe how to construct reinforced walls and masonry elements.

a. Explain how to cut and bend rebar.
b. Explain how to place rebar in reinforced walls.
c. Explain how to install bond beams and bond-beam lintels.
d. Explain how to install precast lintels.
e. Explain how to install piers, pilasters, and columns.

Performance Tasks

Construct shoring for a masonry lintel.
Build a masonry lintel out of CMU.
Build a pier or pilaster.

Trade Terms

Column: A vertical reinforced masonry element designed to support a load, the width of which never exceeds three times its thickness and the height of which always exceeds four times its thickness.

Lintel: The horizontal member or beam over an opening that carries the weight of the masonry above the opening.

Pier: A vertical reinforced masonry element that is typically shorter than a column or pilaster. Piers may be designed to carry loads, or may be purely ornamental.

Using steel reinforcement in masonry walls allows them to withstand greater stress and carry greater loads than unreinforced masonry walls of the same size. The most commonly used reinforcement is steel bar. Steel bar is used for vertical reinforcement and for horizontal bond-beam reinforcement.

Reinforcing steel bar, or rebar, is made of new or recycled steel. It comes in a smooth or a deformed style. The deformed bar is most commonly used in the United States. The deformations are ridges that increase the surface area for bonding.

Each bar has identification stamped on it, as shown in *Figure 10*. ASTM specifications require that the manufacturer stamp each bar with the following information:

- A letter or symbol to indicate the manufacturer's mill
- A number corresponding to the size number of the bar (*Table 3*)
- A symbol or marking indicating the type of steel (*Table 4*)
- A marking to designate the grade (*Table 5*)

Bar sizes are measured in eighths of an inch. The size number corresponds to the number of eighths of an inch of the nominal diameter of the bars. Under this system, a size 5 bar is ⅝ of an inch in diameter. A size 10 is ¹⁰⁄₈ or 1¼ inches in diameter. *Table 3* shows size and weight information for several standard sizes of rebar.

Rebar comes in several grades, which are tested for resistance to pressure in thousands of pounds per square inch (refer to *Table 5*). Rebar may be coated with epoxy or plastic to increase corrosion resistance. As with grade, deformation, and size, coating is part of the project specifications. The location and spacing of the rebar is also part of the project specifications.

It is important to keep rebar and all other reinforcement items clean and dry. Dirt and moisture will prevent a complete bond of the metal to the grout or mortar. They may also set up long-term decay or corrosion inside the grouted cavity itself.

3.1.0 Cutting and Bending Rebar

Rebar can be cut using a portable band saw or abrasive chop saw. Rods up to size 4 (½ inch in diameter) can be cut quickly and easily using a bolt cutter (*Figure 11*). Portable power shears are also used to cut rebar, especially on job sites where a great deal of cutting must be done. Powered rebar cutters can be used to cut rebar up to size 7 (⅞ inch in diameter, *Figure 12*).

Rebar should be bent cold. Rebar may be bent by hand using a tool called a hickey bar or other type of rebar bender (*Figure 13*). Hand benders can be used on rebar up to size 8 (1 inch in diameter). When a great deal of rebar must be bent, a jig can be created to ensure that all bends are uniform.

Fabrication shops use power benders. There are portable models that may be found on larger job sites. Power benders are usually able to bend any size rebar to any shape. Only workers trained in their use should use a power bender.

> **WARNING!**
>
> Powered shears or rebar cutters can be dangerous. Read the operating manual or get training in safe operation of any power equipment before using it.

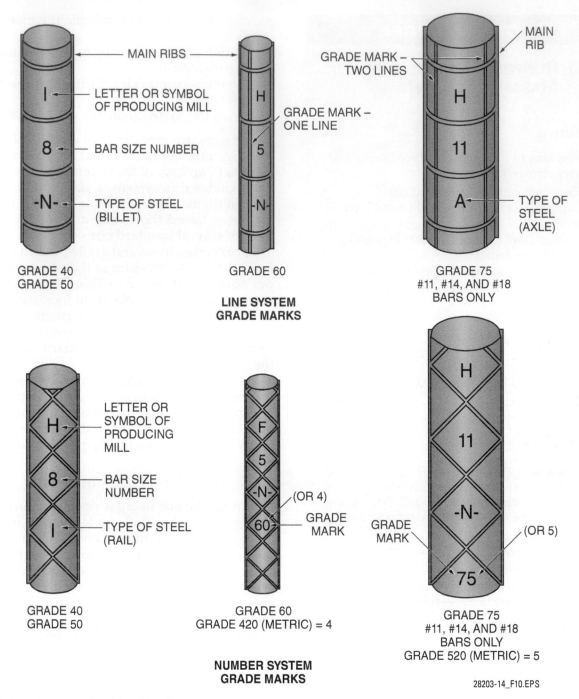

Figure 10 Reinforcement-bar identification.

3.2.0 Placing and Tying Rebar in Reinforced Walls

When working with rods in a reinforced cavity wall, masons first place short steel dowel rods into the concrete footing before it hardens. Two wythes of masonry are then laid. Rebar is placed according to the plans or specifications. It must be held in place so it will not shift during normal construction or grout placement. The bar must be aligned, supported, and rigidly fastened in place before the grout is poured. This is done by using a bar spacer. *Figure 14* shows some forms of bar spacers.

It is often necessary to secure rebar to another piece of rebar, such as when vertical and horizontal reinforcement cross. Soft-annealed iron wire is normally used for tying rebar. The ends of the wires are twisted together with pliers (*Figure 15*). It is important to bend the ends of the tie toward the rebar. Otherwise air pockets may form during grouting. *Table 6* shows the sizes and properties of wire used to secure rebar.

Table 3 Reinforcing Bar Properties

Bar Size, No.	Weight, lb/ft	Diameter, in	Cross-sectional Area, in²	Perimeter, in
3	0.376	0.375	0.11	1.178
4	0.668	0.500	0.20	1.571
5	1.043	0.625	0.31	1.963
6	1.502	0.750	0.44	2.356
7	2.044	0.875	0.60	2.749
8	2.670	1.000	0.79	3.142
9	3.400	1.128	1.00	3.544
10	4.303	1.270	1.27	3.990
11	5.313	1.410	1.56	4.430

Table 4 Reinforcing Bar Types

Symbol/ Marking	Type of Steel
A	Axle (ASTM A617)
S or N	Billet (ASTM A615)
I or IR	Rail (ASTM A616)
W	Low alloy (ASTM A706)

Table 5 Reinforcing Bar Grades

Grade	Identification	Minimum Yield Strength
40 and 50	None	40,000 to 50,000 psi (40 to 50 ksi)
60	One line or the number 60	60,000 psi (60 ksi)
75	Two lines or the number 75	75,000 psi (75 ksi)
420	The number 4	60,000 psi (60 ksi)
520	The number 5	75,000 psi (75 ksi)

Several common ties are shown in *Figure 16*. Each has a particular application.

- *Snap tie* – The simplest and most basic. This tie is normally used in flat, horizontal work to prevent the bars from moving during concrete or grout placement.
- *Wrap-and-snap tie* – Normally used when tying rebar placed in walls to keep the horizontal bars from moving. Also called the wall tie.
- *Saddle tie* – Often used to hold hooked ends of bars in position when tying footing or other mats.

Deformed rebar must be overlapped for a length that is at least 40 times its diameter. Smooth rebar must be lapped a greater distance. The lapping is secured with soft-annealed iron wire ties. Make sure the tie ends are flush to the bar to prevent air pockets in the grout.

28203-14_F11.EPS

Figure 11 Manual rebar cutter and bender.

28203-14_F12.EPS

Figure 12 Portable rebar cutter.

As the wall increases in height, the vertical rods are wired to horizontal rods for stabilizing and bracing, as well as for extra strength. *Figure 17* shows how vertical reinforcement should be positioned. A hook is added to the vertical reinforcement in high-wind areas. Grout is poured between the rods as the courses are laid.

> **WARNING!**
> OSHA standard *29 CFR 1926.701(b)* requires that all protruding reinforcing steel onto and into which workers could fall shall be guarded to eliminate the hazard of impalement.

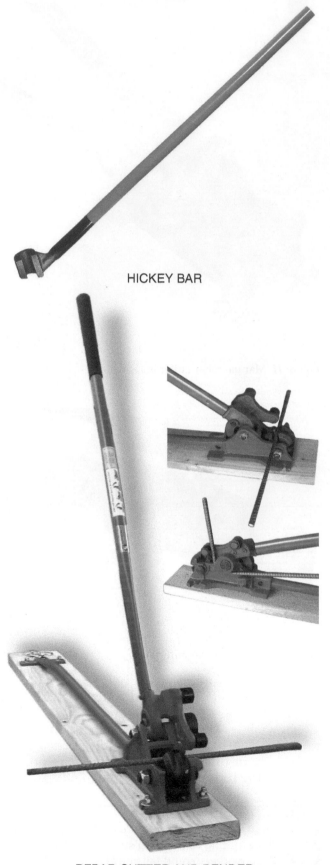

HICKEY BAR

REBAR CUTTER AND BENDER

28203-14_F13.EPS

Figure 13 Hickey bar and rebar bender.

3.3.0 Installing Bond Beams and Bond-Beam Lintels

A bond beam is a grouted, reinforced horizontal or vertical collection of masonry units. The bond beam is spaced through a structure to carry extra loads. A vertical bond beam is often used to strengthen nonbearing structures such as garden walls. A vertical bond beam or internal pilaster can be easily constructed in a hollow masonry wall.

This structure requires placing reinforcement, then framing in the sides of the pilaster. The reinforcement can be metal lath, fine hardware cloth, or cut pieces of masonry. Wood should never be used for this purpose. The beam can be grouted using high-lift techniques. Temporary bracing may or may not be necessary, depending on the age and hardness of the mortar.

Bond beams can be spaced through an otherwise unreinforced wall to add lateral strength. When the core is underneath the bond beam, the grout stop is interrupted. If solid-bottomed bond beams are used, the cells beneath it will need to be grouted first.

The bond beam is constructed of a cut or specially shaped CMU, with rebar cradled horizontally in the course of CMUs. To prevent grout from leaking, wire lath or metal screening must be put over the cells that are not to be grouted to keep the grout out. The rods are put into the beam block, as shown in *Figure 18*, and then grouted in place.

A horizontal bond beam can also be used as a lintel, which is the horizontal loadbearing member or beam over an opening. Such a bond beam, as you would expect, is called a bond-beam lintel. Bond-beam lintels are usually made of the same brick or block as the wall. The inside of the brick or block is cut out, or a special precut block is used. *Figure 19* shows some preshaped bond-beam block that are commonly used for lintels. These block have closed bottoms. *Figure 20* shows a bond-beam block cut from a standard block. The bottom of the unit is left in to hold the reinforcement and grout.

The bond-beam lintel may be built in place. Use dimension lumber to build a brace to support the lintel on the bottom and sides. Then lay out the cut masonry on the bracing and leave it in place until the lintel cures. A layer of concrete or grout is then poured into the cutout masonry. Steel reinforcement rods are laid in the cutouts and covered with another layer of concrete or grout.

3.3.1 Building a Bond-Beam Masonry Lintel

Begin by building the supporting wall to the proper height. Ensure that the supporting masonry jambs are level, plumb, ranged, and that the

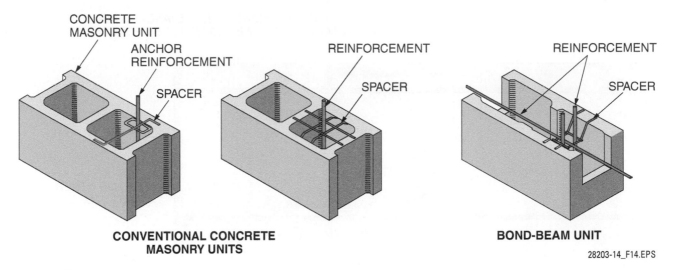

CONVENTIONAL CONCRETE
MASONRY UNITS

BOND-BEAM UNIT

28203-14_F14.EPS

Figure 14 Rebar spacers.

28203-14_F15.EPS

Figure 15 Pliers for tying rebar.

Table 6 Sizes and Properties of Masonry Wire

Wire Size	Nominal Diameter[1], in	Nominal Area, in	Nominal Perimeter, in
W1.1	0.121	0.011	0.380
W1.7	0.148	0.017	0.465
W2.1	0.162	0.020	0.509
W2.8	0.187	0.027	0.587
W4.9	0.250	0.049	0.785

[1] ASTM A82 (ref. 15) permits variation of ±0.003 in. from diameter shown.

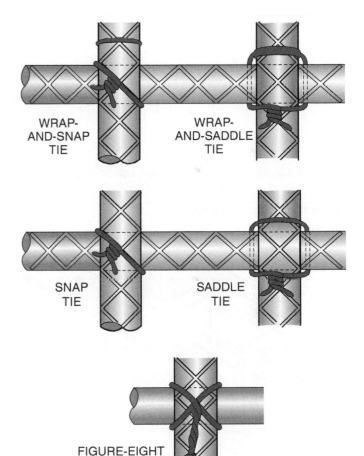

28203-14_F16.EPS

Figure 16 Types of ties.

jambs are the correct height. Next, install temporary support, called shoring. Wood, steel angles, and metals studs can be used to construct shoring. To facilitate the removal of the shoring once the bond-beam masonry lintel is finished, the length of the shoring should be at least ½ inch less than the masonry opening and shimmed to the desired height. Set the top of the support so that the first course of masonry units can be laid dry over the support so that their tops will align with the adjacent units in the supporting wall. Verti-

cal supports can be fastened with cut nails to the masonry jambs. Supports should be placed at a height that will allow the header to be shimmed up to the proper opening height.

Place vertical supports at each end of the lintel, or at regular intervals for larger openings. As a

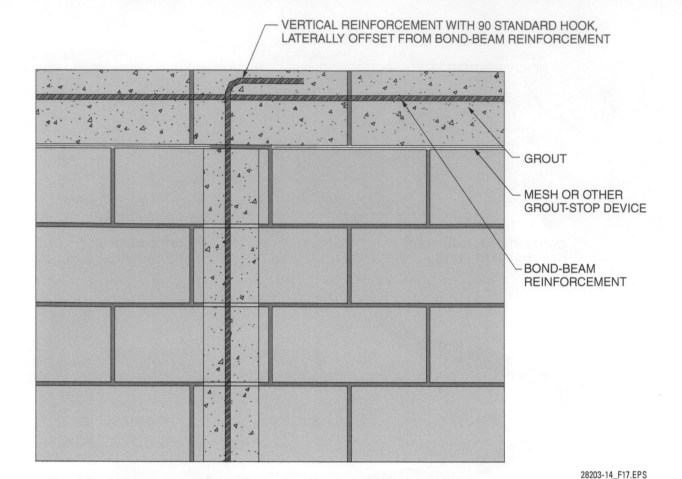

VERTICAL REINFORCEMENT WITH 90 STANDARD HOOK, LATERALLY OFFSET FROM BOND-BEAM REINFORCEMENT

GROUT

MESH OR OTHER GROUT-STOP DEVICE

BOND-BEAM REINFORCEMENT

28203-14_F17.EPS

Figure 17 Vertical reinforcement.

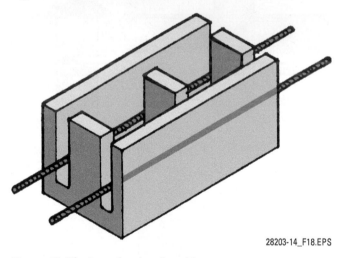

28203-14_F18.EPS

Figure 18 Placing rebar in a bond beam.

rule of thumb, place a vertical support every 24 inches. The opening height should also be at least ½ inch less than the masonry opening.

Set the first course of masonry over the shoring. Units in this course should be buttered carefully, then set on the support and slid against adjacent units to avoid trapping mortar between the units and the temporary support. Backer rod can be used in the bottom joint between these units to prevent mortar and grout from oozing out of the joint. Parge the inside of the head joints on all three sides.

Next, install the lower reinforcing bars. Ensure that the bars have a minimum clearance of ½ inch between the bar and the inside of the masonry unit. Use small pieces of block or tie wire to separate the bars and to secure them in place. Some lintels will require additional courses with horizontal reinforcing bars tied together by vertical reinforcing bars, commonly called stirrups.

Once the lower reinforcing bars have been installed, the remaining masonry lintel courses can be installed. Typically, all courses within a masonry lintel will require open-end units. Care must be taken to minimize mortar droppings within the cells and to parge the inside of the joints to reduce the chance of a blowout. These courses must be installed with care so as not to disrupt the course bearing on the temporary support.

When the mortar has set sufficiently, grout the masonry lintel. With the engineer's permission, lintels may be poured in two pours, one half at a time. Do not allow excessive vibration of the lintel during the grouting process, as this can lead to a blowout.

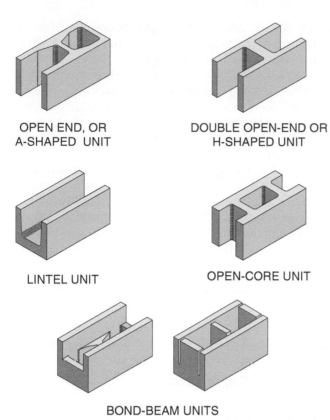

OPEN END, OR
A-SHAPED UNIT

DOUBLE OPEN-END OR
H-SHAPED UNIT

LINTEL UNIT

OPEN-CORE UNIT

BOND-BEAM UNITS

28203-14_F19.EPS

Figure 19 Bond-beam shapes.

28203-14_F20.EPS

Figure 20 Bond-beam block.

When the grout has achieved sufficient strength, remove the temporary shoring. Be careful to avoid damaging the finished masonry faces, especially if they are to be left exposed. Finally, fill and finish the joints on the underside of the lintel and any affected jamb joints. This may require grinding or chiseling out mortar and grout if the inside of any head joint was not parged properly during the initial construction.

Brush and dampen the joints prior to filling with mortar. Tool the joints to the proper finish, taking care to address the vertical return of each joint.

Note that there is another course of reinforced brick on top of the lintel. This type of structure is called a bond beam when it is not directly over an opening and is simply used to reinforce masonry. Bond beams are used where codes or specifications require reinforced masonry walls to stand up to earthquakes or high winds.

3.4.0 Installing Precast Lintels

Precast reinforced concrete lintels are often used over openings in concrete block walls and foundations. They have rectangular cross sections, as shown in *Figure 21*, or one corner may have a recess to fit over a shaped block. Precast concrete lintels may be one-piece or may be split into two thinner sections. This way, you can make up the

lintel cross section that is needed without having a special precast lintel made.

A solid, one-piece lintel is used to support only the weight of block above the span of an opening. An example of a one-piece lintel is shown in *Figure 22*. A one-piece lintel will typically have a height of 7⅝ inches and be the same thickness as the masonry wall. This size will fit in with the normal coursing of the wall block. To provide increased tensile strength to the lintel, steel reinforcing rods may be cast into the lintel. The size of the rod will depend on the load and the length of the lintel.

To place a precast concrete lintel, apply mortar to the masonry on either side of the frame. Care-

28203-14_F21.EPS

Figure 21 Precast concrete lintel.

fully set the lintel into its proper position and press down until the mortar joint is the same thickness as all other joints. Then check that the lintel is level and plumb. Remember that the lintel needs a minimum bearing surface on each end (4 inches for brick, 8 inches for block) unless otherwise specified.

3.5.0 Installing Piers, Pilasters, and Columns

Masons are also responsible for installing other types of reinforced masonry elements in addition to lintels. Three common types of reinforced masonry elements that you will encounter on the job are piers, pilasters, and columns. A structural pier is an isolated (hidden inside a wall) shaft designed to support a concentrated load. Non-isolated, freestanding hollow piers that do not bear structural loads can be built to serve purely decorative functions. A pilaster is a strengthened section of a masonry wall that provides lateral stability for the wall. When completed, it resembles a column projecting from a wall. A column is an isolated vertical structure designed to support a load such as a girder, truss, or beam. In this section, you will learn how to install piers, pilasters, and columns in accordance with the standards used by the masonry industry.

3.5.1 Installing Piers

Piers (*Figure 23*) are typically shorter than pilasters or columns. According to *ACI 530/ASCE 5/ TMS 402, Building Code Requirements for Masonry Structures*, foundation piers must have a minimum nominal thickness of 8 inches. The *International Building Code®* requires that the nominal heights of solidly grouted foundational piers be limited to no more than 10 times their nominal thickness; the nominal height of piers that are not solidly grouted may not exceed four times the pier's nominal thickness.

Structural engineers design solid, reinforced piers that are designed to carry concentrated loads. Piers require reinforcement when carrying flexural loads (loads that can cause the pier to bend) such as the thrust of an arch, a steel beam, a roof truss, or a floor joist. Some piers are designed with bearing plates set in a thin mortar bed to help the pier bear the load placed on it (see *Figure 24*). When installing a narrow pier, use care in selecting brick to ensure that they do not vary in length. Otherwise, both sides of the pier will not be able to be kept plumb for the pier's full length.

3.5.2 Installing Pilasters

A pilaster is a vertical section of thickened wall built with and bonded to the wall. This vertical section of masonry is tied directly to the wall by the use of wall ties or by bonding the masonry units directly with mortar. Pilasters project from the wall on one or both sides. If the pilaster projects on a single side, it is called a flush pilaster. When they appear as ornaments, pilasters are often called buttresses, and may be finished at the top with a sloping face. The portion of the double-

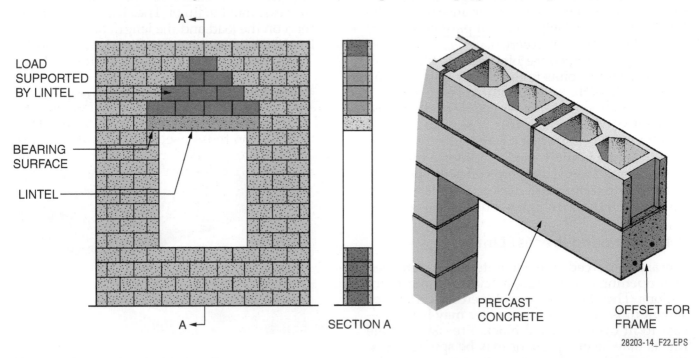

Figure 22 Concrete lintel in block wall.

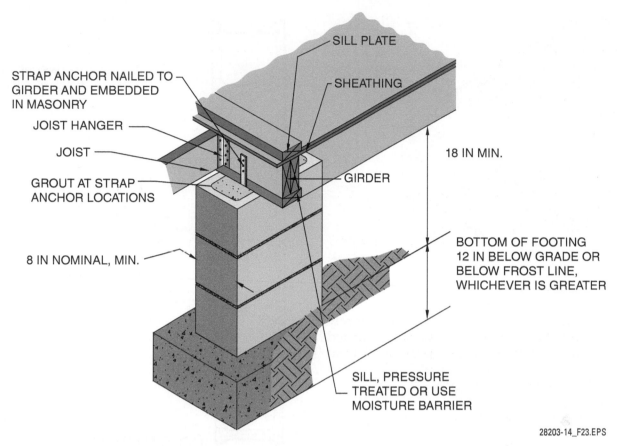

STRAP ANCHOR NAILED TO
GIRDER AND EMBEDDED
IN MASONRY

JOIST HANGER

JOIST

GROUT AT STRAP
ANCHOR LOCATIONS

8 IN NOMINAL, MIN.

SILL PLATE

SHEATHING

18 IN MIN.

GIRDER

BOTTOM OF FOOTING
12 IN BELOW GRADE OR
BELOW FROST LINE,
WHICHEVER IS GREATER

SILL, PRESSURE
TREATED OR USE
MOISTURE BARRIER

28203-14_F23.EPS

Figure 23 Example of a concrete masonry pier.

sided pilaster that shows on the outside of a structure is also typically referred to as the buttress. If the pilaster also projects inside the wall, the inside projection is called a counterfort. Pilasters may also be called interior or exterior pilasters according to the side of the wall that shows the projection. *Figure 25* shows a typical pilaster construction for a concrete block wall.

Although most pilasters reinforce or support a wall, they can serve a number of other functions. They may be used to support heavy loads, such as vertical roof or floor loads. They may also provide a decorative effect.

In reinforced masonry construction, pilasters are sometimes used as recesses to hide heating or air-conditioning ducts. Pilasters also serve to hide vertical steel beams and offer some protection from fire damage. In the case of fire resistance, the pilaster wall is usually only 4 inches thick since it is acting as a fire screen and not to support loads. Finally, pilasters may be used to give additional lateral support to masonry walls. By using pilasters properly, the industry has been able to design and build higher, thinner walls with plain masonry units.

Concrete masonry unit pilasters – Requirements for the placement of concrete masonry pilasters can be calculated by engineering design or may

be determined by requirements specified in local building codes. A general rule of thumb for sizing pilasters is illustrated in *Figure 26*. These guidelines can be used to judge the size of pilasters and the size of the projection:

- The pilaster projection should be no less than $\frac{1}{12}$ the pilaster height between supports.
- The pilaster width should be no less than $\frac{1}{10}$ the wall length between pilasters or other vertical supports.

In laying up pilasters you can make use of special concrete masonry units (*Figure 27*) or plain units similar to those used for the wall itself. If hollow units are used, they can be grouted and may contain embedded reinforcements. Grouted pilasters have all vertical joints fully mortared to keep the grout from flowing into adjacent cells.

Pilasters appear rectangular or square in shape and are classified according to their connection to the wall. Bonded pilasters are constructed as an integral part of the wall (*Figure 28*). Unbonded pilasters (*Figure 29*) are used where control joints are needed for crack control. Lateral support is provided through the use of suitable connections.

Brick pilasters – Brick pilasters are used for the same purpose as concrete masonry pilasters. They may be tied to the wall by use of wall ties or

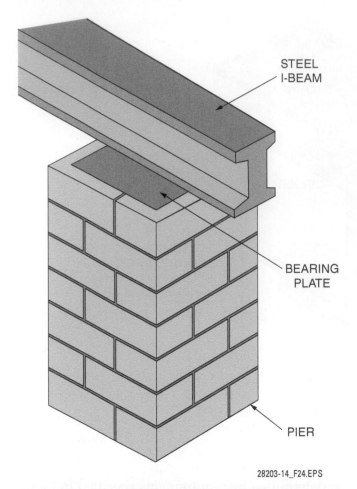

STEEL I-BEAM

BEARING PLATE

PIER

28203-14_F24.EPS

Figure 24 Pier with bearing plate.

28203-14_F25.EPS

Figure 25 Concrete block pilaster.

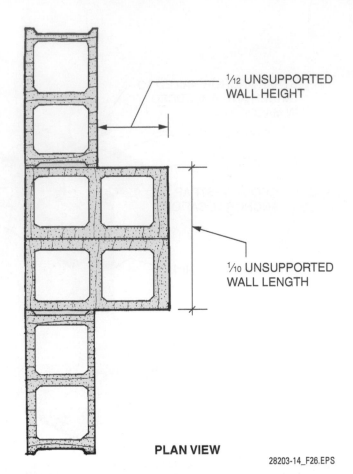

1/12 UNSUPPORTED WALL HEIGHT

1/10 UNSUPPORTED WALL LENGTH

PLAN VIEW

28203-14_F26.EPS

Figure 26 General requirements for pilaster size.

by bonding the masonry units to the wall. Some building codes require that pilasters be tied into the masonry wall a minimum of 4 inches and extend out from the wall 4 to 8 inches. You should check your local codes and specifications to determine these requirements.

Figure 30 shows a section of masonry wall with two pilasters on one side. These pilasters are bonded to the wall according to the plan shown at the bottom of the figure. The units in the first course are laid in one direction. In the second course, the pattern is reversed. This creates the needed bond with the wall and allows use of bats to fill in behind the face brick. Since the bats are half size, they can be used interchangeably to fill in the void as required. Following standard practice, ties are placed every sixth course.

> **NOTE**
>
> Always check your local building codes to determine the size and spacing of pilasters.

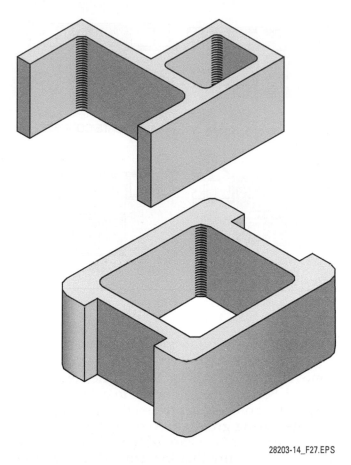

Figure 27 Pilaster unit.

28203-14_F27.EPS

3.5.3 Installing Columns

Columns (see *Figure 31*) are similar in design and function to piers, but they are generally taller and thinner. According to *ACI 530/ASCE 5/TMS 102, Building Code Requirements for Masonry Structures*, a column's width is never greater than three times its thickness, and its height is always greater than four times its thickness. Columns, unlike piers, are always solid or reinforced. Columns may be built inside a wall (isolated), or as freestanding structures. When designing columns, structural engineers must take into account the column's effective height, which is the clear height between supports. The effective height is determined by the risk of movement or twisting when the column is under load. Codes require that off-center axial loads, lateral loads, and out-of-plumb columns (collectively referred to as eccentricity) be factored into a design. The default allowance for eccentricity is 0.1 times each side dimension. If the actual eccentricity exceeds that minimum, it will be noted in the design.

Codes require columns to be vertically and laterally reinforced (see *Figure 32*). A minimum of four vertical bars is required for column reinforcement. When installing vertical ties at the base or below the lowest horizontal reinforcement in a beam, girder, slab, or drop panel, the spacing between ties is halved. Lateral ties enclose the vertical reinforcements and provide support for them. They are placed to prevent the reinforcement from buckling when the column is under compression. Lateral ties may be placed in mortar or grout, although placement in grout is considered to be more effective. The maximum lateral spacing between ties is 16 vertical bar diameters, or 48 diameters of lateral tie bars or wires, or the least cross-sectional dimension of the member.

Columns are sometimes built with cleanouts in the base (refer to *Figure 31*). The cleanouts serve three functions. First, they allow access for the

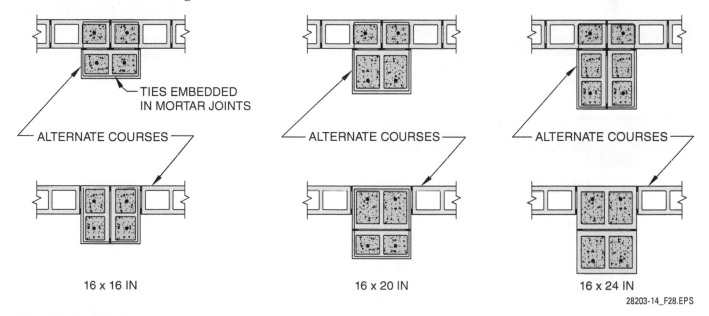

TIES EMBEDDED IN MORTAR JOINTS

ALTERNATE COURSES

ALTERNATE COURSES

ALTERNATE COURSES

16 x 16 IN

16 x 20 IN

16 x 24 IN

28203-14_F28.EPS

Figure 28 Bonded pilaster.

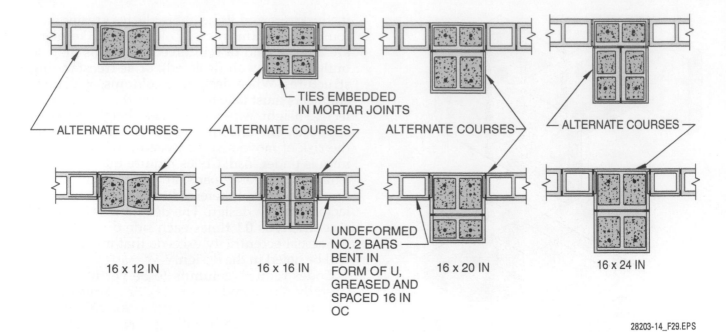

ALTERNATE COURSES

TIES EMBEDDED IN MORTAR JOINTS

ALTERNATE COURSES

ALTERNATE COURSES

ALTERNATE COURSES

UNDEFORMED NO. 2 BARS BENT IN FORM OF U, GREASED AND SPACED 16 IN OC

16 x 12 IN

16 x 16 IN

16 x 20 IN

16 x 24 IN

28203-14_F29.EPS

Figure 29 Unbonded pilaster.

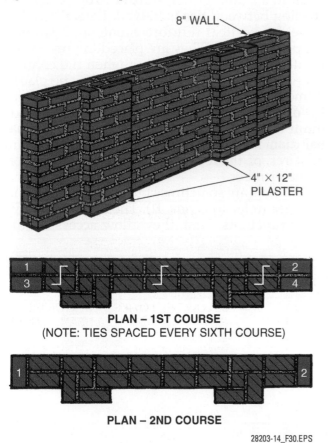

8" WALL

4" × 12" PILASTER

PLAN – 1ST COURSE
(NOTE: TIES SPACED EVERY SIXTH COURSE)

PLAN – 2ND COURSE

28203-14_F30.EPS

Figure 30 Brick wall with pilasters.

removal of fallen mortar before the column is grouted. Second, they allow air to escape during the grouting process, to prevent the forming of air pockets during grouting. And finally, they permit inspection of the base of the column during grouting to ensure that the grout has filled the space. Install a cleanout in each cell of the base course when using full concrete block. Use a vibrator when grouting a column, to ensure all air has been evacuated and that all cavities have been filled.

Wall columns, as their name suggests, are columns that have walls built between them, but which are not structural elements of the wall. When wall columns are used, they are constructed before the surrounding walls. They may incorporate a vertical steel pipe or I-beam at their core. If the local code provides seismic requirements for columns, be sure to follow those requirements during construction.

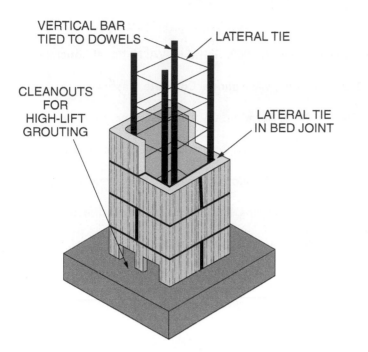

VERTICAL BAR TIED TO DOWELS

LATERAL TIE

CLEANOUTS FOR HIGH-LIFT GROUTING

LATERAL TIE IN BED JOINT

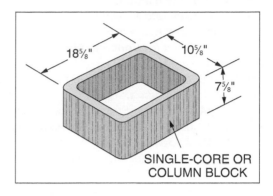

18⅝"

10⅝"

7⅝"

SINGLE-CORE OR COLUMN BLOCK

28203-14_F31.EPS

Figure 31 Masonry column.

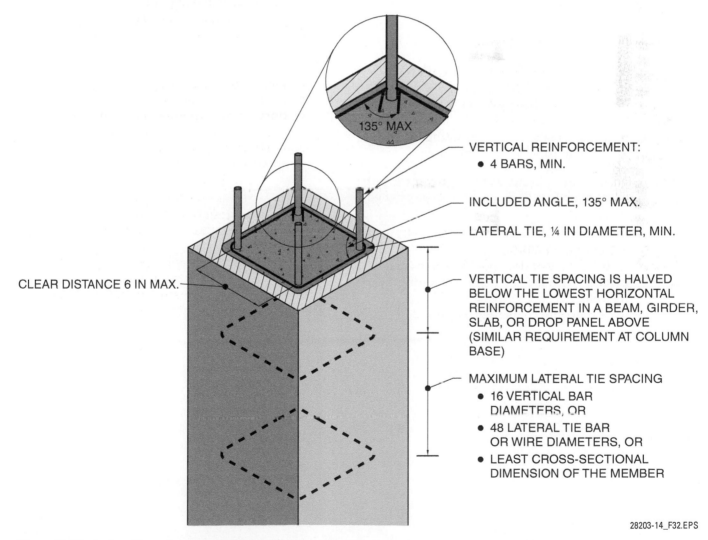

135° MAX

VERTICAL REINFORCEMENT:
• 4 BARS, MIN.

INCLUDED ANGLE, 135° MAX.

LATERAL TIE, ¼ IN DIAMETER, MIN.

CLEAR DISTANCE 6 IN MAX.

VERTICAL TIE SPACING IS HALVED BELOW THE LOWEST HORIZONTAL REINFORCEMENT IN A BEAM, GIRDER, SLAB, OR DROP PANEL ABOVE (SIMILAR REQUIREMENT AT COLUMN BASE)

MAXIMUM LATERAL TIE SPACING
• 16 VERTICAL BAR DIAMETERS, OR
• 48 LATERAL TIE BAR OR WIRE DIAMETERS, OR
• LEAST CROSS-SECTIONAL DIMENSION OF THE MEMBER

28203-14_F32.EPS

Figure 32 Vertical and lateral reinforcement of a concrete column.

Additional Resources

TEK 2-1A, Typical Sizes and Shapes of Concrete Masonry Units. 2002. Herndon, VA: National Concrete Masonry Association. **www.ncma.org**

TEK 3-2A, Grouting Concrete Masonry Walls. 2005. Herndon, VA: National Concrete Masonry Association. **www.ncma.org**

TEK 5-3A, Concrete Masonry Foundation Wall Details. 2003. Herndon, VA: National Concrete Masonry Association. **www.ncma.org**

TEK 9-4A, Grout for Concrete Masonry. 2005. Herndon, VA: National Concrete Masonry Association. **www.ncma.org**

TEK 12-4D, Steel Reinforcement for Concrete Masonry. 2011. Herndon, VA: National Concrete Masonry Association. **www.ncma.org**

TEK 17-3A, Allowable Stress Design of Concrete Masonry Columns. 2001. Herndon, VA: National Concrete Masonry Association. **www.ncma.org**

TEK 17-4B, Allowable Stress Design of Concrete Masonry Pilasters. 2000. Herndon, VA: National Concrete Masonry Association. **www.ncma.org**

3.0.0 Section Review

1. Rebar should be bent _____.
   a. when cold
   b. after placement
   c. when heated
   d. no more than 90 degrees

2. A bar spacer is used to align, support, and rigidly fasten rebar in place _____.
   a. once grout has been poured
   b. before grout is poured
   c. only when using low-lift grouting
   d. only when using high-lift grouting

3. When installing bond beams, prevent grout from leaking into cells that are not to be grouted by _____.
   a. installing a bearing plate over the cells with a thin layer of mortar
   b. using CMU that does not have holes
   c. placing wire lath or metal screening over the cells
   d. use grout with coarse aggregate

4. The thickness of a one-piece lintel will typically be _____.
   a. twice the nominal height of the masonry units used in the wall
   b. thicker than the masonry wall
   c. thinner than the masonry wall
   d. the same thickness as the masonry wall

5. A buttress is another term for a _____.
   a. pilaster that does not project from the wall
   b. double-sided pilaster
   c. portion of a double-sided pilaster that shows on the outside of a structure
   d. portion of a double-sided pilaster that shows on the inside of a structure

SUMMARY

Grout is made of cementitious material, water, admixtures, and aggregate. It is looser than mortar and has slightly different adhering and setting properties. It adds strength to structural units and can be used alone or in combination with steel reinforcement bars. Admixtures can be added to the mix to modify its properties.

Grout is usually delivered ready-mixed because it needs to be agitated continuously after mixing. It is pumped or poured into enclosed spaces in lifts or layers. The receiving space must be free of mortar protrusions and debris. The grout must be rodded or vibrated to clear air bubbles and fill voids. Care must be taken to prevent the grout from splashing on the masonry, as it is hard to remove.

Grouting techniques vary depending upon the height of the lift. Low-lift technique can be used for lifts of less than 5 feet. High-lift grouting is complicated because of the way the wall must be built. Cleanout openings must be made at the bottom of the wall when high-lift grouting is used.

Along with grout, reinforcing steel bar, or rebar, is used in reinforced walls and in reinforced masonry elements such as bond beams, lintels, piers, pilasters, and columns, to allow them to withstand greater stress and carry greater loads than unreinforced masonry walls or elements of the same size. Reinforced masonry elements are designed to provide vertical or horizontal support in a structure. They can be enclosed in a wall, freestanding, or built with walls running between them without contributing to the structural strength of the wall. Proper grouting technique is essential for reinforced masonry structures to ensure that they provide sufficient strength to support the loads placed on them.

1. A reinforced wall contains _____.
   a. only grout
   b. both grout and steel reinforcement
   c. voids filled with rubble
   d. only steel reinforcement

2. When making fine grout, the only acceptable aggregate is _____.
   a. lime
   b. gravel
   c. sand
   d. cement

3. As it cures, grout shrinks _____.
   a. no more than 5 percent
   b. 5 to 10 percent
   c. 10 to 15 percent
   d. 15 to 20 percent

4. An admixture used to replace some of the portland cement and provide greater slump with less water is _____.
   a. fly ash
   b. shrinkage compensator
   c. retarder
   d. antifreeze compound

5. Grout will need additional water if the air is _____.
   a. cool and dry
   b. hot and damp
   c. cool and damp
   d. hot and dry

6. Laboratory tests are used to establish the compressive strength of grout after it has cured for _____.
   a. 12 hours
   b. 28 days
   c. 48 hours
   d. 30 days

7. Grout mixed in large quantities must meet the requirements of ASTM specification _____.
   a. C94
   b. C107
   c. C324
   d. C404

8. When grout is mixed on the job site, the water and dry ingredients should be blended for a minimum of _____.
   a. three minutes
   b. five minutes
   c. eight minutes
   d. 10 minutes

9. The amount of grout placed in one uninterrupted operation is called a _____.
   a. layer
   b. placement
   c. lift
   d. flight

10. Use rodding to consolidate grout if the lift is _____.
    a. 36 inches or more
    b. 12 inches or more
    c. 36 inches or less
    d. 12 inches or less

11. High-lift grouting for a cavity wall that is 8 feet in height should be completed in _____.
    a. a single lift
    b. two lifts
    c. three lifts
    d. four lifts

12. The condition that occurs when grout mounds over a reinforcement element and creates a void is called _____.
    a. bridging
    b. separation
    c. voiding
    d. blowout

13. Mortar protruding into a cavity will prevent the grout from _____.
    a. curing properly
    b. consolidating
    c. forming a complete bond
    d. segregating

14. Turn off mechanical vibrators _____.
    a. after removing the vibrator from the grout
    b. as the vibrator is being removed from the grout
    c. as a surface skin begins to form over the grout
    d. when the grout level rises slightly

15. Reinforcing bar (rebar) is made from new or recycled _____.
    a. cast iron
    b. aluminum
    c. steel
    d. fiberglass

16. Grade 520 reinforcing bar has a minimum yield strength of _____.
    a. 40,000 psi
    b. 52,000 psi
    c. 60,000 psi
    d. 75,000 psi

17. A device used for hand-bending of rebar is called a _____.
    a. rebar jig
    b. hickey bar
    c. bend former
    d. hockey stick

18. Pieces of rebar are secured together by using ties made from _____.
    a. soft-annealed iron wire
    b. nylon
    c. brass wire
    d. stainless steel wire

19. A horizontal loadbearing member placed over an opening is called a _____.
    a. stile
    b. spreader beam
    c. strongback
    d. lintel

20. Steel reinforcing rods may be cast into a one-piece lintel to increase its _____.
    a. deflection resistance
    b. tensile strength
    c. lateral stability
    d. compressive strength

Trade Terms Quiz

Fill in the blank with the correct term that you learned from your study of this module.

1. A hollow masonry structure other than a wall in which the voids in the masonry units are filled with grout and reinforcing bar is called a(n) _____.

2. _____ is the mounding of grout or cement over an obstruction, creating a void under the obstruction.

3. A hollow masonry wall in which the voids are filled with grout but not reinforcing bar is called a(n) _____.

4. A(n) _____ is the horizontal member or beam over an opening that carries the weight of the masonry above the opening.

5. A recess or groove in one placement of grout or concrete that is later filled with a new placement of grout or concrete so that the two lock together in a tongue-and-groove configuration is called a(n) _____.

6. _____ involves poking the grout with a rod in order to consolidate it.

7. One continuous placement of grout or cement without interruption, equivalent to one layer, is called a(n) _____.

8. A(n) _____ is a vertical reinforced masonry element that is typically shorter than a column or pilaster.

9. The swelling or rupture of a cavity wall from too much pressure caused by pouring liquid grout into the cavity is called _____.

10. _____ is a reinforcing bar embedded in concrete, mortar, or grout in such a manner that it acts together with the other components to resist loads.

11. A vertical reinforced masonry element designed to support a load, the width of which never exceeds three times its thickness and the height of which always exceeds four times its thickness, is called a(n) _____.

12. A(n) _____ is a vertical reinforced masonry element consisting of a thickened section of a wall to which it is structurally bonded.

13. A course of masonry units with steel rebar inserted and held in place by a solid fill of grout or mortar, used as a lintel or reinforcement beam to distribute stress, is called a(n) _____.

14. A(n) _____ is a hollow masonry wall in which the voids in the masonry units are filled with grout and reinforcing bar.

15. Consolidating grout with the use of a mechanical vibrator is called _____.

Trade Terms

Blowout
Bond beam
Bridging
Column
Grouted wall

Key
Lift
Lintel
Pier
Pilaster

Rebar
Reinforced masonry element
Reinforced wall
Rodding
Vibrating

Trade Terms Introduced in This Module

Blowout: The swelling or rupture of a cavity wall from too much pressure caused by pouring liquid grout into the cavity.

Bond beam: A course of masonry units with steel rebar inserted and held in place by a solid fill of grout or mortar; used as a lintel or reinforcement beam to distribute stress.

Bridging: The mounding of grout or cement over an obstruction, creating a void under the obstruction.

Column: A vertical reinforced masonry element designed to support a load, the width of which never exceeds three times its thickness and the height of which always exceeds four times its thickness.

Grouted wall: A hollow masonry wall in which the voids are filled with grout but not reinforcing bar.

Key: A recess or groove in one placement of grout or concrete that is later filled with a new placement of grout or concrete so that the two lock together in a tongue-and-groove configuration.

Lift: One continuous placement of grout or cement without interruption, equivalent to one layer.

Lintel: The horizontal member or beam over an opening that carries the weight of the masonry above the opening.

Pier: A vertical reinforced masonry element that is typically shorter than a column or pilaster. Piers may be designed to carry loads, or may be purely ornamental.

Pilaster: A vertical reinforced masonry element consisting of a thickened section of a wall to which it is structurally bonded.

Rebar: Reinforcing bar embedded in concrete, mortar, or grout in such a manner that it acts together with the other components to resist loads.

Reinforced masonry element: A hollow masonry structure other than a wall in which the voids in the masonry units are filled with grout and reinforcing bar.

Reinforced wall: A hollow masonry wall in which the voids in the masonry units are filled with grout and reinforcing bar.

Rodding: Poking the grout with a rod in order to consolidate it.

Vibrating: Consolidating grout with the use of a mechanical vibrator.

Additional Resources

This module presents thorough resources for task training. The following resource material is suggested for further study.

ACI 530/ASCE 5/TMS 402, Building Code Requirements for Masonry Structures, Latest Edition. Reston, VA: American Society of Civil Engineers.

ACI 530.1/ASCE 6/TMS 602, Specifications for Masonry Structures, Latest Edition. Reston, VA: American Society of Civil Engineers.

ASTM C94, Specifications for Ready-Mixed Concrete, Latest Edition. West Conshohocken, PA: ASTM International.

ASTM C404, Standard Specification for Aggregates for Masonry Grout, Latest Edition. West Conshohocken, PA: ASTM International.

ASTM C476, Standard Specification for Grout for Masonry, Latest Edition. West Conshohocken, PA: ASTM International.

Bricklaying: Brick and Block Masonry. 1988. Brick Industry Association. Orlando, FL: Harcourt Brace & Company.

Concrete Masonry Handbook for Architects, Engineers, Builders, Fifth edition. 1991. W. C. Panerese, S. K. Kosmatka, and F. A. Randall, Jr. Skokie, IL: Portland Cement Association.

Technical Note 11E, *Guide Specifications for Brick Masonry, Part 5, Mortar and Grout*. 1991. Reston, VA: The Brick Industry Association. **www.gobrick.com**

TEK 2-1A, Typical Sizes and Shapes of Concrete Masonry Units. 2002. Herndon, VA: National Concrete Masonry Association. **www.ncma.org**

TEK 3-2A, Grouting Concrete Masonry Walls. 2005. Herndon, VA: National Concrete Masonry Association. **www.ncma.org**

TEK 5-3A, Concrete Masonry Foundation Wall Details. 2003. Herndon, VA: National Concrete Masonry Association. **www.ncma.org**

TEK 9-4A, Grout for Concrete Masonry. 2005. Herndon, VA: National Concrete Masonry Association. **www.ncma.org**

TEK 12-4D, Steel Reinforcement for Concrete Masonry. 2011. Herndon, VA: National Concrete Masonry Association. **www.ncma.org**

TEK 17-3A, Allowable Stress Design of Concrete Masonry Columns. 2001. Herndon, VA: National Concrete Masonry Association. **www.ncma.org**

TEK 17-4B, Allowable Stress Design of Concrete Masonry Pilasters. 2000. Herndon, VA: National Concrete Masonry Association. **www.ncma.org**

Figure Credits

Section Review Answers

Answer	Section Reference	Objective
Section One		
1. b	1.1.0	1a
2. d	1.2.0	1b
3. d	1.3.0	1c
4. a	1.4.0	1d
5. d	1.5.0	1e
6. c	1.6.0	1f
Section Two		
1. a	2.1.0	2a
2. b	2.2.1	2b
3. a	2.3.0	2c
4. c	2.4.0	2d
Section Three		
1. a	3.1.0	3a
2. b	3.2.0	3b
3. c	3.3.0	3c
4. d	3.4.0	3d
5. c	3.5.2	3e

NCCER CURRICULA — USER UPDATE

NCCER makes every effort to keep its textbooks up-to-date and free of technical errors. We appreciate your help in this process. If you find an error, a typographical mistake, or an inaccuracy in NCCER's curricula, please fill out this form (or a photocopy), or complete the online form at **www.nccer.org/olf**. Be sure to include the exact module ID number, page number, a detailed description, and your recommended correction. Your input will be brought to the attention of the Authoring Team. Thank you for your assistance.

Instructors – If you have an idea for improving this textbook, or have found that additional materials were necessary to teach this module effectively, please let us know so that we may present your suggestions to the Authoring Team.

NCCER Product Development and Revision

13614 Progress Blvd., Alachua, FL 32615

Email: curriculum@nccer.org
Online: www.nccer.org/olf

❏ Trainee Guide   ❏ Lesson Plans   ❏ Exam   ❏ PowerPoints   Other _____

Craft / Level: _____   Copyright Date: _____

Module ID Number / Title: _____

Section Number(s): _____

Description: _____

_____

_____

_____

_____

Recommended Correction: _____

_____

_____

_____

_____

Your Name: _____

Address: _____

_____

Email: _____   Phone: _____

28204-14

Masonry Openings and Metal Work

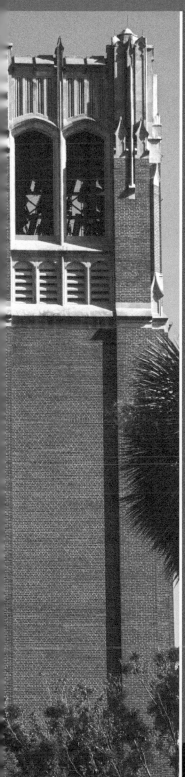

Many masonry structures require the installation of reinforcements to enable the structure to withstand cracking under load. Reinforcements are installed around openings such as door frames, windowsills, lintels, chases, and recesses. They are also embedded in the mortar of masonry wythes to tie a single masonry wythe together, to tie two masonry wythes together, and to tie a masonry wythe and a structural element. This module provides instruction on the methods and materials used to install masonry openings, tie wythes together, and tie wythes to structural elements. This module discusses standards for the various types of openings and reinforcements, and steps and procedures for installation.

Module Four

Objectives

When you have completed this module, you will be able to do the following:

1. Describe the methods and materials used to install masonry openings.
   a. Describe how to use and install door and window frames.
   b. Describe how to use and install windowsills.
   c. Describe how to use and install steel lintels.
   d. Describe how to use and install chases and recesses.
2. Describe the methods and materials used to tie a single masonry wythe together.
   a. Describe how to use and install ladder and truss joint reinforcement.
   b. Describe how to use and install seismic reinforcements.
3. Describe the methods and materials used to tie two masonry wythes together.
   a. Describe how to use and install flexible anchors.
   b. Describe how to use and install horizontal anchors.
4. Describe the methods and materials used to tie a masonry wythe to structural elements.
   a. Describe how to use and install rigid ties and bolts.
   b. Describe how to use and install bearing plates.
   c. Describe how to use and install saddles.
   d. Describe how to use and install strap ties.

Performance Tasks

Under the supervision of your instructor, you should be able to do the following:

1. Install a hollow metal door frame.
2. Install a sill and a lintel.
3. Install a bearing plate.
4. Install a strap tie.

Trade Terms

Anchor	Horizontal joint reinforcement	Sill
Beam pocket	Jamb	Skew
Bond beam	Panel	Wick
Chase	Pintle	
Fastener	Reveal	

Industry-Recognized Credentials

If you're training through an NCCER-accredited sponsor, you may be eligible for credentials from NCCER's Registry. The ID number for this module is 28204-14. Note that this module may have been used in other NCCER curricula and may apply to other level completions. Contact NCCER's Registry at 888.622.3720 or go to www.nccer.org for more information.

Code Note

Codes vary among jurisdictions. Because of the variations in code, consult the applicable code whenever regulations are in question. Referring to an incorrect set of codes can cause as much trouble as failing to reference codes altogether. Obtain, review, and familiarize yourself with your local adopted code.

Contents

Topics to be presented in this module include:

Figures and Tables

1.0.0 MASONRY OPENINGS

Objective

Describe the methods and materials used to install masonry openings.

a. Describe how to use and install door and window frames.
b. Describe how to use and install windowsills.
c. Describe how to use and install steel lintels.
d. Describe how to use and install chases and recesses.

Performance Tasks One and Two

Install a hollow metal door frame.

Install a sill and a lintel.

Trade Terms

Anchor: A metal assembly used to attach masonry to a structural support.

Bond beam: A course of masonry filled with steel reinforcing rods and grout that serves as a lintel or reinforcement beam designed to strengthen a wall.

Chase: A continuous recess built into a wall to receive pipes, wires, or heating ducts.

Fastener: A metal assembly used to attach building parts to masonry.

Jamb: The side of an opening, or the vertical framing member on the side of the opening, usually for door and window frames.

Reveal: The side of an opening in a wall for a window or door. This is the part of the masonry jamb around a window or door frame that can be seen from the frame to the face of the masonry wall.

Sill: A horizontal member under a door or window. Slip sills fit inside the door or window frame; lug sills extend beyond the frame and into the masonry on the jamb sides of the frame.

Skew: The condition when two parts come together at an angle that is not 90 degrees, or perpendicular, to each other.

Wick: A bundle of fibers that are loosely twisted or braided and woven together to form a cord that can carry water away from an area by capillary action, which occurs as long as the drip end is lower than the absorption end.

Every structure requires openings of some type. Most buildings have doors and windows that fit into the masonry walls. Some masonry walls also have more elaborate detailing of the surface, as shown in *Figure 1*. There may also be a need for special structural treatments, such as pilasters or bracing.

When accurate planning and design have been done according to modular planning techniques, the various kinds of openings in the walls should create no particular difficulties for the mason. However, you must be fully aware of the exact position of all openings and be ready to apply some special techniques where these openings occur. You can easily make a mistake at this point if the location of the opening is not marked on the footing or wall base before you begin to lay masonry units.

Specialized block is available to construct window and door openings (*Figure 2*). It is important to be aware of the masonry skills used in placing sills, lintels, and jambs. In placing chases, or recesses, you must be aware of the effect that these openings have on the strength of the wall.

Hollow metal frames are more durable than wood, and factory finishes can give hollow metal frames an attractive appearance. More important, improved welding techniques enable mass production, substantially lowering costs.

The four parts of any opening, such as a window, are the sill, the two jambs, and the header or lintel (see *Figure 3*). The sill is the horizontal bottom of the opening; the jambs are the vertical sides; and the lintel is the horizontal top of the opening. For doors, sills may be set directly on the foundation or on a few courses of masonry. For windows, sills are set on top of masonry courses.

28204-14_F01.EPS

Figure 1 Detailing on masonry walls.

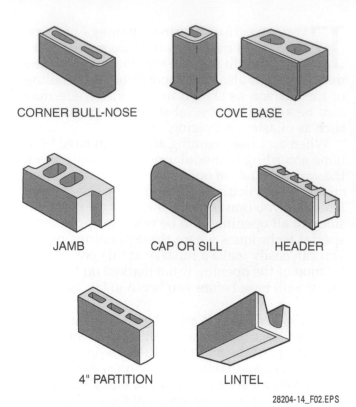

CORNER BULL-NOSE COVE BASE

JAMB CAP OR SILL HEADER

4" PARTITION LINTEL

28204-14_F02.EPS

Figure 2 Specialized block.

When working with block, the most efficient window and door openings are sized to accommodate the block both horizontally and vertically, with enough room available for masonry sills. Any change from this results in openings requiring a number of block pieces and, of course, increased labor costs. The location and size of doors in concrete block walls require the same considerations necessary for windows.

When cutting brick to fit against a jamb of a window or door, it is easier and quicker to make all necessary cuts before laying any units. All cuts made should be the same size. Small differences in the wall space can be made up by adjusting the thickness of the head joints.

Openings in a brick wall are not as difficult to work out as they may first appear. Modular and standard measurements allow opening frames to have common measurements with brick. Window and door frames are generally available in sizes that will fit in brick openings without the need for excessive cutting of brick. These standard sizes permit the brick coursing to fit evenly with the opening height, and the opening width is based on the modular dimensions of the brick.

When installing brick openings, planning is crucial. The mason must work out the bond before the wall is built. This will ensure that there will be no irregular lap conditions when the bond stops at the frame. Calculations should be made so that any cutting will be reduced to a minimum.

HEADER OR LINTEL

JAMB

SILL

28204-14_F03.EPS

Figure 3 Parts of a window.

If this has been attended to competently, you will only have to read the plans and correctly lay out the bond.

It is generally the responsibility of the carpenter working on the job to set the windows the mason must lay to. Before beginning, check the windows to be certain they are level, plumb, and the same elevation. In some cases, however, the window may not have been set at all. The carpenter crews may be behind schedule, a window unit may have been broken and re-ordered, or it may not have been delivered to the job site at all. If everything else is ready for the brickwork, the masonry contractor cannot afford to hold up work; check with your supervisor to determine what action to take.

1.1.0 Installing Door and Window Frames

Aluminum and steel are used for hollow metal frames, with steel the most popular for door frames. *Figure 4* shows a typical hollow metal door frame, with both a welded and a tab-locked corner

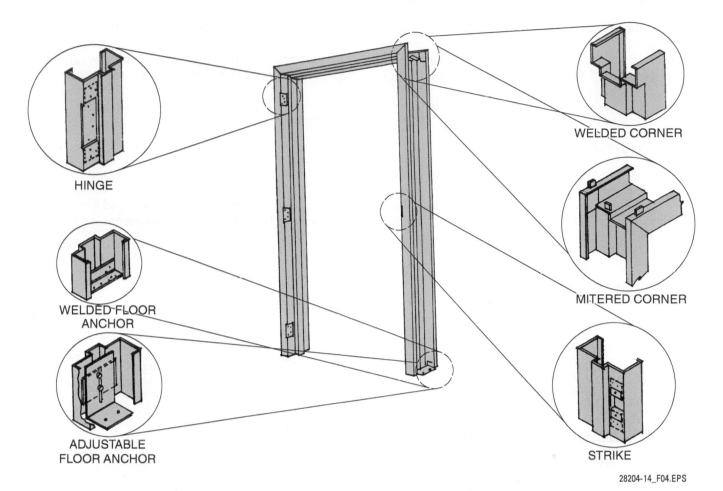

HINGE

WELDED FLOOR ANCHOR

ADJUSTABLE FLOOR ANCHOR

WELDED CORNER

MITERED CORNER

STRIKE

28204-14_F04.EPS

Figure 4 Standard hollow metal door frame.

shown. When using metal frames, you must set the frames in place before the opening is laid up with square-cut jamb block. Remember that all joints around the jamb must be well filled with mortar to avoid an air or moisture leak. The cores closest to the jamb should be grouted to provide additional strength around the door frames.

Steel frames may be made of 12- through 18-gauge steel. Different gauges of steel are used for different applications and for different degrees of durability. Hollow metal steel frames come with steel ties or fasteners, usually a minimum of three for each jamb. Door frames also have floor knees for anchorage to the floor. The door frames are reinforced at all critical points, including the corners, and at the hinge, the strike (the plate that surrounds the recess in the frame that receives the door's latch or bolt when the door is shut), or closer locations on the frame. Steel frames are often zinc coated for outdoor use.

Hollow metal aluminum frames are popular for windows, as shown in *Figure 5*, and also for doors. They come in a variety of finishes and are quite durable. Typically, an aluminum frame will be less than $\frac{1}{10}$ inch thick, with the anchorage ar-

eas slightly thicker. All screws, nuts, washers, and other installation materials may be of aluminum, stainless steel, or other noncorrosive material. If weatherproofing or soundproofing is desired,

28204-14_F05.EPS

Figure 5 Aluminum windows on a commercial building.

vinyl seals are used between the frame and the wall.

Both steel and aluminum frames can be ordered assembled (set-up) or unassembled (knock-down). Unassembled frames offer savings in transportation and handling costs. The jambs and stretchers are usually individually wrapped to prevent scratching. The corners are mitered and

Planning Saves Time

When laying block to openings, it is very important that the block be laid out and planned so units do not have to be cut or at least cut infrequently. Note how many block must be cut in the first example. This is time consuming and wasteful. The second figure shows good planning. Standard full-length or half-length block can be used. Special jamb block can be used around the window and door openings. The jamb blocks have a recess cut into them so the window and door frames can be set easily.

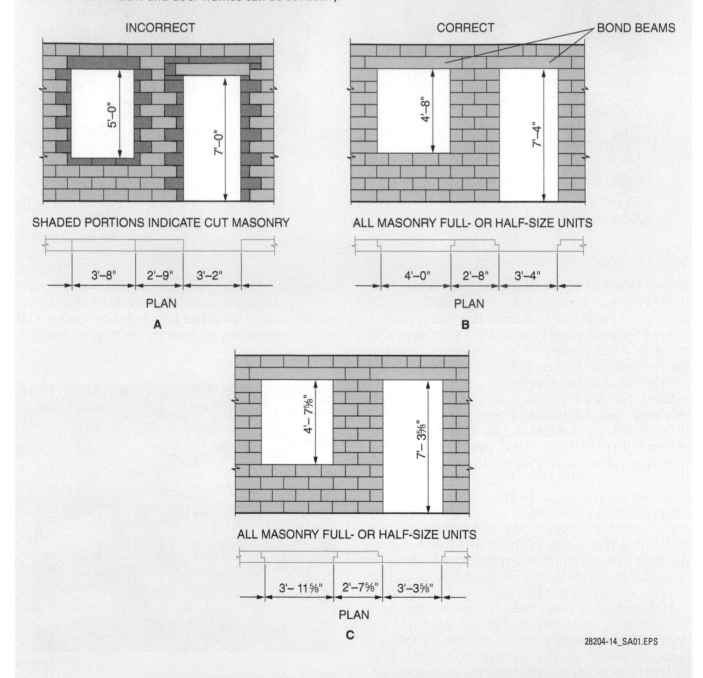

INCORRECT

SHADED PORTIONS INDICATE CUT MASONRY

5'-0"
7'-0"

3'-8"   2'-9"   3'-2"

PLAN

A

CORRECT   BOND BEAMS

ALL MASONRY FULL- OR HALF-SIZE UNITS

4'-8"
7'-4"

4'-0"   2'-8"   3'-4"

PLAN

B

ALL MASONRY FULL- OR HALF-SIZE UNITS

4'-7⅝"
7'-3⅝"

3'-11⅝"   2'-7⅝"   3'-3⅝"

PLAN

C

28204-14_SA01.EPS

reinforced, and have lock tabs for rigid joining. A factory-assembled frame has the corners welded and ground smooth at the factory.

Most manufacturers offer the three basic kinds of hollow metal frames: standard, special, and drywall.

- A standard frame simply uses a standard design and dimensions. As such, it will fit most wall conditions. A standard frame may be ordered either knockdown or set-up.
- Special frames are manufactured to order and are usually delivered assembled. Unlike standard frames, they may have fasteners welded in place. Special frames often call for special wall conditions, so it is important to check the drawings before installing them.
- Drywall frames are designed for installation after the drywall partition is in place. The drywall partitions can be placed, and even finished, before installing this frame. Drywall frames are available for wood- or steel stud partitions, and can be used for certain exterior stud and masonry applications. Drywall frames are always furnished knockdown. Installation is simple; many manufacturers claim installation times of less than five minutes per frame.

The frame itself can be fastened to the wall by using **anchors** in the mortar joints or by driving tempered nails through the frames and directly into the mortar joints. Anchors are discussed in detail in the section titled *Tying Two Masonry Wythes Together*. Either way, the frame should be positioned the proper distance back from the face of the wall. This position depends on the thickness of the wall.

The distance from the face of the frame to the face of the masonry wall is known as the **reveal**. The carpenter has the responsibility of setting the frame in a level and plumb position. The mason has the responsibility of laying the jamb block in such a manner that this can be achieved. The mason should always check the frame to be sure that it is level and plumb before finishing the wall.

Fasteners or ties give standard frames their design flexibility. Ties for stud walls and for existing structures normally come welded in place on the jambs. For masonry walls, most fasteners are of the snap-in variety, and can be ordered to accommodate most wall sizes. Door frames may have adjustable floor ties that may be set flush or extend 1 inch below the finish floor. *Figure 6* shows an adjustable frame tie that snaps onto the jamb.

Follow the manufacturer's instructions when installing any type of hollow metal frame. There are two different methods of installation, one for knockdown frames and the other for set-up frames. The following instructions are given for door frames; it is assumed that a slip sill will be installed after the framework is complete.

1.1.1 Knockdown-Frame Installation

Knockdown frames come from the manufacturer unassembled. The components must be bolted or welded together before they are installed in the wall. The following steps are general guidelines for installing a knockdown frame. The instructions are for masonry construction, with information for frame construction shown in parentheses.

Step 1   Check the frame to ensure that it is for the correct opening and that all parts are present.

Step 2   Push the top of the hinge jamb against the masonry (or over the frame) wall. Hold the top of the jamb in place and push the bottom against the masonry (or over the frame) wall at the base.

Step 3   Position the frame head against (or over) the wall. Align the head tabs with the jamb slots and slide the head towards the jamb, inserting the tabs into the slots.

Step 4   Push the top of the strike jamb against (or over) the wall, aligning the jamb slots and head tabs as before. When the tabs engage, push the bottom of the frame against the wall at the base.

Step 5   Check the level of the frame head and the reveal. Adjust as necessary.

Step 6   Adjust all jamb fasteners until all are in contact with the supporting structure if the wall is built up.

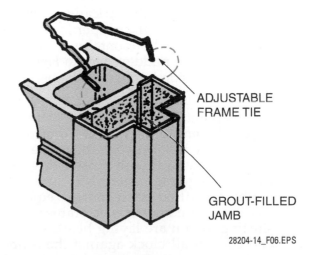

ADJUSTABLE FRAME TIE

GROUT-FILLED JAMB

28204-14_F06.EPS

Figure 6 Adjustable frame tie.

Step 7 Plumb the hinge jamb and secure. Check the level of the head again. Plumb the strike jamb and secure. At a minimum, fasten at the top, bottom, and middle of each jamb.

Step 8 Check the square, skew, and plumb of the frame and secure permanently according to the manufacturer's recommendations.

Step 9 If the wall is not complete around the frame, continue laying masonry. Completely fill the void area between the masonry and the frame with mortar or grout. If you are laying block, fill the end cells of all block against the jamb with grout.

Step 10 Clean any smeared mortar from the metal parts of the frame.

1.1.2 Set-Up Door Frame Installation

The following steps apply to set-up frames. This is the type of hollow metal frame that is most commonly found on masonry jobs. *Figure 7* shows this type of frame set in concrete masonry units (CMU, commonly called block).

Step 1 Ensure that the frame is manufactured correctly and that the frame is for the correct opening.

Step 2 Mark the frame location, set the frame, and brace it in position with plumb braces. Refer to *Figure 7*.

Step 3 Make sure the door will open on the correct side and in the correct direction. Check the reveal to make sure the door is set back the proper distance.

Step 4 Check the frame for plumb, square, and skew. Adjust as necessary.

Step 5 Brace the frame both vertically and horizontally to secure the proper location of the frame. Brace horizontally at the bottom and in the middle (refer to *Figure 7*). Scrap dimension lumber can be used for braces.

Step 6 At this point, begin laying the masonry units adjacent to the frame. Completely fill the void area between the frame and the masonry unit with mortar or grout as the laying progresses. Make sure both jambs are filled with mortar equally as courses are laid to keep frames from shifting. If you are laying block, fill the end cells of all block against the frame jamb with grout.

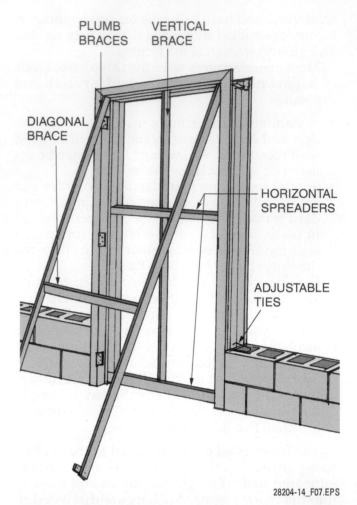

28204-14_F07.EPS

Figure 7 Braced assembled door frame.

Step 7 Install at least three fasteners at each jamb as the work progresses. Follow the manufacturer's recommendations on the type and location of the fasteners. At a minimum, fasten at the top, bottom, and middle of each jamb close to the hinges.

Constantly check the frame location and plumb as the wall is erected.

1.1.3 Bracing Jigs

Setting assembled door and window frames usually requires lumber to be cut for use as bracing. Several manufacturers offer ready-made proprietary bracing jigs for this work. Some of these bracing jigs consist of telescoping poles with universal clamps and adjustment tensioners on each end (*Figure 8*).

These jigs are designed so that one end clamps to the frame and the other end clamps to ductwork, bar joists, I-beams, or other fixed structural members already in place. Because the telescoping poles are long, the securing member does not need to be close at hand. The poles can be clamped to ceiling beams, so the floor around the door frame can be left clear.

Other types of bracing jigs are available. Manufacturers provide installation instructions with these products.

1.2.0 Installing Sills

The primary function of a sill is to divert water away from the building. Sills can also provide a decorative accent to windows (*Figure 9*). Generally, the slope of the sill should be at least 15 degrees below the horizontal, sloping away from the window, and the sill should extend at least 1 inch beyond the face of the wall at the sill's closest point to the wall. Cored block and brick may be used in sills, although uncored units may be required at either end of the sill to prevent exposure of the cores for aesthetic reasons as well as to prevent the collection of moisture and dirt.

Sills typically consist of either a single unit or multiple units. They can be built in place or prefabricated. Various materials can be used, such as stone, wood, metal, or concrete, depending on the project design. Typically, single-unit sills are classified as either slip sills or lug sills, as shown in *Figure 10*. Slip sills are the same length as the window or door opening, and lug sills extend into the masonry on either side of the opening. The slip sill may be left out when the opening is being laid and set at a later time. The lug sill, however, must be set when the masonry is up to the bottom of the window or door opening.

Windowsills and doorsills are made of concrete, brick, or cut stone. They function to support window and door frames so that the frames will not sag, crack, or tilt. They also serve to prevent the entry of water, wind, and dirt and add a finished appearance to the opening. One-piece sills are designed so that they have a slight degree of downward slope from the edges to the center of the sill. This is designed to facilitate water runoff.

SINGLE-UNIT SILL

MULTIPLE-UNIT SILL

28204-14_F09.EPS

Figure 9 Decorative sill.

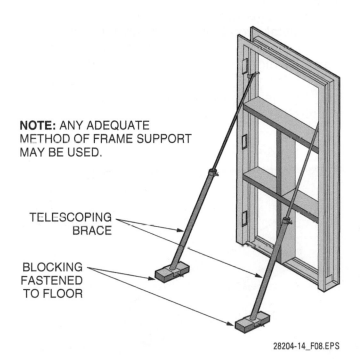

NOTE: ANY ADEQUATE METHOD OF FRAME SUPPORT MAY BE USED.

TELESCOPING BRACE

BLOCKING FASTENED TO FLOOR

28204-14_F08.EPS

Figure 8 Telescoping brace.

Sills cover the full widths of openings and are supported by either the block under the opening or by the slab or foundation directly below the opening. Although the usual practice is to use precast sills, which can be placed in an opening after the wall or foundation has been laid up, cast-in-place or combination sills can be used.

Sills of different types or from different manufacturers may have slight variations that affect their method of installation. The mason should always check the plans and the sill construction to determine the proper method of installation before beginning.

The following steps can be used, in general, for sill placement:

Step 1 From the plans, determine the position of the opening according to the number of courses under the sill and the number of block on either side.

Step 2 Lay up the wall to the height of the masonry unit immediately under the sill.

Step 3 Apply a full bed of mortar to the masonry unit directly under the sill. The width of all joints around the sill should remain constant.

Step 4 Set the sill gently in place, and press it down until the mortar oozes from the joints.

Step 5 Remove the excess mortar, and smooth the joint surfaces on both sides of the wall.

Step 6 As always, check all measurements carefully. Measure joint thickness, and check the level of the sill by placing a level along the horizontal line of the sill. If the sill is not level, press down at either end. If the sill cannot be leveled by this slight adjustment, remove the sill, apply fresh mortar, and reset.

> **WARNING!**
> Concrete sills and lintels can be extremely heavy. Lift the sills properly, using two people and proper lifting techniques. Be sure sills and lintels are properly supported. If they fall, they could cause injuries and damage materials.

Flashing and weepholes are required beneath the sills. The flashing should extend through the brick to the wall's exterior face at the lower end of the flashing. Drip edges are optional. Extend the sill flashing beyond the sill ends to the first head joint outside of the jamb of the opening, and turn the ends up and outward for at least 1 inch. This will allow the moisture to drain away from the sill and to the adjacent wall. Place weepholes on top of the flashing. The maximum spacing of weepholes should be 16 inches if wicks are used, and 24 inches if no wicks are used. All sills require a drip. Sills with long runs may require the installation of anchors to support them; refer to the construction drawings for details.

> **NOTE**
> The term *drip* refers to a cut or casting on a masonry unit, such as a sill, that is designed to let water drip off the masonry unit without running down the wall. The term *drip edge* refers to the bent edge of a piece of flashing that is also designed to prevent water from running down a wall.

When laying block lug sills, apply the mortar near the ends of the sill only. After the wall has been completed, mortar can be forced under the sill. When sills are installed during the construction of the wall or foundation, they must be protected against breakage and staining. Wrap the sill in kraft paper or polyethylene to prevent mortar droppings from falling on the sill.

After the block sill has been properly laid, stretch a taut line and lay the block course by course, including the jamb block. Take special care to keep all the blocks level, plumb, and in alignment. Fill all joints around the jamb well in order to prevent leakage of moisture or air.

Windowsills in brick buildings can be of brick, stone, metal, wood, or precast concrete. Stone and precast concrete require additional labor and are generally more expensive than brick. Brick is the

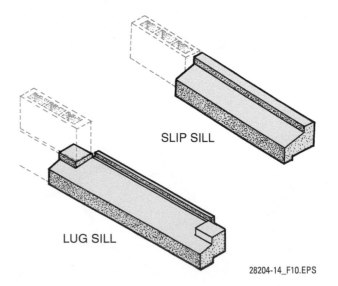

28204-14_F10.EPS

Figure 10 Lug and slip sills.

preferred choice for windowsills. It adds to the appearance of the building, is inexpensive, and can be placed by the same workers who lay up the wall.

Brick for sills should be laid on edge, rowlock style, and sloped at least 15 degrees or ¾ inch to shed water and produce a visually appealing effect (*Figure 11*).

Brick windows begin a full number of courses above the beginning of the wall. Using the 4-inch modular system makes the coursing easier, but a common mistake is forgetting that three courses equal 8 inches. At sill height, lay the sloped rowlock course to occupy the height of two normal courses.

As soon as the mortar is firmly set, place the window frame and brace it so that it will remain level, plumb, and square. Carpenters usually set the window frames; however, the mason may set frames when carpenters are not available and the work must proceed. The bracing remains in place until the wall is built over the top of the frame. If the top course does not come within ¼ inch of the top framing piece, the difference can be adjusted with mortar.

On brick veneer walls, the brick is butted flush against the window jambs and the frame is nailed into the framing of the wall. Regardless of how frames are fastened to the masonry wall, remember that the frame should be set the proper distance back from the face of the wall. The position depends upon the thickness of the wall.

1.2.1 One-Piece Slip and Lug Sills

For block construction, both window and door sills are usually one piece of concrete or cut stone. They support window and door frames so that the frames will not sag, crack, or tilt. They also prevent the entry of water, wind, and dirt.

Sills cover the full widths of openings; they are supported either by the masonry under the opening or by the slab or foundation directly below the opening. The usual practice is to use precast slip sills that can be placed in an opening after the wall

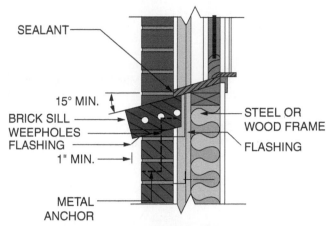

BRICK SILL IN FRAME/BRICK VENEER CONSTRUCTION

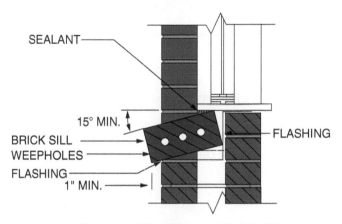

BRICK SILL IN CAVITY WALL CONSTRUCTION

28204-14_F11.EPS

Figure 11 Brick windowsills.

or foundation has been laid up. A detailed cross section of sill placement is shown in *Figure 12*.

Slip sills are the same width as the window or door opening, but lug sills extend into the masonry on either side of the opening. The slip sill can be omitted when the opening is laid and set at a later time. The lug sill, however, must be set when the masonry is up to the bottom of the window or door opening.

Checking Plumb and Level on Door Frames

Take care to ensure that a metal frame is absolutely plumb and square. An improper installation will affect the performance of the door and make fitting the door difficult or impossible. A little extra time spent installing the frame will make door installation much easier.

Use a level to plumb a door jamb. Check for accuracy before and during the process. One way to check squareness is by measuring diagonally from top to bottom at the corners. Equal measurements indicate that the frame is square.

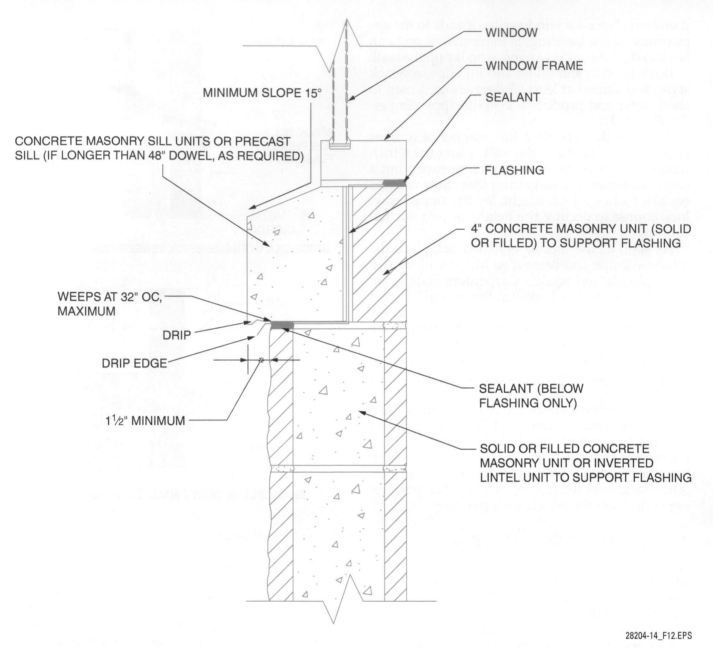

WINDOW

WINDOW FRAME

MINIMUM SLOPE 15°

SEALANT

CONCRETE MASONRY SILL UNITS OR PRECAST
SILL (IF LONGER THAN 48" DOWEL, AS REQUIRED)

FLASHING

4" CONCRETE MASONRY UNIT (SOLID
OR FILLED) TO SUPPORT FLASHING

WEEPS AT 32" OC,
MAXIMUM

DRIP

DRIP EDGE

SEALANT (BELOW
FLASHING ONLY)

$1\frac{1}{2}$" MINIMUM

SOLID OR FILLED CONCRETE
MASONRY UNIT OR INVERTED
LINTEL UNIT TO SUPPORT FLASHING

28204-14_F12.EPS

Figure 12 Typical detail for a slip sill.

1.2.2 Laying One-Piece Sills

Sills of different types, or from different manufacturers, may need different installation methods. Always check the plans and manufacturer's documents before installing a sill. The following is a general outline for placing a one-piece sill:

Step 1 Use the plans to determine the position of the opening according to the number of courses under the sill and the number of masonry units on either side of it.

Step 2 Lay up the wall to the height of the masonry immediately under the sill.

Cut All the Brick First

When cutting brick to fit against a jamb of a window or door, it is good practice to make all necessary cuts to build to the top before actually laying any units. Cutting brick one at a time for each individual course is very time consuming. All cuts made should be the same size. Any difference in the wall space should be compensated for in the header joints rather than in the size of the cuts.

Step 3 Apply a full bed of mortar to the masonry directly under the sill. The width of all joints around the sill should remain constant.

Step 4 Set the sill gently in place and press it down until the mortar oozes from the joints.

Step 5 Remove excess mortar and smooth the joint surfaces on both sides of the wall.

Step 6 As always, check all measurements carefully. Measure joint thickness and check the level of the sill by using the level along the horizontal line of the sill. If the sill is not level, press down at either end. If the sill cannot be leveled by this slight adjustment, remove the sill, apply fresh mortar, and reset.

When lug sills are laid, the mortar should be applied near the ends of the sill only. After the wall has been completed, mortar can be forced under the sill.

1.2.3 Brick Sills

Windowsills in brick buildings can be of brick as well as metal, stone, or precast concrete. Brick for sills are usually laid on edge, rowlock style, and sloped about 15 degrees or ¾ inch to shed water. The slope is maintained by thickening the mortar bed, and the laying angle is maintained by a line.

Figure 13 shows this type of sill, which is flush with the edges of the window opening. Note that raising the edge of the rowlock produces a slight lip. The window frame can be set flush against this lip to hold it tightly in place and to minimize any collection of dirt.

Most brick sills are slip sills in that they do not extend past the window frame. Sometimes, specifications call for lug sills that need to be set when the masonry is up to the bottom of the window opening.

1.2.4 Lintels

When the masonry courses have been completed to the height of the jamb, you are ready to set the lintel. This is the final step in completing the construction of any opening. A lintel provides support over windows, doors, or other openings. It ensures that no loads are placed on the door or window frame underneath. The building design and the load to be supported determine the type of lintel to be used. Steel, precast concrete, lintel block, reinforced brick, or metal straps can be used as lintel material. Be sure the masonry un-

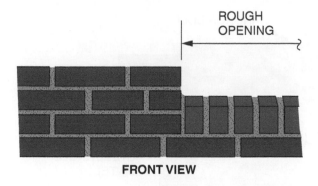

FRONT VIEW

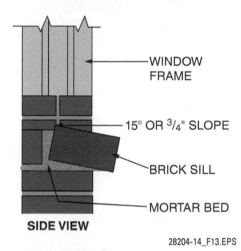

SIDE VIEW

28204-14_F13.EPS

Figure 13 Rowlock brick windowsill.

derneath that supports the lintel has set up before placing the lintel. You learned how to install bond beam lintels and precast lintels in the module titled *Reinforced Masonry*. You will learn how to install steel lintels in the section of this module titled *Installing Steel Lintels*.

The ends of the lintel transfer the load to the underlying masonry. The ends should extend into the masonry on either side a minimum of 8 inches for block and 4 inches for brick to distribute the weight more evenly. *Figure 14* shows a properly installed brick lintel. Insets vary according to project specifications. Larger spans require the lintel to extend even deeper into the wall. The lintel should be strong enough to support the load and resist bending. Regardless of the unit width, ⅔ of the unit must be resting on the steel angle lintel.

Window Frames

Window frames have a sill plate as well as a frame header. The sill plate is installed first, then both jambs, then the header. Detailed instructions for assembly are usually provided in the manufacturer's literature.

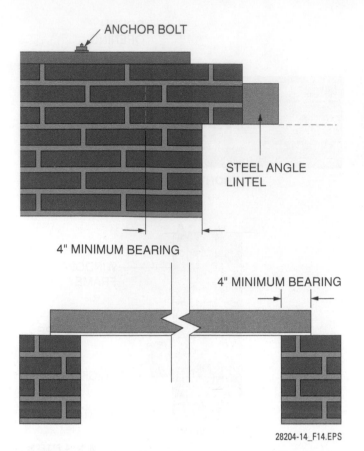

4" MINIMUM BEARING

STEEL ANGLE LINTEL

ANCHOR BOLT

4" MINIMUM BEARING

28204-14_F14.EPS

Figure 14 A 4-inch minimum bearing is required for each end of brick lintel.

Slight movement often occurs at the location of lintels. For this reason, control joints are often located at the ends of the lintel to allow it to move slightly in response to temperature changes without cracking the underlying joint. A noncorrosive metal plate is set under the supported end of the lintel at the side of the opening. A full mortar bed is added to the plate and then raked out to a depth of ¾ inch when it has set up but is still soft enough to tool. The opening in the mortar joint is later filled with caulking. The lintel can then move on the plate without causing movement in the masonry.

1.3.0 Installing Steel Lintels

Masons use two types of steel lintels: angle irons and I-beams. Angle irons, which are shaped like the capital letter L when viewed in cross section, are one of the most common types of steel lintels (*Figure 15*). Angle iron should have a thickness of

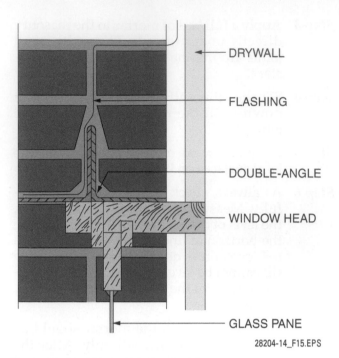

DRYWALL

FLASHING

DOUBLE-ANGLE

WINDOW HEAD

GLASS PANE

28204-14_F15.EPS

Figure 15 Angle irons used as lintels.

at least ¼ inch and a width not less than 3½ inches to support 4-inch masonry units. These lintels should be supported on either side by a distance of at least 4 inches.

Use a single angle iron for 4-inch walls, two angle irons placed back to back for 8-inch walls, and three angle irons for 12-inch walls. *Figure 16* shows two angle irons placed back to back in a brick wall. The ends should extend 4 to 8 inches into the brickwork to more evenly distribute the weight. Steel lintels should have a minimum thickness of ¼ inch and a minimum width of 3 inches to support the standard 4-inch masonry unit.

I-beam lintels, which as their name suggests resemble the capital letter I in cross section, are used to support masonry that spans great lengths or that carries highly concentrated loads (see *Figure 17*). Corrugated ties are welded to the faces of I-beams to attach masonry veneers or face shells that completely encase the beam. A continuous plate is welded to the bottom of the beam to support the weight of the masonry. The plate must be sized to sufficiently support the masonry. The plates are in turn connected to the supporting masonry jambs with steel bearing plates, anchor bolts, or rebar. Weldable rebar is attached to the top of the lintel to connect the wall above.

Protect Sills from Damage

When sills are installed during construction, they must be protected against breakage and staining. Wrap the sill in kraft paper or polyethylene to keep mortar droppings off the sill.

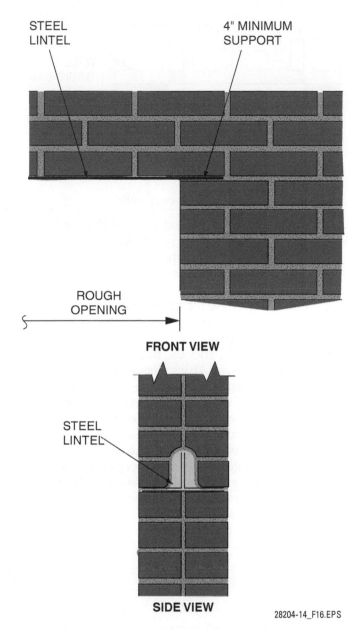

FRONT VIEW

SIDE VIEW

28204-14_F16.EPS

Figure 16 Steel angle iron lintels for brick.

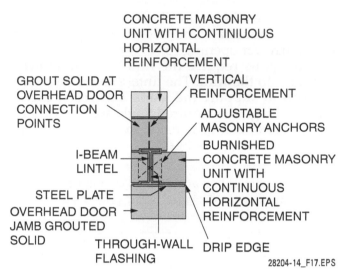

Figure 17 I-beam used as a lintel.

28204-14_F17.EPS

To install I-beams, follow these steps:

Step 1 Build the supporting wall to the proper height. Ensure that the supporting masonry jambs are level, plumb, ranged, and to the correct height. Consider the length of the studs, bolts, or rebar that will be used for the connector prior to grouting the jamb to bearing height.

Step 2 Install the required connector. Ensure that the connectors are centered on the wall, set to the proper elevation, and placed the proper distance apart.

Step 3 Place the I-beam lintel. Ensure that the welded or bolted connections are made as specified. It is common for the steel erection contractor to perform this step; if so, be sure to coordinate the work so it takes place without disrupting other masonry work. Place at least one temporary vertical support midspan along the lintel. Long spans may require more supports at regular intervals.

Step 4 Encase the I-beam in masonry face shells or veneer as specified. Be very careful when installing face shells, as they are very unstable prior to the setting of the mortar. Use light taps to adjust the shells gently while the mortar is still fresh. Heavy tapping will likely break the mortar bond and can potentially disrupt the previously installed units. Add mortar behind the face shells as you progress in order to provide additional stability. Engage the ties and install the remaining courses in the same manner.

Step 5 Set the first course of masonry over the steel lintel. Units in this course should be buttered carefully, set on the support, and slid against adjacent units to avoid trapping mortar between the units and the lintel.

Step 6 Install reinforcement and grout if required. It is common for the first few courses over an I-beam lintel to be reinforced bond beams in order to provide additional strength.

Step 7 Once the masonry structure has achieved its design strength, remove the temporary vertical support(s).

Steel lintels that are installed over large openings, or that carry heavier than normal loads, should have a larger loadbearing surface at each end and should also be thicker than those used over narrower openings or to carry normal loads. Larger spans require the lintel to extend even deeper into the wall. The lintel should be strong enough to support the loads and resist excessive bending. The required angle size for various lengths of openings is listed in *Table 1* for openings up to 16 feet in length.

> **NOTE**
>
> Steel lintels that are installed over large openings are designed by the project's design team.

1.4.0 Installing Chases and Recesses

Chases and recesses are horizontal or vertical spaces left in a wall for the purpose of containing plumbing, heating ducts, electrical wiring, or other equipment. Often, specifications require these me-

Table 1 Steel Angle Size for 4-Inch Masonry Walls

Span	Size of Angle
0'–5'	3" × 3" × ¼"
5'–9'	3½" × 3½" × ⁵⁄₁₆"
9'–10'	4" × 4" × ⁵⁄₁₆"
10'–11'	4" × 4" × ³⁄₈"
11'–15'	6" × 4" × ³⁄₈"
15'–16'	6" × 4" × ½"

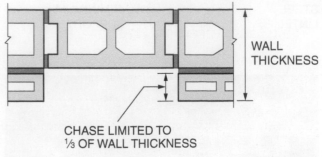

WALL THICKNESS

CHASE LIMITED TO ⅓ OF WALL THICKNESS

28204-14_F18.EPS

Figure 18 Typical chase in a masonry wall.

chanical items to be placed behind the finished surface of the wall. If plans call for chases or recesses, they should be formed as the wall is built. Otherwise, the other trades must go through the expensive and time-consuming process of chiseling them out after the block is in place.

Chases, shown in *Figure 18*, are generally formed on the inside of the wall and vary in width from 4 to 12 inches. Because these openings reduce the wall thickness and its strength, their depth is limited to ⅓ of the wall's thickness; that is, a 12-inch wall should never have a chase more than 4 inches deep unless special structural considerations are made. Bond the masonry units that form the chase directly to the rest of the wall. Stack joints are generally not acceptable for such bonding because the joints are weak. Make certain that the chase is plumb and that all excess mortar is cut off so that pipes, electrical wires, or ductwork will fit properly.

Residential Windows

Decorative windows are often specified in high-end brick homes. They come in many shapes and sizes. A skilled mason must be able to construct openings to fit these windows and the decorative accents around them.

28204-14_SA02.EPS

28204-14_SA03.EPS

28204-14_SA04.EPS

28204-14_SA05.EPS

Additional Resources

Bricklaying: Brick and Block Masonry. 1988. Brick Industry Association. Orlando, FL: Harcourt Brace & Company.

Concrete Masonry Handbook, Fifth edition. W. C. Panerese, S. K. Kosmatka, and F. A. Randall, Jr. Skokie, IL: Portland Cement Association.

Technical Note 31B, *Structural Steel Lintels*. 1987. Reston, VA: The Brick Industry Association. **www.gobrick.com**

Technical Note 36, *Brick Masonry Details, Sills, and Soffits*. 1988. Reston, VA: The Brick Industry Association. **www.gobrick.com**

1.0.0 Section Review

1. The parts of a masonry opening do *not* include the _____.

   a. reveal
   b. sill
   c. jamb
   d. lintel

2. The primary function of a sill is to _____.

   a. support the weight of the load over the masonry opening
   b. divert water away from the building
   c. serve as a step or platform
   d. be a decorative element

3. The minimum thickness of a steel lintel should be _____.

   a. ⅛ inch
   b. ¼ inch
   c. ⅓ inch
   d. ½ inch

4. The depth of a chase should never exceed _____.

   a. ⅓ of the wall's thickness
   b. ½ of the wall's thickness
   c. the thickness of the wall
   d. 1½ times the wall's thickness

2.0.0 TYING A SINGLE WYTHE TOGETHER

Objective

Describe the methods and materials used to tie a single masonry wythe together.

a. Describe how to use and install ladder and truss joint reinforcement.
b. Describe how to use and install seismic reinforcements.

Trade Terms

Horizontal joint reinforcement: A system of connected steel wire that provides added structural integrity to a masonry wythe by distributing lateral loads evenly and by tying the masonry elements in the structure together mechanically.

Panel: A section of wall between control joints, wall ends, or a control joint and wall end.

Moisture and temperature changes can cause the materials in a masonry wall or other masonry structure to expand and contract at different rates, which in turn can cause them to crack. The structural integrity of a single-wythe masonry structure can be maintained through the use of horizontal joint reinforcement. Horizontal joint reinforcement contributes to structural integrity by distributing lateral loads evenly and by tying the masonry elements in the structure together mechanically. This has the effect of increasing the resistance of masonry structure to cracking. In addition to tying the wythe together, horizontal joint reinforcement reinforces masonry structures in the following ways:

- Bonding intersecting walls
- Limiting movement between different materials in a wall
- Giving horizontal structural reinforcement

Horizontal joint reinforcement is usually made of galvanized steel wire. As with vertical rebar, horizontal joint reinforcement comes in different gauges of steel. The basic types of metal horizontal joint reinforcement are discussed in the following sections.

2.1.0 Installing Ladder and Truss Joint Reinforcement

Continuous horizontal joint reinforcement for single-wythe walls is manufactured in ladder (rectangular) and truss (triangular) styles (see *Figure 19*). Continuous horizontal joint reinforcement is made of galvanized steel to prevent corrosion. They are available in different wire sizes. Standard widths are approximately 1⅝ inches less than the wall thickness. This sizing ensures good mortar coverage on both faces of the wall. Cross wires are welded diagonally or perpendicular to the longitudinal wires, usually at 16-inch spacings. In single-wythe block walls, the simple two-wire ladder type gives good protection against cracking. The truss type is stronger, but interferes with easy placing of vertical rebar.

The usual distance for joint reinforcement spacing is every 8, 16, or 24 inches vertically. The distance between wall ends or between control joints is known as a **panel**.

Joint reinforcement should be placed in the first and second bed joints immediately above and below wall openings. It should extend at least 24 inches beyond the openings, or to the end of the panel. In addition, it should be placed two or three courses immediately below the top of a wall. Continuous joint reinforcement should be cut at control joints and expansion joints.

All ties must be firmly embedded in, and bonded to, the mortar. Additionally, any exposed metal will corrode, as well as detract from the appearance of the masonry. Ties must be embedded at least ⅝ inch from the exposed face of the wall.

The steps for installing ties of this type are as follows:

Step 1 Place the reinforcement directly on the masonry course. To ensure continuity, overlap continuous wire or hardware cloth for at least 6 inches.

Step 2 If you are using tab ties, stagger the tabs so they do not form a continuous vertical line.

Step 3 Spread a full mortar bed over the reinforcement.

Step 4 Check for and fill any voids or pockets to ensure that no moisture will collect around the metal.

Step 5 Maintain consistent joint size so that the reinforced joints will be no larger than all other joints.

Step 6 Lay the next course or courses of masonry.

Step 7 When working with two wythes, always check that they are perfectly level with one another.

Continuous joint reinforcement can be shaped, cut, and bent to fit corners. It must be overlapped at least 6 inches to provide good load transfer. When cutting reinforcement, remember to leave enough space for a good mortar joint to bury the ends. Prefabricated corners and tees are available for virtually any wall configuration.

> **WARNING!**
>
> Continuous joint reinforcement can cause a safety hazard. If steel wire protrudes past the end of the course, it can injure anyone passing by. Avoid accidents by bending down any surplus continuous reinforcement, being sure to bend the wire up carefully when preparing to lay the intersecting block. If it is not possible to bend the wire immediately, hang a warning flag from the end.

2.2.0 Installing Seismic Reinforcements

Seismic reinforcements are a special type of continuous horizontal reinforcement designed for use in areas that are prone to earthquakes (see *Figure 20*). In the event of an earthquake, seismic reinforcements enable masonry wythes to maintain their structural integrity. Because they are designed to function by connecting two adjacent masonry wythes, they are a hybrid between the single-wythe reinforcement systems discussed in this section and the double-wythe anchor systems discussed in the section titled *Tying Two Masonry Wythes Together*. Special seismic retrofit anchors have been developed to anchor existing masonry walls to floors and roofs. Manufacturers' data sheets give instructions for installing these systems.

Seismic reinforcement systems are designed with grooves, slots, or indentations through which a continuous wire is strung, which allows the continuous wire to serve as an integral component of the tie system (refer to *Figure 20*). When the joint reinforcement and the continuous wire are secured in the mortar joints on the veneer wythe, in the event of an earthquake the veneer wall will be able to move as a single unit while maintaining its structural integrity and its connection to the backing wythe. Seismic reinforcements can also protect veneer walls against problems caused by thermal expansion and contraction. Local building codes specify the applicable requirements for using seismic reinforcements in masonry structures.

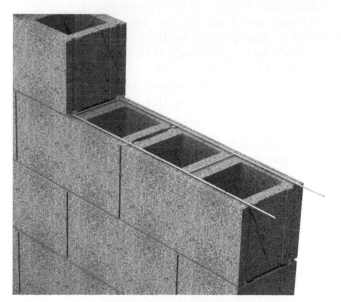

LADDER (RECTANGULAR)
CROSS BRACING

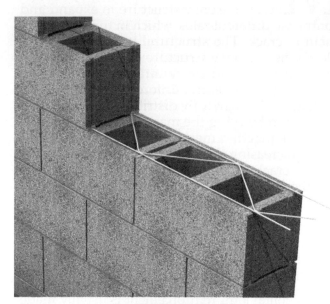

TRUSS (TRIANGULAR)
CROSS BRACING

28204-14_F19.EPS

Figure 19 Ladder- and truss-type continuous joint reinforcement for tying single-wythe walls together.

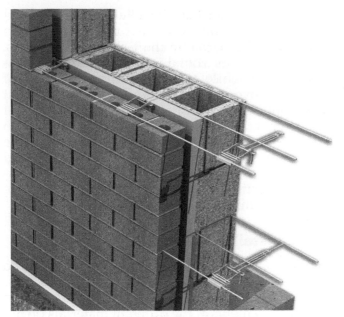

LADDER (RECTANGULAR) TYPE

TRUSS (TRIANGULAR) TYPE

28204-14_F20.EPS

Figure 20 Seismic reinforcements showing continuous wire on the veneer wythe.

Additional Resources

ASTM A82, *Standard Specification for Steel Wire, Plain, for Concrete Reinforcement*, Latest Edition. West Conshohocken, PA: ASTM International.

ASTM A185, *Standard Specification for Steel Welded Wire Reinforcement, Plain, for Concrete*, Latest Edition. West Conshohocken, PA: ASTM International.

ASTM A496, *Standard Specification for Steel Wire, Deformed, for Concrete Reinforcement*, Latest Edition. West Conshohocken, PA: ASTM International.

ASTM A951, *Standard Specification for Steel Wire for Masonry Joint Reinforcements*, Latest Edition. West Conshohocken, PA: ASTM International.

TEK 12-4D, *Steel Reinforcement for Concrete Masonry*. 2011. Herndon, VA: National Concrete Masonry Association. **www.ncma.org**

2.0.0 Section Review

1. The typical spacing for cross wires in continuous joint reinforcement is _____.

   a. 12 inches
   b. 14 inches
   c. 16 inches
   d. 18 inches

2. Seismic reinforcement systems are designed to tie an adjacent masonry wythe through the use of a _____.

   a. pintle-and-eye system
   b. series of bolts
   c. tongue-and-groove system
   d. continuous wire

SECTION THREE

3.0.0 TYING TWO MASONRY WYTHES TOGETHER

Objective

Describe the methods and materials used to tie two masonry wythes together.
  a. Describe how to use and install flexible anchors.
  b. Describe how to use and install horizontal anchors.

Trade Term

Pintle: A type of masonry fastener that allows a horizontal reinforcement to maintain a joint while allowing the wall to have some freedom of vertical movement.

Anchors and their fasteners connect a masonry veneer to an interior masonry wythe in a cavity wall. Anchors may be of the continuous (also called three-wire) joint-reinforcement type, or they may be individual anchors attached to the backing wall separately. Use continuous joint reinforcement for multiwythe walls of one type of masonry. Either truss or ladder type is acceptable (see *Figures 21* and *22*). The tab type of continuous joint reinforcement is used for tying block and brick wythes together. These rigid reinforcements transfer loads horizontally and vertically. The tab configuration is particularly well suited for transferring lateral and ver-

tical loads. Notice that the adjustable ladder and truss reinforcements in the figures are shown with *pintles* inserted in the T-shaped extensions. Pintles allow horizontal reinforcements to maintain a joint while allowing limited freedom of vertical movement of the wall due to expansion and contraction, vibration, and other sources of movement. *Figure 23* shows pintles inserted into an eye-wire truss, which is a type of adjustable continuous joint reinforcement that is designed specifically for use with pintles. The eyes are extensions with holes at the ends into which the pintles are inserted.

3.1.0 Installing Flexible Anchors

Some special types of anchors are adjustable to allow some movement. These anchors tie materials with different rates of heat expansion so that small movements of the different materials do not cause cracks. A common type of flexible anchor is the veneer anchor, as shown in *Figure 24*.

Seismic anchors are a special type of flexible anchor (see *Figure 25*). They serve the same purpose as seismic reinforcements discussed in the section *Installing Seismic Reinforcements*, but instead of being connected to each other as part of a continuous joint reinforcement system, seismic anchors are connected individually to the structural backing. Their installation requirements are the same as those of other types of flexible anchors.

3.2.0 Installing Horizontal Anchors

Horizontal anchors, also called unit ties, do not transfer as much vertical load as continuous ties. Horizontal anchors are useful for connecting

THREE-WIRE LADDER TYPE

LADDER TAB TYPE

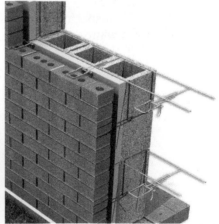
ADJUSTABLE LADDER TYPE

28204-14_F21.EPS

Figure 21 Ladder-type continuous joint reinforcement for tying two wythes together.

    NCCER – *Masonry Level Two*   28204-14

THREE-WIRE TRUSS TYPE

TRUSS TAB TYPE

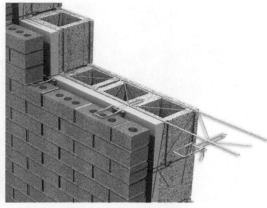
ADJUSTABLE TRUSS TYPE

28204-14_F22.EPS

Figure 22 Truss-type continuous joint reinforcement for tying two wythes together.

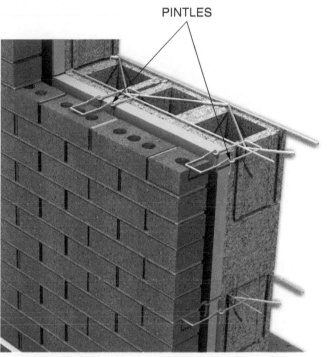

PINTLES

28204-14_F23.EPS

Figure 23 Pintles inserted into eye-wire adjustable continuous joint reinforcement.

28204-14_F24.EPS

Figure 24 Veneer anchor.

composite wythes in a solid wall. Rigid unit ties are available in three types: box (also called rectangular), Z-type (*Figure 26*), and corrugated. Brick manufacturers often recommend certain types of wall ties for use with their products based on their experience with installing them.

Some rectangular or Z ties may have crimps or dips in them. These are designed to allow condensation inside a masonry cavity to collect and drip down. However, because crimps can weaken the load transfer strength of the tie, some building codes do not permit their use. Newer rectan-

gular or Z ties have plastic drip rings on the wire instead of a crimp.

Shaped metal ties and hardware cloth are also used for joint reinforcement or to tie materials together. *Figure 27A* shows some commonly used shaped ties. Some have dovetail anchors, which fit into special slots. Slots are made in the concrete to receive the shaped end of the anchor. On some jobs, the shaped metal bars may be called anchors. *Figure 27B* shows a piece of wire-mesh wall tie. The rectangular strips of mesh are used alone or with a metal tie bar to tie nonbearing walls to intersecting walls.

28204-14_F25.EPS

Figure 25 Seismic anchors.

Because they allow some limited movement, adjustable anchors are often installed as individual unit ties. *Figure 28* shows some adjustable unit ties for attaching masonry to a backup structure of concrete, wood or metal studs, or block. These units consist of two parts, with movement allowed between the parts. The plate part is attached to the backup structure and the wire part is mortared to the masonry unit. Another kind of adjustable assembly is shown in *Figure 29*. These adjustable assemblies are used to tie masonry face wythes to a masonry backup. The two parts of these assemblies allow contraction and expansion

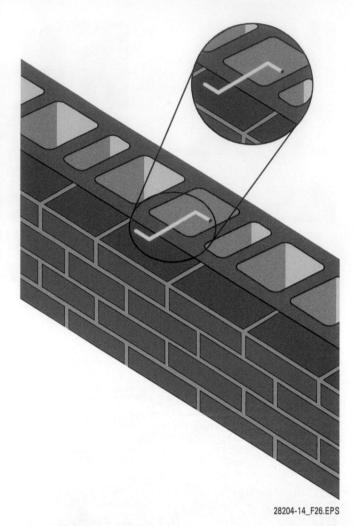

28204-14_F26.EPS

Figure 26 Z-type unit tie.

between the two wythes. Both of these types of individual adjustable horizontal ties maintain position but do not transfer load.

Adjustable Unit Ties

Adjustable unit ties (anchors) are installed to secure a multiwythe wall. The metal base is secured to the wall. As the mason lays additional courses, the metal wire is attached to the base and anchored in the brick.

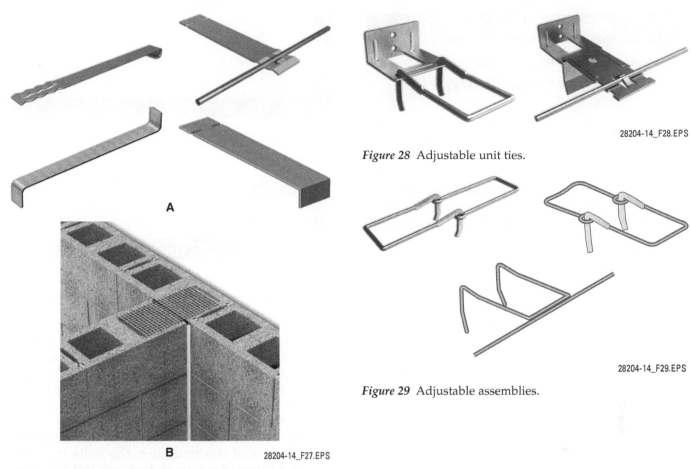

Figure 28 Adjustable unit ties.

28204-14_F28.EPS

Figure 29 Adjustable assemblies.

28204-14_F29.EPS

B

28204-14_F27.EPS

Figure 27 Typical metal anchors and mesh wall tie.

Additional Resources

ASTM E488, Standard Test Methods for Strength of Anchors in Concrete Elements, Latest Edition. West Conshohocken, PA: ASTM International.

ASTM E754, Standard Test Method for Pullout Resistance of Ties and Anchors Embedded in Masonry Mortar Joints, Latest Edition. West Conshohocken, PA: ASTM International.

TEK 12-1B, Anchors and Ties for Masonry. 2011. Herndon, VA: National Concrete Masonry Association. **www.ncma.org**

3.0.0 Section Review

1. A common type of flexible anchor is the _____.
   a. corrugated anchor
   b. veneer anchor
   c. three-wire truss
   d. tab tie

2. Individual adjustable horizontal ties do not _____.
   a. require attachment to the wythe
   b. maintain position
   c. allow for vertical movement
   d. transfer load

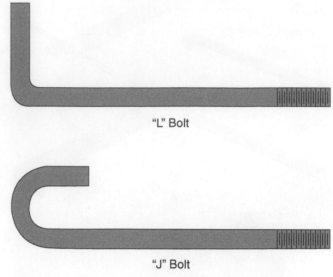

"L" Bolt

"J" Bolt

28204-14_F30.EPS

Figure 30 Simple rod tie bolts.

SECTION FOUR

4.0.0 TYING A MASONRY WYTHE TO A STRUCTURAL ELEMENT

Objective

Describe the methods and materials used to tie a masonry wythe to structural elements.

    a. Describe how to use and install rigid ties and bolts.
    b. Describe how to use and install bearing plates.
    c. Describe how to use and install saddles.
    d. Describe how to use and install strap ties.

Performance Tasks 3 and 4

    3. Install a bearing plate.
    4. Install a strap tie.

Trade Term

Beam pocket: An opening in a vertical masonry wall that allows an intersecting structural beam to bear on or pass through the wall.

To ensure the structural integrity of masonry structures, masons often install devices that provide structural support to masonry elements by tying masonry wythes to the structural elements of the wall or building. Like other forms of horizontal joint reinforcement, these devices are embedded in the mortar or grout during construction of the masonry structure. Commonly used devices include rigid ties and bolts, bearing plates, saddles, and hurricane ties. The following sections discuss the applications and installation procedures for each type of embedded item. Always refer to the local applicable code and specifications and to the project plans when installing embedded items.

4.1.0 Installing Rigid Ties and Bolts

Rigid masonry wall ties resist compression and shear stresses, but flexible ties do not resist shear stresses. Rigid ties include galvanized metal strips and galvanized bolts. The bolts have different heads for different purposes. The strip tie is commonly used for attaching loadbearing partition walls.

Tie-bolt devices can be simple threaded rods as shown in *Figure 30*. They can also be complex sleeve-and-tie devices, with various head types. These heavy-gauge connectors usually fasten a beam, ledge, sill plate, or other structural member to a masonry wall. Different ties may be set in mortar or epoxy, or they may be power driven, according to the engineering specifications.

When the plans call for bolts to anchor other components of the structure, the bolts must be mortared into place when they are tightened so they will not come loose. They are spaced according to the plans or drawings. Bolts are normally set in the center of a wall unless otherwise specified.

Check the drawings before starting the wall, because you may have to notch the brick so that the angled end of the bolts will fit snugly. Use a line to ensure that all the bolts are set the same distance from the face of the wall. Also, check the drawings for the height the bolts must protrude above or beyond the wall, as this will vary from job to job.

Once you have set the locations, form a base under each bolt with mortar. Use masonry scraps as a base if the bolt protrudes into a core. Fill mortar around the bolt tightly, then use your trowel or a stick to pack the mortar. Make sure the bolt shank is plumb or level, depending on the location. Try to keep the mortar off the threaded part of the bolt. When you finish, be sure to clean the threads so the nut will easily thread onto the bolt.

Installing bolts for steel beams or plates calls for absolute spacing, as the steel is predrilled with no room for adjustments. The construction superintendent or carpenter usually checks the placement of the bolts for these jobs.

4.2.0 Installing Bearing Plates

Another type of embedded item is a bearing plate. Bearing plates serve as connection points between masonry walls and steel structural members such as beams or joists. The bearing plate is welded or bolted to the structural member and embedded into the masonry by means of welded studs, anchor bolts, welded rebar, or a combination of these. Made of steel, often galvanized, bearing plates come in various thicknesses and sizes.

Project specifications may call for bearing plates under the ends of **lintels**, arches, beams, or other concentrations of vertical load. Bearing plates are also used on top of columns, pilasters, or other load concentrators.

To install a bearing plate, follow these steps:

Step 1 At least one lift prior to the installation of the bearing plate, lay out and locate the bearing plate. Refer to design drawings to determine the lateral location and the top of the plate's elevation, and to ensure that the proper type of plate is being used for the installation. Mark the center of the plate on the wall.

Step 2 Determine the reinforcing requirements for the bearing plate (if any) and install masonry accordingly. Note that some plates require a continuous bond beam, while others require a 4-foot bond beam centered on the plate or else require one or two solid grouted cells under the plate.

Step 3 Build the masonry wall to the bearing height. Include any required bond-beam units and reinforcement. Determine if the vertical reinforcement will conflict with the plate and cut as needed to avoid conflict. Use open-end or center-web units as needed to accommodate plate anchors.

Step 4 Plug the sides of the beam pocket to contain the grout within the spaces to be grouted. This may require the use of small cut pieces of masonry and mortar to separate the pocket from the remainder of the wall.

Step 5 Build masonry around the pocket. Ensure the size of the pocket is sufficient to accommodate the structural member and to allow room for the connection. To prevent grout from seeping into the pocket, cross-web all masonry units that are adjacent to the pocket.

Step 6 Apply double-bond-beam mesh to ensure that the grout does not hang into the pocket and interfere with installation.

Step 7 Confirm that the bearing plate fits easily into and out of the pocket and that it can be easily installed to the proper center and height marks. Mark the top of the plate height adjacent to the pocket to ensure proper placement.

Step 8 Place grout and set the bearing plate according to the center and height marks.

4.3.0 Installing Saddles

Saddles are welded metal hangers used in block masonry structures (see *Figure 31*). They are installed to provide support for structural beams that intersect a masonry wall. Saddles are an alternative to beam pockets, which are openings in a vertical masonry wall that allow an intersecting structural beam to bear on or pass through the wall. The procedure for installing a saddle is the same as that used for bearing plates.

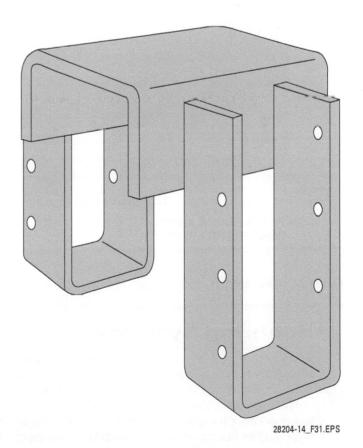

28204-14_F31.EPS

Figure 31 Masonry saddle.

4.4.0 Installing Strap Ties

Masons use two kinds of straps in masonry construction: hurricane ties and purlin anchors, also called PA straps. As you learned in *Masonry Level One*, a hurricane tie is a special kind of tie that is used to tie the frame to the foundation (*Figure 32*). Hurricane ties are used to provide vertical reinforcement in masonry structures, but are not used to tie veneer.

Purlin anchors provide horizontal reinforcement for masonry structures. They are used to fasten or connect wooden framing structures to masonry support walls (*Figure 33*). These straps are flat galvanized metal straps with bends designed to engage masonry construction on one end and with holes designed to allow for bolted and/or nailed connections to wooden structural framing elements on the other end.

To install a PA strap, follow these steps:

Step 1 At least one lift prior to the installation of the PA strap, determine the strap locations and mark the locations on the wall.

Step 2 Place a center mark at each strap location on the top course of the lift below.

Step 3 Straps are usually used in conjunction with wood ledgers so the height will often match the top of ledger. Depending on the detail, the height may also be raised the thickness of the deck above the top of ledger to allow the strap to rest over the deck. Be sure to check the details to ensure accuracy of placement.

Step 4 Build the wall to the course below the height of the strap. Make sure to take into consideration any bond-beam courses, grouted cells, and/or reinforcement. Place bond-beam mesh as needed to contain grout in the necessary cells. Transfer the center mark from below to the location of the strap.

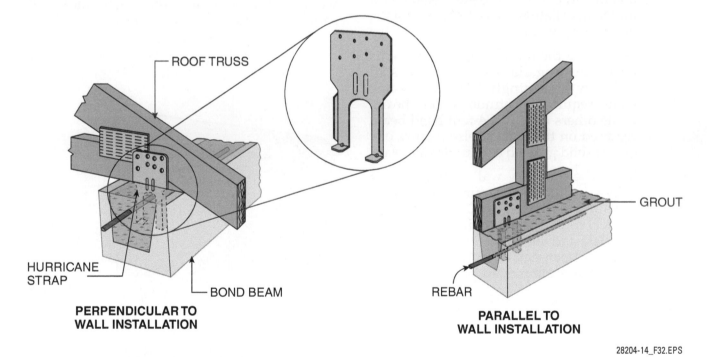

28204-14_F32.EPS

Figure 32 Hurricane ties.

Anchoring Systems

Masonry anchoring systems are designed to retrofit veneer anchors for concrete, block, and wooden backings. They work by means of mechanical expansion, screws, or epoxy adhesive. These anchoring systems are used either in repairs where ties were not originally installed, to replace a failed system, or in new construction. Manufacturers' data sheets usually provide installation instructions.

Wedge anchors have a stainless-steel split expansion ring, threaded stud-bolt body, nut, washer, and expander. They expand to fit securely into holes drilled into masonry units.

Step 5 Determine whether cuts will be necessary for the strap installation due to interference from webs or masonry-unit heads. Bond-beam cuts may be adequate if the strap height aligns with a bed joint. If the strap height does not align with a bed joint, a combination of center webs, open ends, or other cuts may be required.

Step 6 Mark, make, and install the necessary cuts to facilitate strap installation. Set the cut units using full head and bed joints. Parge the inside of cut units for added strength and to help prevent blowouts.

Step 7 If the strap is to be placed as the work progresses, ensure that the strap is installed accurately on the center mark, embedded to the proper depth in the masonry, and extended perpendicularly to the masonry wall. The proper depth is usually marked on each strap with a groove noted as the wall face. This method can be beneficial if, for example, the height of the straps will make them difficult to reach after the lift is completed, or if the wall is to be grouted the same day.

Step 8 If the straps will be placed at the same time as the grout is placed, the grout must be placed and consolidated first. Then, chip a small neat hole into the face of the masonry and work the strap into the hole to the proper depth while ensuring that it remains perpendicular to the wall. Take care to avoid damaging the masonry units and to avoid damaging the strap or bending it excessively during the installation. Finally, ensure that the strap has

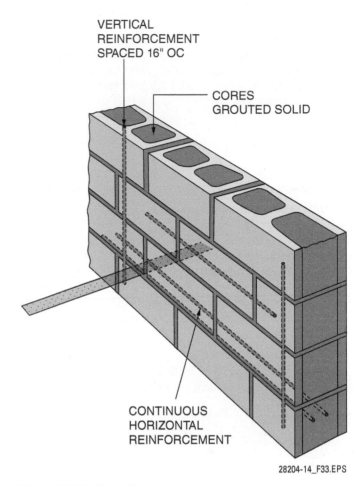

VERTICAL REINFORCEMENT SPACED 16" OC

CORES GROUTED SOLID

CONTINUOUS HORIZONTAL REINFORCEMENT

28204-14_F33.EPS

Figure 33 Purlin anchors.

been encased solidly in grout. The grout should ooze slightly from the hole around the strap, and the strap should feel snug in the wall. If the strap is loose or if the grout has been compromised, use a mechanical vibrator to work grout back into the voids and pack it solid around the strap.

Additional Resources

ASTM E754, *Standard Test Method for Pullout Resistance of Ties and Anchors Embedded in Masonry Mortar Joints*, Latest edition. West Conshohocken, PA: ASTM International.

TEK 12-1B, *Anchors and Ties for Masonry*. 2011. Herndon, VA: National Concrete Masonry Association.

www.ncma.org

4.0.0 Section Review

1. Loadbearing partition walls are often attached using _____.

   a. strip ties
   b. seismic ties
   c. bearing plates
   d. corrugated ties

2. To distribute the vertical load over an arch, the project specifications may call for the installation of _____.

   a. seismic anchor
   b. bearing plates
   c. continuous joint reinforcement
   d. saddle

3. As a method of supporting intersecting beams, saddles are an alternative to _____.

   a. corrugated anchors
   b. purlin anchors
   c. beam pockets
   d. bearing plates

4. When installing PA straps, the strap locations should be determined and marked on the wall at least _____.

   a. one lift prior to installation
   b. two lifts prior to installation
   c. three lifts prior to installation
   d. four lifts prior to installation

SUMMARY

This module introduced the wide range of metal reinforcements that masons use to ensure that masonry structures are able to withstand cracking when under load. Reinforcements are installed around openings and embedded in the mortar of masonry wythes. Aluminum and steel are used for the hollow metal frames used in doors. Prefabricated and built-in-place windowsills may be single unit or multiple units that are made from brick, stone, wood, metal, or concrete. Structural steel lintels, also called angle irons, may be fabricated in one or two pieces and thick enough to support the weight of the masonry units above them.

Continuous joint reinforcement, anchors, and ties are used to bond masonry wythes into a single structural element that resists cracking. Reinforcements, anchors, and ties are used to tie a single masonry wythe together, to tie two masonry wythes together, and to tie masonry wythes to structural elements such as steel beams and reinforced concrete. Ladder and truss joint reinforcement and seismic reinforcements are common types of reinforcement that tie single wythes together. To tie two adjacent masonry wythes together, masons can use flexible anchors, seismic anchors, and horizontal ties. When connecting masonry wythes to structural elements, masons may choose from rigid ties and bolts, bearing plates, saddles, and hurricane ties.

1. Before laying masonry units, you should mark the location of wall openings on the _____.
   a. sill
   b. floor plan
   c. footing
   d. door schedule

2. The horizontal top element of a window is the _____.
   a. lintel
   b. jamb
   c. sill
   d. muntin

3. To accommodate openings in a brick wall, the bond must be worked out by _____.
   a. the mason
   b. the architect
   c. an engineer
   d. the window installer

4. On each jamb, hollow steel door frames are usually provided with a minimum of three _____.
   a. floor knees
   b. tab locks
   c. steel ties
   d. bearing plates

5. Most fasteners used for standard frames in masonry walls are _____.
   a. welded in place on jambs
   b. installed after the drywall partition
   c. set-ups
   d. snap-ins

6. A set-up door frame installation should be braced horizontally _____.
   a. at the bottom and in the middle
   b. at the middle only
   c. at the bottom only
   d. at the top and at the bottom

7. For proper drainage, sills should slope below the horizontal by at least _____.
   a. 5 degrees
   b. 7.5 degrees
   c. 15 degrees
   d. 33 degrees

8. If wicks are not used with sill flashing, a spacing of 24 inches should be used for _____.
   a. rigid ties
   b. weepholes
   c. bearing plates
   d. perimeter drains

9. For windowsills in brick buildings, the preferred choice of material is _____.
   a. stone
   b. metal
   c. precast concrete
   d. brick

10. Control joints are often installed at each end of a lintel to _____.
    a. create a stronger bond
    b. allow slight movement
    c. provide tensile strength
    d. level the lintel

11. The width of an angle-iron lintel that is used to support 4-inch masonry units should not be less than _____.
    a. 3½ inches
    b. 4 inches
    c. 4½ inches
    d. 5 inches

12. One problem with using a truss-type joint reinforcement is that it _____.
    a. is difficult to cut to size
    b. is more expensive than ladder-type joint reinforcement
    c. interferes with easy placement of vertical rebar
    d. is hard to keep flat

13. Compared to the thickness of the wall in which it is installed, the standard width of continuous horizontal joint reinforcement is approximately _____.
    a. 1⅝ inches greater
    b. 1⅝ inches less
    c. 1¾ inches greater
    d. 1¾ inches less

14. Ties must be bedded in mortar at least ⅝ inch from the exposed face of a wall to prevent _____.

    a. corrosion
    b. cracking
    c. condensation
    d. lateral movement

15. Tab-type reinforcements transfer _____.

    a. only vertical loads
    b. both lateral and vertical loads
    c. only lateral loads
    d. only lateral and diagonal loads

16. Flexible anchors connect materials with different rates of heat expansion to prevent _____.

    a. veneer separation
    b. windstorm damage
    c. formation of cracks
    d. lateral movement

17. Corrugated, Z-type, and box ties are examples of _____.

    a. continuous ties
    b. horizontal anchors
    c. flexible ties
    d. seismic reinforcements

18. Slots made in concrete receive the dovetail anchors of _____.

    a. shaped metal ties
    b. tab-type reinforcements
    c. J-bolts
    d. slip lugs

19. Bearing plates are installed to handle concentrations of _____.

    a. lateral pressure
    b. vertical load
    c. horizontal load
    d. shear stress

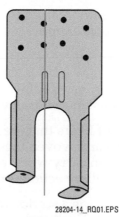

28204-14_RQ01.EPS

Figure 1

20. The device shown in *Review Question Figure 1* is a _____.

    a. rigid-tie assembly
    b. seismic anchor
    c. adjustable anchor
    d. hurricane tie

Trade Terms Quiz

Fill in the blank with the correct term that you learned from your study of this module.

1. A course of masonry filled with steel reinforcing rods and grout that serves as a lintel or reinforcement beam designed to strengthen a wall is called a(n) _____.

2. A(n) _____ is a metal assembly used to attach building parts to masonry.

3. A metal assembly used to attach masonry to a structural support is called a(n) _____.

4. A(n) _____ is a horizontal member under a door or window that fits inside the door or window frame.

5. A bundle of fibers that are loosely twisted or braided and woven together to form a cord that can carry water away from an area by capillary action is called _____.

6. A(n) _____ is a section of wall between control joints, wall ends, or a control joint and wall end.

7. A type of masonry fastener that allows a horizontal reinforcement to maintain a joint while allowing the wall to have some freedom of vertical movement is called a(n) _____.

8. The _____ is the side of an opening, or the vertical framing member on the side of the opening, usually for door and window frames.

9. A system of connected steel wire that provides added structural integrity to a masonry wythe by distributing lateral loads evenly and by tying the masonry elements in the structure together mechanically is called _____.

10. A(n) _____ is the side of an opening in a wall for a window or door that can be seen from the frame to the face of the masonry wall.

11. A continuous recess built into a wall to receive pipes, wires, or heating ducts is called a(n) _____.

12. A(n) _____ is an opening in a vertical masonry wall that allows an intersecting structural beam to bear on or pass through the wall.

13. The condition when two parts come together at an angle which is not 90 degrees or perpendicular to each other is called _____.

Trade Terms

Anchor
Beam pocket
Bond beam
Chase
Fastener
Horizontal joint reinforcement
Jamb

Panel
Pintle
Reveal
Sill
Skew
Wick

Trade Terms Introduced in This Module

Anchor: A metal assembly used to attach masonry to a structural support.

Beam pocket: An opening in a vertical masonry wall that allows an intersecting structural beam to bear on or pass through the wall.

Bond beam: A course of masonry filled with steel reinforcing rods and grout that serves as a lintel or reinforcement beam designed to strengthen a wall.

Chase: A continuous recess built into a wall to receive pipes, wires, or heating ducts.

Fastener: A metal assembly used to attach building parts to masonry.

Horizontal joint reinforcement: A system of connected steel wire that provides added structural integrity to a masonry wythe by distributing lateral loads evenly and by tying the masonry elements in the structure together mechanically.

Jamb: The side of an opening, or the vertical framing member on the side of the opening, usually for door and window frames.

Panel: A section of wall between control joints, wall ends, or a control joint and wall end.

Pintle: A type of masonry fastener that allows a horizontal reinforcement to maintain a joint while allowing the wall to have some freedom of vertical movement.

Reveal: The side of an opening in a wall for a window or door. This is the part of the masonry jamb around a window or door frame that can be seen from the frame to the face of the masonry wall.

Sill: A horizontal member under a door or window. Slip sills fit inside the door or window frame; lug sills extend beyond the frame and into the masonry on the jamb sides of the frame.

Skew: The condition when two parts come together at an angle that is not 90 degrees, or perpendicular, to each other.

Wick: A bundle of fibers that are loosely twisted or braided and woven together to form a cord that can carry water away from an area by capillary action, which occurs as long as the drip end is lower than the absorption end.

Additional Resources

This module presents thorough resources for task training. The following resource material is suggested for further study.

ASTM A82, Standard Specification for Steel Wire, Plain, for Concrete Reinforcement, Latest Edition. West Conshohocken, PA: ASTM International.

ASTM A185, Standard Specification for Steel Welded Wire Reinforcement, Plain, for Concrete, Latest Edition. West Conshohocken, PA: ASTM International.

ASTM A496, Standard Specification for Steel Wire, Deformed, for Concrete Reinforcement, Latest Edition. West Conshohocken, PA: ASTM International.

ASTM A951, Standard Specification for Steel Wire for Masonry Joint Reinforcements, Latest Edition. West Conshohocken, PA: ASTM International.

ASTM E488, Standard Test Methods for Strength of Anchors in Concrete Elements, Latest Edition. West Conshohocken, PA: ASTM International.

ASTM E754, Standard Test Method for Pullout Resistance of Ties and Anchors Embedded in Masonry Mortar Joints, Latest Edition. West Conshohocken, PA: ASTM International.

Bricklaying: Brick and Block Masonry. 1988. Brick Industry Association. Orlando, FL: Harcourt Brace & Company.

Concrete Masonry Handbook, Fifth edition. W. C. Panerese, S. K. Kosmatka, and F. A. Randall, Jr. Skokie, IL: Portland Cement Association.

Technical Note 31B, *Structural Steel Lintels.* 1987. Reston, VA: The Brick Industry Association. **www.gobrick.com**

Technical Note 36, *Brick Masonry Details, Sills, and Soffits.* 1988. Reston, VA: The Brick Industry Association. **www.gobrick.com**

TEK 12-1B, Anchors and Ties for Masonry. 2011. Herndon, VA: National Concrete Masonry Association. **www.ncma.org**

TEK 12-4D, Steel Reinforcement for Concrete Masonry. 2011. Herndon, VA: National Concrete Masonry Association. **www.ncma.org**

Figure Credits

Section Review Answers

Answer	Section Reference	Objective
Section One		
1. a	1.0.0	1
2. b	1.2.0	1b
3. b	1.3.0	1c
4. a	1.4.0	1d
Section Two		
1. c	2.1.0	2a
2. d	2.2.0	2b
Section Three		
1. b	3.1.0	3a
2. d	3.2.0	3b
Section Four		
1. a	4.1.0	4a
2. b	4.2.0	4b
3. c	4.3.0	4c
4. a	4.4.0	4d

NCCER CURRICULA — USER UPDATE

NCCER makes every effort to keep its textbooks up-to-date and free of technical errors. We appreciate your help in this process. If you find an error, a typographical mistake, or an inaccuracy in NCCER's curricula, please fill out this form (or a photocopy), or complete the online form at **www.nccer.org/olf**. Be sure to include the exact module ID number, page number, a detailed description, and your recommended correction. Your input will be brought to the attention of the Authoring Team. Thank you for your assistance.

Instructors – If you have an idea for improving this textbook, or have found that additional materials were necessary to teach this module effectively, please let us know so that we may present your suggestions to the Authoring Team.

NCCER Product Development and Revision

13614 Progress Blvd., Alachua, FL 32615

Email: curriculum@nccer.org
Online: www.nccer.org/olf

❏ Trainee Guide   ❏ Lesson Plans   ❏ Exam   ❏ PowerPoints   Other _____

Craft / Level: _____   Copyright Date: _____

Module ID Number / Title: _____

Section Number(s): _____

Description: _____

_____

_____

_____

Recommended Correction: _____

_____

_____

_____

Your Name: _____

Address: _____

Email: _____   Phone: _____

28205-14

Advanced Laying Techniques

After mastering the basic techniques of laying concrete masonry units and brick in courses and wythes, you need to develop the knowledge and skill required to build structures that accomplish some specific purpose. This module explains the construction of walls, control and expansion joints, and corners and intersections. Typically, the architect or engineer will determine specific construction procedures based on various factors. These include wind loading, lateral force, soil condition, and the function of the structure. However, on jobs such as small garden walls and retaining walls, you may be required to recommend the best approach to the client or owner. As a mason, you must be prepared to determine safe and proper practices by knowing the fundamental requirements, as well as where to look for design and structural details.

Module Five

Trainees with successful module completions may be eligible for credentialing through NCCER's National Registry. To learn more, go to **www.nccer.org** or contact us at **1.888.622.3720**. Our website has information on the latest product releases and training, as well as online versions of our *Cornerstone* magazine and Pearson's product catalog.

Your feedback is welcome. You may email your comments to **curriculum@nccer.org**, send general comments and inquiries to **info@nccer.org**, or fill in the User Update form at the back of this module.

This information is general in nature and intended for training purposes only. Actual performance of activities described in this manual requires compliance with all applicable operating, service, maintenance, and safety procedures under the direction of qualified personnel. References in this manual to patented or proprietary devices do not constitute a recommendation of their use.

Objectives

When you have completed this module, you will be able to do the following:

1. Identify the structural principles and fundamental uses of basic types of walls.
   a. Identify the structural principles and fundamental uses of solid masonry walls.
   b. Identify the structural principles and fundamental uses of hollow masonry walls.
   c. Identify the structural principles and fundamental uses of cavity walls.
   d. Identify the structural principles and fundamental uses of composite walls.
   e. Identify the structural principles and fundamental uses of anchored veneer walls.
   f. Identify the structural principles and fundamental uses of retaining walls.
   g. Identify the structural principles and fundamental uses of freestanding walls.
2. Identify the requirement for and function of control joints and expansion joints.
   a. Identify the effects of temperature and moisture on control joints and expansion joints.
   b. Identify the uses of control joints.
   c. Identify the uses of expansion joints.
3. Lay out and construct various corners and intersections.
   a. Lay out and construct toothing.
   b. Lay out and construct corbeling.
   c. Lay out and construct intersecting walls.
   d. Lay out and construct angled corners.

Performance Tasks

Under the supervision of your instructor, you should be able to do the following:

1. Lay out and construct a composite wall with control joints and expansion joints.
2. Lay out and construct intersections.
3. Lay out and construct angled corners.

Trade Terms

Breaking the bond	Empirically designed	Segmental retaining wall (SRW)
Cap	Humored	Toothing
Coping	Pencil rod	

Industry-Recognized Credentials

If you're training through an NCCER-accredited sponsor, you may be eligible for credentials from NCCER's Registry. The ID number for this module is 28205-14. Note that this module may have been used in other NCCER curricula and may apply to other level completions. Contact NCCER's Registry at 888.622.3720 or go to **www.nccer.org** for more information.

Code Note

Codes vary among jurisdictions. Because of the variations in code, consult the applicable code whenever regulations are in question. Referring to an incorrect set of codes can cause as much trouble as failing to reference codes altogether. Obtain, review, and familiarize yourself with your local adopted code.

Contents

Topics to be presented in this module include:

Figures and Tables ───────────

Figures and Tables (continued)

1.0.0 STRUCTURAL PRINCIPLES AND USES OF WALLS

Objective

Identify the structural principles and fundamental uses of basic types of walls.

a. Identify the structural principles and fundamental uses of solid masonry walls.
b. Identify the structural principles and fundamental uses of hollow masonry walls.
c. Identify the structural principles and fundamental uses of cavity walls.
d. Identify the structural principles and fundamental uses of composite walls.
e. Identify the structural principles and fundamental uses of anchored veneer walls.
f. Identify the structural principles and fundamental uses of retaining walls.
g. Identify the structural principles and fundamental uses of freestanding walls.

Trade Terms

Breaking the bond: Starting a course with a cut so as to center the header unit over the head joints in the course below.

Cap: Masonry units laid on top of a finished wall.

Coping: The materials or masonry units used to form a cap or finish on top of a wall, pier, chimney, or pilaster to protect the masonry below from water penetration. Coping is usually projected from both sides of the wall to provide a protective covering as well as an ornamental design.

Empirically designed: Design based on the application of physical limitations learned from experience or on observations gained through experience, but not based on structural analysis.

Humored: A slang term for gradually adjusting a section of masonry in order to correct alignment issues.

Segmental retaining wall (SRW): A wall made of segmental block stacked on top of each other without mortar bonding.

There are many reasons why masonry walls are popular. These include sustainability, durability, ease of maintenance, design flexibility, attractive appearance, and competitive cost. Masonry walls range from simple, single-wythe walls with no reinforcement to highly reinforced multiwythe walls. They can be plain foundation walls or highly ornate and decorative walls (*Figure 1*). The various wall types are classified according to seven basic systems:

- Solid wall
- Hollow wall
- Cavity wall
- Composite wall
- Anchored veneer wall
- Retaining wall
- Freestanding wall

Masonry walls can be reinforced by adding reinforcing steel or grout. Reinforcing techniques are presented in the module *Reinforced Masonry*. Two types of walls that are built to provide something other than an enclosure are retaining walls and freestanding, or garden, walls. There are many designs and materials for both types. While freestanding walls are not subjected to the pressure that is applied to retaining walls, reinforcement and bonding are very important in both. Proper foundation construction is also an important part of both types of structures.

1.1.0 Identifying the Structural Principles and Fundamental Uses of Solid Masonry Walls

Solid masonry walls are common. They are used for either loadbearing or nonbearing construction. Solid masonry walls are typically built of solid masonry units laid up with full mortar joints. Masonry units are considered solid if they have less than 25 percent void area.

The simplest solid masonry wall is a single-wythe wall. They can be built using brick or uncored concrete masonry units (CMU, commonly known as block). The single-wythe brick wall shown in *Figure 2* can be used for low perimeter walls, partition walls, or for other nonbearing applications.

28205-14_F01.EPS

Figure 1 Ornate decorative wall.

Figure 2 Single-wythe solid brick wall.

sistance and sound insulation. Solid block is more difficult to work with than hollow block; therefore, it is rarely used.

Multiwythe solid walls of brick or block like those shown in *Figure 4* are used in loadbearing construction. Multiwythe walls can be used to offer a finished wall surface from either side. Alternatively, a cheaper backup wythe and a more expensive facing wythe can be combined. This wall system offers structural capabilities at a lower cost.

The two wythes of masonry are bonded together. Masonry headers, horizontal anchors, or continuous joint reinforcement secures the wythes to each other. The wall system is masonry bonded, and the masonry units overlap in alternate courses. To completely bond the two wythes of masonry, fill the collar joint with mortar.

> **NOTE**
>
> Always check to make sure which building codes are used in your area. Codes are updated periodically; make sure you are using the latest edition.

A single-wythe solid block wall, shown in *Figure 3*, can be either loadbearing or nonbearing. They can be used for structural purposes to support floor or roof loads. When used as a nonbearing partition wall, they offer excellent fire re-

In 8-inch walls, header courses can be used for the full width of the wall section. In most modern construction, masonry headers have been replaced by metal horizontal anchors or continuous joint reinforcement. They are grouted solidly in

28205-14_F02.EPS

Terminology

As mason, you need to understand basic laying technique terms related to bonding directions, the position of the masonry unit, and the type of mortar joint.

Stretcher, header, and rowlock refer to the basic position of a masonry unit within a wall. The stretcher unit is the normal position of a masonry unit, with its side exposed in the face of the wall. The header unit is laid flat on its widest side; its end is exposed in the wall face. The rowlock is laid on the side with the end exposed in the wall face.

Collar, head, and bed joints refer to the position or types of mortar joints. Head and bed joints refer to the vertical and horizontal mortar joints that are seen in the exposed face of a wall. A collar joint is the mortar joint between the wythes of masonry in a multiwythe wall system.

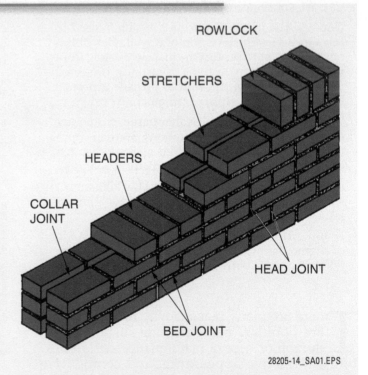

28205-14_SA01.EPS

CONCRETE MASONRY UNIT

28205-14_F03.EPS

Figure 3 Single-wythe solid block wall.

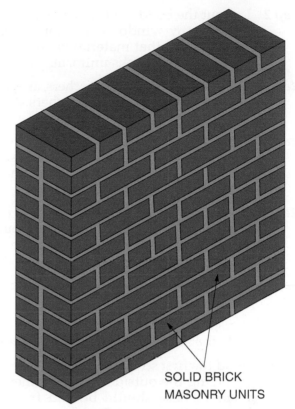

SOLID BRICK
MASONRY UNITS

MULTIWYTHE BRICK WALL

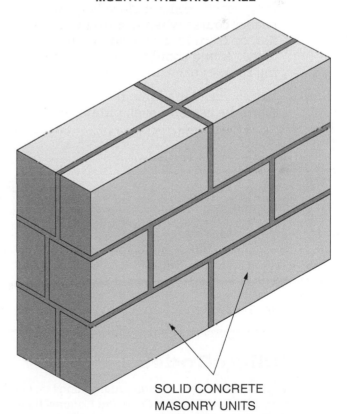

SOLID CONCRETE
MASONRY UNITS

MULTIWYTHE CONCRETE BLOCK WALL

28205-14_F04.EPS

Figure 4 Multiwythe uncored brick and block walls.

the bed joints to tie multiple wythes together. To-day, solid masonry walls do not typically exceed 12 inches in thickness except under special conditions.

There are many different types of bonding patterns and designs for solid brick walls. Load-bearing requirements, location, and cost usually determine the specific design. As an example of a typical solid masonry wall, we will use an 8-inch common bond. Headers are placed every seven courses for a double-wythe wall. The heading course serves as a bond between the inside and outside 4-inch tiers. This design can be laid up fast and has an effective bond.

The recommended procedure for constructing an 8-inch common-bond wall with leads is as follows:

Step 1 Lay out the wall location as with any single-wythe wall. In addition, locate a second line inside the first wall line. The distance between lines is equal to the average length of the brick being used.

Step 2 Lay out the bond to eliminate excess cutting for any windows or doors. This is important so that material is not wasted and the pattern looks uniform.

Step 3 Lay the first course of stretchers, as shown in *Figure 5*. Four or five stretchers should be sufficient.

Step 4 Lay the header course as shown in *Figure 6*. The header course normally starts at the second course. However, it should start wherever specified.

Step 5 Lay successive courses of stretchers until reaching the next specified header course. Level, plumb, and range each course as it is laid.

Step 6 Lay the header course, breaking the bond in the usual manner.

Step 7 Finish laying the first corner, and complete the second corner in the same way.

Step 8 Build the first stretcher course of the wall by laying the outside first, and then the inside. These should be laid from the leads to the center. Level, plumb, and range each course before beginning the header course when working with a partner, or from lead to lead when working alone. A line should be used on all backing courses to ensure a good face.

Step 9 Lay the header course from each lead toward the center. Level, plumb, and range

the header course as you did in Step 8 above.

Step 10 Lay the outside wythe up to the next header course, keeping it level, plumb, and ranged.

Step 11 Lay the inside tier up to the same height. Be careful to keep it level with the outside tier as it is laid up. This is important because you must have a level surface for the header course.

Step 12 Continue the work until the desired height is reached. Tool the joints and brush the wall at the proper time. Retool the joints if necessary.

1.2.0 Identifying the Structural Principles and Fundamental Uses of Hollow Masonry Walls

Hollow masonry walls are built with hollow masonry units. These include brick, block, or structural tile. Cored masonry units contain more than 25 percent void area. Typically, all units are laid with full mortar joints.

Hollow masonry walls may be either loadbearing or nonbearing, depending on the structural capabilities of the masonry units used. They can be single wythe or multiwythe. Construction of hollow masonry walls is similar to construction of solid masonry walls.

Bonding requirements will vary depending on the local conditions and building codes. Pattern

28205-14_F05.EPS

Figure 5 Stretcher course.

28205-14_F06.EPS

Figure 6 Header course.

Building Codes

Prior to 2000 there were several organizations that issued model building codes. The most widely used were the *Southern Standard Building Code*, the *National Building Code*, and the *Uniform Building Code*. In 2000, three organizations merged into the International Code Council. They issued the *International Building Code*® (IBC). This model code has been adopted and is now law throughout most of the United States. The National Fire Protection Association also issued a model code called *NFPA 5000, Building Construction and Safety Code*®. It is used in a few communities and by the state of California.

bond is used on some types of multiwythe walls. Alternatively, horizontal anchors and continuous joint reinforcement are also used to form a structural bond. You learned about horizontal anchors and continuous joint reinforcement in the module titled *Masonry Openings and Metal Work*.

Horizontal anchors (*Figure 7*) and continuous joint reinforcement (*Figure 8*) are used to bond wythes. They are made from corrosion-resistant rods or wire. They are embedded in the horizontal mortar joints and across all wythes. Modern building codes permit the use of rigid steel bonding anchors. They specify that at least one anchor should be used for every 2.67 square feet of wall surface.

Anchor type and spacing is determined by the design team, following the local applicable code. Horizontal anchors are usually placed in alternate courses. They are staggered so that no two anchors form a continuous vertical line. The typical spacing between horizontal anchors in a horizontal position is 32 inches. In a vertical position, horizontal anchors should be installed every 16 inches for best results. There is a high degree of moisture present in all masonry walls. Anchors should be coated to prevent corrosion. The most popular coating method is galvanizing the anchors.

Adjustable horizontal anchors for veneer walls are shown in *Figure 9*. Using adjustable anchors speeds the construction of multiwythe walls. Each wythe can be laid independently of the other instead of laying up both wythes at the same time.

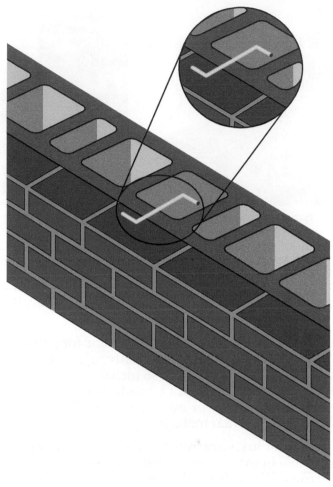

28205-14_F07.EPS

Figure 7 Examples of horizontal anchors.

The Monadnock Building

Builders often use solid masonry brick walls in building construction. They offer strength, stability, and insulating value. These properties increase with the thickness of the wall. Experience shows that empirically designed, unreinforced bearing walls should be large.

The Monadnock Building in Chicago is 16 stories high with unreinforced loadbearing brick walls ranging in thickness from 12 inches at the top to more than 6 feet at the ground. At the time the Monadnock Building was built, wall wythes were bonded together with masonry headers.

28205-14_SA02.EPS

THREE-WIRE TRUSS TYPE

TRUSS TAB TYPE

ADJUSTABLE TRUSS TYPE

28205-14_F08.EPS

Figure 8 Examples of truss-type continuous joint reinforcement.

Continuous joint reinforcement consists of metal rods or wires with attached cross wires. It serves three functions:

- It acts as horizontal reinforcement for single-wythe masonry walls, usually with vertical spacing of no more than 16 inches.
- It acts as reinforcement to control cracking.
- It bonds wythes in multiwythe walls without using individual metal anchors.

Continuous joint reinforcement is generally available in many widths. Select a width 2 inches narrower than the wall. Sections are usually 10 feet long. If the masonry face is exposed to the weather, the joint reinforcement must have a mortar cover of at least ⅝ inch.

1.3.0 Identifying the Structural Principles and Fundamental Uses of Cavity Walls

Cavity walls are built in two wythes that are completely separated by an unbroken airspace of at least 2 inches. The inner and outer walls are tied together by rigid metal anchors or continuous joint reinforcement embedded in the mortar joints. The face wythe and the backup wythe may be of similar materials, as in the case of the brick cavity wall or the block cavity wall shown in *Figure 10*. The two wythes may also be of different materials.

The thickness of the face wythe is generally 4 inches (nominal dimension), while the backup wythe can be 4, 6, or 8 inches in thickness, depending on the structural requirements for the wall. However, the general rule is that the thickness of the interior wythe should be equal to or greater than the thickness of the outer wythe.

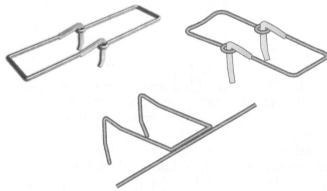

28205-14_F09.EPS

Figure 9 Adjustable horizontal anchors.

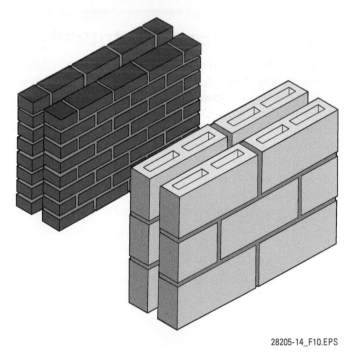

28205-14_F10.EPS

Figure 10 Brick cavity wall and block cavity wall.

In most cases, the inner wythe carries the vertical load of the structure while the outer wythe acts as a weather barrier and as a visual design feature. Considering the wythes together, the nominal thickness of most cavity walls will be 10 to 14 inches if a 2-inch airspace is used.

The cavity wall is designed to offer three major advantages over solid or hollow masonry wall systems:

- It increases the insulating value of the wall. The cavity may be left open; filled with fiber insulating materials such as water-repellent vermiculite or silicone-treated perlite; or filled with insulating mats or rigid insulation.
- When properly constructed, it prohibits the passage of water or moisture across the wall. If proper flashing and weepholes are used in the construction of the wall, rain penetration to the interior wall is greatly reduced. Any moisture that comes through the first wythe of a cavity wall flows to the bottom of the cavity and out through the weepholes. Flashing is installed at the base of the cavity wall and any other necessary locations in the wall as required by the design to direct water to the weepholes.
- A cavity wall also prevents the formation of condensation on interior surfaces. If the cavity is left open, a vapor barrier or damp-proofing is required to prevent interior condensation. Building codes specify the number of anchors or reinforcements required for a cavity wall, as well as their specific locations.

In the past, the masonry wythes of cavity walls were held together with masonry headers or bonders. However, the advantage of a cavity wall system is to have an airspace that is completely free of obstructions for insulation and drainage purposes. To accomplish this, noncorroding horizontal anchors or continuous joint reinforcement is used to bond the wythes of masonry.

1.3.1 Mortar and Mortar Joints

The bond formed between the mortar, the metal anchors or joint reinforcement, and the masonry units is an important single property in cavity-wall construction. Since the cavity wall is constructed of two tiers or wythes of masonry completely separated by an unbroken airspace, the mortar joints must be solid and completely filled if the wythes are to act as a unified wall system.

In areas where a great deal of pressure is expected from high winds, Type S mortar is recommended. For areas where average conditions occur, Type N mortar may be used. In either case, full bed joints should be made, the anchors or continuous joint reinforcement binding the wythes should be completely covered with mortar, and the head joints should be filled completely.

In applying the mortar, you must make certain that the cavity remains as free of mortar droppings as possible. If mortar droppings remain in the cavity, they may create dams which clog the weepholes and prevent moisture from escaping. Excess mortar in the cavity can also create bridges between wall wythes for condensation to cross.

To prevent mortar droppings, use a trowel to flatten or cut any mortar from the cavity, or bevel the bed joints as you lay up each unit. Even with these precautions, some mortar may fall into the cavity. By using a mortar collection device, like that shown in *Figure 11*, any droppings can be separated to allow moisture to exit the wall.

1.3.2 Flashing

Some people believe that if cavity walls are properly designed and built with well-tooled mortar joints, the structure will be weather-tight. However,

Hollow Concrete Walls

Hollow block walls, like the one shown, are used for a variety of applications. These include foundation walls, interior or exterior loadbearing walls, and nonbearing partition walls. The cored block is easily handled, and the wall is economically laid up. However, these units must be protected from exposure to high levels of moisture.

28205-14_SA03.EPS

Figure 11 Mortar collection device.

28205-14_F11.EPS

experience has shown that in a high-wind condition, moisture or water can enter even the best-constructed cavity walls. Any moisture that penetrates the outer wythe of a cavity wall should be free to flow to the bottom of the cavity. From there it should be channeled by flashing materials to the weepholes, where it can flow out of the cavity.

The application of flashing material will help prevent water penetration or provide water drainage. Flashing materials can be thin sheets of stainless steel, thermal plastic, rubberized asphalt, or copper. Copper offers the advantage that it is long lasting. Plastic flashings do not react with mortar ingredients and often eliminate the need for control or expansion joints since the plastic is flexible.

Flashing is especially important in areas with severe weather patterns. Flashing should be used in most building practices including base, sill, roof, and head areas. Examples are shown in *Figure 12* and discussed in the following:

- Base flashing should be placed at the base of the cavity to direct any moisture in the cavity to the weepholes.
- Sill flashing should be placed under and behind all sills. These sills should extend at least 1 inch from the wall to drain water from the sill, to prevent water stains on the masonry units, and to provide a drip rail for the underside of the sill.
- Roof flashing should be placed where roof lines abut with masonry walls. This consists of a base flashing placed under the roofing and turned up against the masonry units, and counterflashing overlapped against the base flashing and extended securely into the mortar joints.
- Head flashing should be placed over all openings except those completely protected by an overhang.

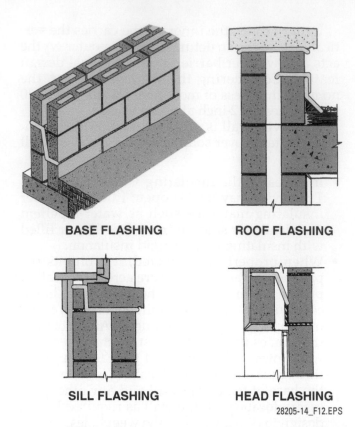

BASE FLASHING   **ROOF FLASHING**

SILL FLASHING   **HEAD FLASHING**

28205-14_F12.EPS

Figure 12 Use of flashing in cavity walls.

When placing any flashing, remember the basic bonding rules. Through-wall flashing should be placed on a thin bed of sealant. The flashing should then be covered with a thin bed of mortar to create the proper bond for the next masonry course. The masonry on which the flashing is bedded should be smooth enough that the flashing is not punctured. Any joints in the flashing must be completely sealed.

Remember that the effect of flashing on wall strength depends on the mortar placement and the mortar-flashing-masonry bond. Flashing does not affect a wall's compressive strength unless the flashing is placed directly in contact with the masonry units. If this occurs, the wall will have no flexural strength at that point.

1.3.3 Weepholes

Flashing concealed in tooled mortar joints is not self-draining. Weepholes must be provided to allow the moisture to escape through the outer wythe, as shown in *Figure 13*. Without these holes, water can be trapped within the wall cavity, cause moisture problems, and even weaken the structure.

Weepholes should be provided every 16 to 32 inches on center in head joints immediately above flashing sheets. Do not place weepholes where they will be covered with soil after the construction is complete.

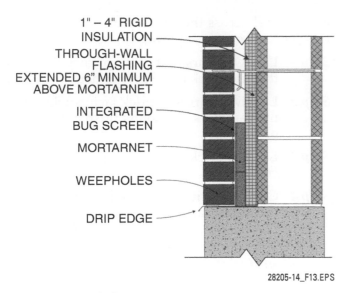

1" – 4" RIGID INSULATION

THROUGH-WALL FLASHING EXTENDED 6" MINIMUM ABOVE MORTARNET

INTEGRATED BUG SCREEN

MORTARNET

WEEPHOLES

DRIP EDGE

28205-14_F13.EPS

Figure 13 Weepholes.

1.4.0 Identifying the Structural Principles and Fundamental Uses of Composite Walls

A composite wall is a multiwythe wall system consisting of wythes of the same or different materials bonded together in such a manner that all imposed loads are distributed over all wythes of the wall system.

The wythes of masonry are bonded together in several ways:

- By using a full vertical collar joint
- By using masonry headers

- By using noncorroding-metal horizontal anchors or continuous joint reinforcement

Composite walls of 8 to 12 inches in thickness are normally used as exterior walls and will usually consist of a brick- or stone-face wythe bonded to a concrete-masonry or structural clay-tile backup wythe. In some parts of the country, this type of construction is incorrectly called a brick veneer. *Figure 14A* shows a typical brick/block composite wall using a ladder-type continuous joint reinforcement. *Figure 14B* shows a different arrangement for bonding using headers to bond the two wythes every six courses. Composite systems are particularly effective in preventing the passage of rain through the wall because of the continuous collar joint.

1.4.1 Bonding for Composite Walls

Composite walls present two problems that the mason must consider. The loadbearing strength of the wall is limited by the strength of the weakest combination of units and mortar, and the strength of one of the wythes. This means that to ensure even distribution of the load, you must properly bond the anchors, the units, and the wythes to one another.

Since the wall is composed of different materials, the different parts of the wall will expand and contract at different rates when there are changes in temperature and moisture. However, such movements are generally not enough to en-

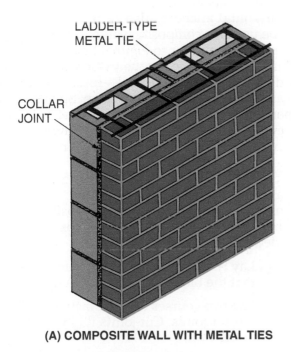

LADDER-TYPE METAL TIE

COLLAR JOINT

(A) COMPOSITE WALL WITH METAL TIES

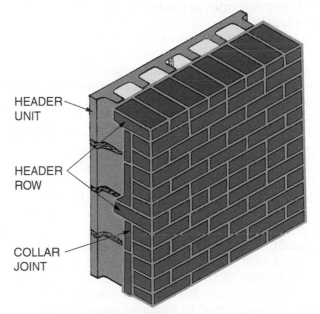

HEADER UNIT

HEADER ROW

COLLAR JOINT

(B) COMPOSITE WALL WITH MASONRY BOND

28205-14_F14.EPS

Figure 14 Composite wall with masonry bond.

danger the stability of the wall if all mortar joints are properly filled and tooled.

The work should have full, solid, and compact head and collar joints, along with smooth bed joints. Although it is difficult to fill collar joints and joints between brick headers solidly, extra effort should be made to do this so that the wall will act as a unified structure that can resist moisture penetration and cracking.

Proper bonding is easiest to achieve when the brick courses are level with the block, particularly at bonding height. For this reason, it is best to carry up the facing and backing together.

Unless it is absolutely necessary, the facing should not be carried up more than five courses ahead of the backing. The practice of carrying up the backing and inserting the wall horizontal anchors or continuous joint reinforcement so that the facing can be laid up later may cause problems if the brick is not level with the block at the height of the horizontal anchors or continuous joint reinforcement. It may not be possible to get the brick under the horizontal anchor or continuous joint reinforcement, or it may be necessary to bend the anchor or reinforcement to embed it against the brick. Both result in a weaker bond.

1.4.2 Using Brick Headers

Using brick headers may present special problems if the headers are not laid correctly. You must take extra care with any cross joint (a mortar joint spread across the full length of the brick header), or moisture penetration could result. Single head joints applied only to the end of the header are not acceptable.

Another common problem presented by headers is the difficulty of making sure that the set of the mortar is not broken when tying between brick and heavy block. This occurs when the block backup unit is placed immediately after a header course of brick. The block is placed on the lower course of block as well as the back half of the brick header. If the block is laid before the mortar under the header stiffens, the header brick will settle unevenly. In this case, a crack may develop in the mortar joint under the header, raising the possibility of moisture penetration. Also, since the header brick lies twisted in the face of the wall, the appearance of the building suffers. To avoid this problem, the brickwork can be built to the height of the header course before laying any backup block. Block work can be humored (gradually adjusted in order to correct alignment issues) to remain in line with the brickwork. The block must also be laid exactly level in height to match the outside brickwork, or the header will not lie level across the two wythes. The steps in this procedure are as follows:

Step 1   Lay two courses of backup block. Be sure that the block is the proper height.

Step 2   Lay five courses of brick facing using full mortar joints. Cut joints flush at the back of the wall.

Step 3   Lay the brick header with a full mortar cross joint.

Step 4   Continue with the remaining courses in the same manner.

1.4.3 Using Continuous Joint Reinforcement and Horizontal Anchors

Continuous joint reinforcement is the most popular type of metal tie used for composite walls (refer to *Figure 8*). Continuous joint reinforcement not only ties the masonry work together but also reduces shrinkage in the mortar joints and helps prevent cracks due to settlement. For a composite wall such as one constructed of block and brick, the reinforcements are usually spaced 16 inches on center.

Continuous joint reinforcement is manufactured in 10-foot lengths. It should overlap at least 6 inches when two pieces are used together. This ensures that the tensile strength will not be interrupted throughout the length of the bed joint. It is extremely important that the reinforcement be embedded and covered completely with mortar before any masonry unit is laid on top of it.

Horizontal anchors, also called unit ties, can also be used for composite wall construction. The Z-type unit tie (refer to *Figure 7*) and the rectangular, or box, unit tie are the two most commonly used. Both are made from $\frac{3}{16}$-inch steel, which may be coated to prevent corrosion. Specifications generally require that one anchor must be in place for every 2.67 square feet of wall surface. A spacing of 16 inches on center vertically and 32 inches on center horizontally satisfies this requirement. The anchor placement should be staggered so that the anchors do not line up vertically.

To build a block-and-brick composite wall using metal anchors, the following steps are recommended:

Step 1   Lay two courses of backup block. Be sure that the block is the proper height.

Step 2   Lay five courses of brick facing using full mortar joints. Cut joints flush at the back of the wall.

Step 3 Place the metal reinforcement and spread the bed joint of mortar. Be sure the reinforcement is fully embedded in the mortar joint.

Step 4 Continue with the remaining courses in the same manner.

It should be noted that Steps 1 and 2 may be reversed. With either sequence, the facing and backup wythes are laid up together at each spacing of the reinforcement.

1.5.0 Identifying the Structural Principles and Fundamental Uses of Anchored Veneer Walls

A veneered wall system consists of a single wythe of masonry as a facing attached to a wood frame, metal frame, or other structural system. The masonry veneer acts as a protective coating for the structure—as a barrier to wind and rain, and as an insulator of sound and heat. The veneer may be brick, stone, tile, or uncored concrete masonry materials (*Figure 15* and *Figure 16*), or a rain screen. The masonry facing is not a loadbearing wythe but is designed to carry only its own weight. Although these masonry units are fastened to the framework, they are not bonded to that backing with masonry or mortar.

The masonry veneer is attached to the structural frame by noncorroding horizontal anchors. Generally, building codes or project specifications specify the spacing of metal anchors. These metal anchors permit some slight horizontal and vertical movement of the veneer parallel to the face of the structural frame.

A minimum airspace of at least 1 inch is provided between the facing and backup. The backup is usually covered with a moisture-resistant material, and flashing and weepholes are included at the bottom of the airspace to eliminate water that may pass through the masonry veneer. Because the veneer wythe is exposed to changing weather conditions, the masonry units and mortar must be capable of withstanding the effects of the weather.

1.5.1 Building Code Requirements

Building codes regulate the design of anchored veneers by setting requirements based on empirical data. For example, *ACI 530/ASCE 5/TMS 402, Building Code Requirements for Masonry Structures,* and *ACI 530.1/ASCE 6/TMS 602, Specifications for Masonry Structures,* limit the use of empirical design to walls subject to wind pressure less than 25 pounds per square foot (psf). Higher wind pressures require analytical design.

Although some codes require only a 1-inch clear cavity, the minimum recommended width of the open cavity between a veneer and the backing is 2 inches. It is more difficult to keep a nar-

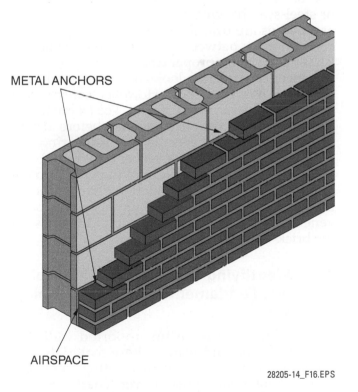

METAL ANCHORS

AIRSPACE

28205-14_F15.EPS

Figure 15 Brick veneer on frame surface.

METAL ANCHORS

AIRSPACE

28205-14_F16.EPS

Figure 16 Brick veneer on masonry surface.

rower cavity clear of mortar during construction. When rigid insulation is installed, the clear distance between the face of the insulation and the back of the facing wythe must be 2 inches. Most codes limit the maximum empirical distance (that is, the maximum distance not based on analysis) between backing and facing to 4½ inches. This limitation is based on the stiffness and load-transfer capability of horizontal anchors and continuous joint reinforcement. Wider airspaces can be employed but must be designed by an engineer.

For height limitations, the *International Building Code®* limits the maximum height of an **empirically designed** wall to 35 feet. *ACI 530/ASCE 5/TMS 402, Building Code Requirements for Masonry Structures* limits anchored veneer backed by wood framing to a height of 30 feet prior to a horizontal expansion joint being installed. Veneer backed by cold-formed steel framing must be supported at every floor above 30 feet. Be sure to check your local codes so that you understand the requirements for supporting and anchoring the veneer to the loadbearing structure.

1.5.2 Brick-Veneer Construction Details

For many years, brick-veneer construction was limited principally to wood-frame houses. Now, it is used on low-rise commercial and institutional construction. It is frequently used for high-rise buildings, especially with concrete-masonry or steel-stud backing systems. Brick-veneer wall assemblies are drainage-type walls. Maintain a clear airspace between the brick veneer and the backing to ensure proper drainage.

Veneer construction looks like cavity-wall construction in its final form. However, construction and function are very different. For example, the supporting structure is completed before attaching the veneer. The veneer is not loadbearing.

The first 30 feet of brick veneer on a frame backing must transfer the weight of the veneer to the foundation. A typical foundation detail is shown in *Figure 17*. The foundation supporting the brick veneer must at least equal the total thickness of the brick-veneer wall assembly.

1.6.0 Identifying the Structural Principles and Fundamental Uses of Retaining Walls

A retaining wall is an unsupported wall that holds back soil and water. These materials constantly press against the wall, creating a pressure that can potentially push it over. Therefore, a retaining wall must be designed and constructed so

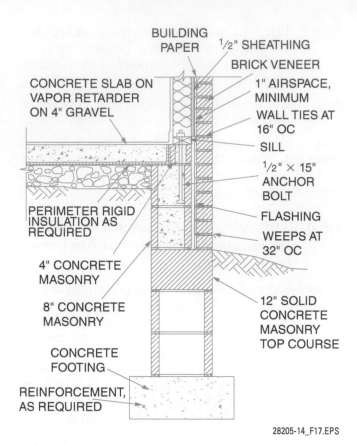

that these pressures will not push it over or cause it to bow out in the middle.

There are five basic types of traditional masonry retaining walls:

- Gravity
- Cantilever
- Segmental
- Counterfort
- Buttress

The different types of retaining walls are shown in *Figure 18*. The counterfort and buttress walls are not used very much today. However, they may be required where the lateral force on the wall is extremely high and there is limited space behind the wall to build a gravity-type structure.

A gravity-type wall depends primarily on its own weight to hold back the soil pressure. A cantilever-type wall uses the weight of the soil, together with its own strength, to get the same results. Reinforced cantilever walls offer the most economical design and are most commonly used. The size of the reinforcing steel used in the wall will determine its overall strength. The concrete footing anchors the stem and resists overturning and sliding due to both vertical and lateral forces.

The services of an engineer are useful when designing retaining walls. The wall and the foot

Figure 17 Brick-veneer details.

Labels in Figure 17: BUILDING PAPER · ½" SHEATHING · BRICK VENEER · CONCRETE SLAB ON VAPOR RETARDER ON 4" GRAVEL · 1" AIRSPACE, MINIMUM · WALL TIES AT 16" OC · SILL · ½" × 15" ANCHOR BOLT · PERIMETER RIGID INSULATION AS REQUIRED · FLASHING · WEEPS AT 32" OC · 4" CONCRETE MASONRY · 8" CONCRETE MASONRY · 12" SOLID CONCRETE MASONRY TOP COURSE · CONCRETE FOOTING · REINFORCEMENT, AS REQUIRED · 28205-14_F17.EPS

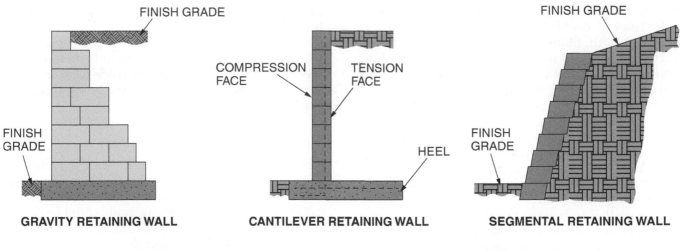

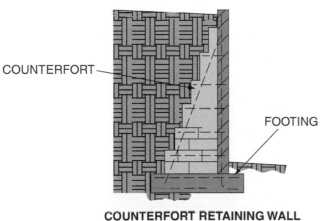

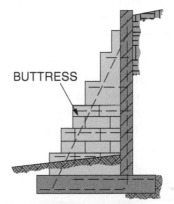

28205-14_F18.EPS

Figure 18 Types of retaining walls.

should be designed so that the computed pressure does not exceed the bearing value of the soil.

Proprietary systems of interlocking concrete masonry units called segmental block are shown in *Figure 19*. They also offer economical and attractive solutions for unreinforced retaining-wall applications of moderate height.

Some of the primary considerations in retaining-wall design should be as follows:

- Use the proper cap or coping to prevent water from collecting or standing on top of the wall.
- If coping consists of masonry units, through-wall flashing should be installed directly beneath the coping.
- Apply a waterproof coating on the back of the wall to prevent saturating the masonry.
- Use permeable backfill behind the wall to collect water and prevent soil saturation and increased hydrostatic pressure.
- Install weepholes or drain lines to drain moisture.
- Use expansion or control joints to permit longitudinal thermal and moisture movement.

28205-14_F19.EPS

Figure 19 Interlocking-block retaining wall.

1.6.1 Reinforcement

Reinforcement for retaining walls can be accomplished in several different ways. Externally, the wall can be reinforced with pilasters and buttresses placed along the wall face. Internally, re-inforcement consists of steel reinforcing bar or grout or both. Steel reinforcement can be placed vertically and tied into the footing, or placed horizontally with grout to form a bond beam at different elevations along the face of the wall.

If the wall being built is a cavity wall, vertical steel reinforcement is embedded in the footing before the wythes are laid. As the wall increases in height, the horizontal rods are periodically wired to the vertical rods for extra strength. The stress the wall will bear determines the size of the reinforcing rods used.

The walls can be structurally bonded by grout, which is poured into the cavity between the wythes of masonry. The grout core seals the space between the wythes and bonds the reinforcing steel. Full bed joints should be used for reinforced walls. Grouting techniques are detailed in *Reinforced Masonry*. It should also be noted that grouting techniques for reinforced masonry walls vary in different parts of the country. You should check the plans, specifications, and local building codes to determine the specific requirements for your project. The requirements for the size and place-ment of reinforcing steel may also vary according to the local codes.

Figure 20 shows a cross section of a typical cantilever retaining wall constructed of reinforced block. A general specification guide for reinforced concrete masonry walls of 8- or 12-inch thickness is given in *Table 1*.

1.6.2 Construction Techniques

The cantilever retaining wall uses the weight of the soil pushing down on the footing along with the strength of the vertical wall to resist the soil and water pressure pushing against the back face of the wall. For these types of walls, the same principles of good construction should be followed as in building any other reinforced masonry wall. Note in *Figure 20* that the wall is properly capped to prevent water intrusion from the top.

Provide standard moisture protection. Build weepholes and proper drainage to allow water to flow through the wall.

Single-wythe walls are constructed using cored masonry units laid with face-shell mortar bedding. The vertical cores are aligned to form

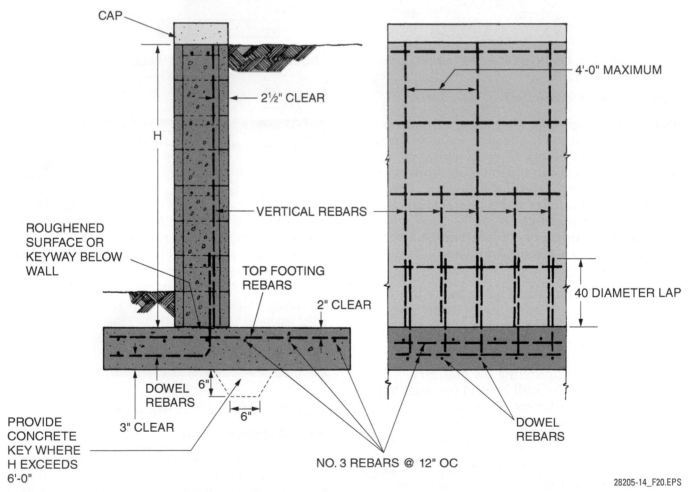

Figure 20 Cantilever retaining wall.

Table 1 Example Specifications for Reinforced Concrete Masonry Retaining Wall

Block Width	Height of Wall	Distance to Wall Face	Width of Footing	Thickness of Footing	Size and Spacing of Vertical Rods in Wall	Size and Spacing of Horizontal Rods in Footing
8"	3'-4"	12"	2'-8"	9"	No. 3 @ 32" OC	No. 3 @ 27" OC
	4'-0"	12"	3'-0"	9"	No. 4 @ 32" OC	No. 3 @ 27" OC
	4'-8"	12"	3'-3"	10"	No. 5 @ 32" OC	No. 3 @ 27" OC
	5'-4"	14"	3'-8"	10"	No. 4 @ 16" OC	No. 4 @ 30" OC
	6'-0"	15"	4'-2"	12"	No. 6 @ 24" OC	No. 4 @ 25" OC
12"	3'-4"	12"	2'-8"	9"	No. 3 @ 32" OC	No. 3 @ 27" OC
	4'-0"	12"	3'-0"	9"	No. 3 @ 32" OC	No. 3 @ 27" OC
	4'-8"	12"	3'-3"	10"	No. 4 @ 32" OC	No. 3 @ 27" OC
	5'-4"	14"	3'-8"	10"	No. 4 @ 24" OC	No. 3 @ 25" OC
	6'-0"	15"	4'-2"	12"	No. 4 @ 16" OC	No. 4 @ 30" OC
	6'-8"	16"	4'-6"	12"	No. 6 @ 24" OC	No. 4 @ 22" OC
	7'-4"	18"	4'-10"	12"	No. 7 @ 32" OC	No. 5 @ 26" OC
	8'-0"	20"	5'-4"	12"	No. 7 @ 24" OC	No. 5 @ 21" OC
	8'-8"	22"	5'-10"	14"	No. 7 @ 16" OC	No. 6 @ 26" OC
	9'-4"	24"	6'-4"	14"	No. 8 @ 8" OC	No. 6 @ 21" OC

continuous vertical spaces that can receive grout. Two-core block is generally preferred over three-core block because of the ease of placing grout. Cavity walls have wythes spaced 1 to 6 inches apart and can be constructed of cored or uncored units, depending on the building requirements. Grouting is usually done as the wall is built.

Concrete footings for retaining walls should be placed on firm undisturbed soil. In areas subject to freezing, they should also be placed below the frost line to avoid heave and possible damage to the wall. If the soil under the footing is soft or silty clay, it may be necessary to place compacted fill before pouring the concrete. If the footing elevation must vary, steps must be located adjacent to a grout pocket with the pocket extending to the top of the lower footing.

Failure to drain the backfill area behind retaining walls causes a buildup of hydrostatic pressure, which can quickly become critical if rainfall is heavy. In mild climates, weepholes at the base of the wall should be provided at 4- to 8-foot intervals. In areas where precipitation is heavy or poor drainage conditions exist, a continuous longitudinal drain of perforated pipe should be placed near the base with discharge areas located beyond the ends of the wall. Two methods for ensuring adequate drainage are shown in *Figure 21*.

Waterproofing requirements for the back face of the retaining wall will depend on the climate, soil conditions, and type of masonry units used. Seepage through a brick wall can cause efflorescence if soluble salts are present, but a waterproof membrane will prevent this water movement.

Walls of porous concrete units should always receive waterproof backing because of the excessive expansion and contraction that accompanies variable moisture content. In areas subject to freezing, a waterproof membrane can prevent freeze-thaw cycle damage to the units.

Segmental retaining walls (SRWs) are one of the newest developments in the masonry industry. These dry-stacked, interlocking-block retaining walls are offered by a variety of manufacturers in many different designs and configurations. An example of this type of wall is shown in *Figure 22*.

The units are stepped back slightly in each course, or battered, toward the embankment. Some units interlock simply by their shape, while others use pins or dowels to connect successive courses. Because they are dry-stacked without mortar, interlocking retaining walls are simple and fast to install. The open joints allow free drainage of soil moisture, and the stepped-back designs reduce overturning stresses.

Because they are dry-stacked, segmental retaining walls are relatively flexible and can tolerate movement and settlement without distress. The footings for SRWs are typically supported on flexible aggregate leveling-beds. Supplemental drainage can be provided by gravel backfill and collection pipes at the base of the wall if necessary.

The maximum height that can be constructed using a single-unit-depth SRW is directly proportional to its weight, width, and vertical batter for any given soil type and site geometry conditions. The height can be increased by using multiple

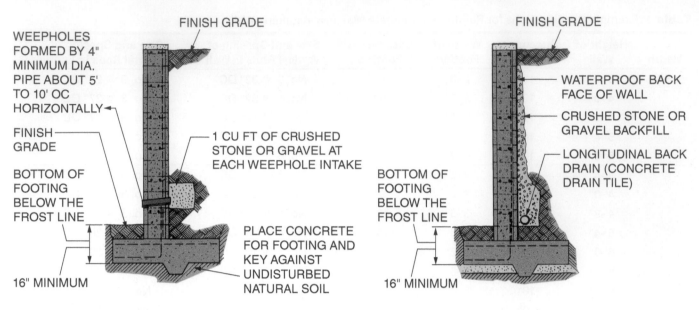

WITH PERMEABLE BACKFILL

WEEPHOLES FORMED BY 4" MINIMUM DIA. PIPE ABOUT 5' TO 10' OC HORIZONTALLY

FINISH GRADE

BOTTOM OF FOOTING BELOW THE FROST LINE

16" MINIMUM

FINISH GRADE

1 CU FT OF CRUSHED STONE OR GRAVEL AT EACH WEEPHOLE INTAKE

PLACE CONCRETE FOR FOOTING AND KEY AGAINST UNDISTURBED NATURAL SOIL

WITH IMPERMEABLE BACKFILL

FINISH GRADE

WATERPROOF BACK FACE OF WALL

CRUSHED STONE OR GRAVEL BACKFILL

LONGITUDINAL BACK DRAIN (CONCRETE DRAIN TILE)

BOTTOM OF FOOTING BELOW THE FROST LINE

16" MINIMUM

28205-14_F21.EPS

Figure 21 Drainage for retaining wall.

unit depths. Before beginning to build a segmental retaining wall, consult the manufacturer's design specifications to determining the maximum allowable height and the required batter for the type of block being used.

1.7.0 Identifying the Structural Principles and Fundamental Uses of Freestanding Walls

Freestanding walls include garden walls, fences, and screens of brick, tile, and concrete masonry units. They can take many forms. Screen-wall masonry units like those shown in *Figure 23* are often used for privacy fences.

Fences and walls should be able to safely withstand wind loads of at least 5 psf, with many codes specifying a minimum resistance of 20 psf. In hurricane areas, higher wind-pressure resistance is needed. *Table 2* shows pressures in psf and corresponding wind gust velocities in miles per hour (mph).

Without reinforcement, high and straight garden walls or fences lack vertical tensile strength and are unstable in strong winds. It is recommended that for 10-psf wind pressure, the height above grade should not exceed ¾ of the wall thickness squared. For example, an 8-inch-thick wall could be a maximum height of 48 inches (¾ × 8 × 8 = 48). This formula does not depend on a bond between the foundation and the wall. Therefore, reinforcing will greatly increase the height of the wall for a given thickness.

28205-14_F22.EPS

Figure 22 Segmental retaining wall.

Besides the basic straight wall made with a single-wythe or composite system, several other designs are popular in various parts of the country. A pier-and-panel wall is composed of a series of relatively thin panels, usually less than 4 inches thick, that are braced by masonry piers. This type of wall is relatively easy to build and is economical because of the reduced panel thickness. It is also well suited for uneven terrain. Foundations are only required for the piers.

Figure 23 Screen wall.

1.7.1 Reinforcement

Local building codes will determine how high garden walls and fences can be built and whether they must be reinforced or not. Building a garden wall or fence is the same as building other types of solid or composite masonry walls. Reinforcing requirements will depend on the specifications for wind loading, earthquake resistance, and soil conditions.

For walls that require some type of reinforcement, the main concern is the embedding of the reinforcing steel. The reinforcing steel must be embedded into the footer or pier enough to anchor the wall firmly to the foundation and to resist any lateral pressure.

1.7.2 Construction Techniques

Screen walls and garden walls are normally constructed the same way as interior partition walls.

Table 2 Pressure on Walls from Wind Gusts

Wind Gust Velocity (mph)	Pressure (psf)
40	5
57	10
69	15
80	20

Screen walls are governed by the same height-to-thickness (h/t) ratio lateral-support requirements as interior partitions, but walls with interrupted bed joints should be designed more conservatively because of reduced flexural strength and lateral load resistance.

Concrete masonry screen wall units should meet the minimum requirements of *ASTM (American Society for Testing and Materials) C90, Standard Specification for Loadbearing Concrete Masonry Units*. Brick should meet *ASTM C216, Standard Specification for Facing Brick (Solid Masonry Units Made from Clay or Shale)*, or *C652, Standard Specification for Hollow Brick (Hollow Masonry Units Made From Clay or Shale)*, Grade SW. Clay tile units should meet *ASTM C530, Standard Specification for Structural Clay Nonloadbearing Screen Tile*, Grade NB. The mortar used for exterior screen walls is typically Type N. Type S can be used in high-wind areas.

Many patterns can be applied to uncored brick garden walls. The final design will depend on the function of the wall and the availability of colors and sizes.

Perforated walls (*Figure 24*) can also be built with any of the standard brick types. These walls are constructed by omitting the mortar from head joints and separating the units to form voids. The walls may be laid up in single- or double-wythe construction. In double-wythe walls, separate header or rowlock courses alternate with stretcher courses to form different patterns.

Figure 24 Perforated wall.

Double-wythe walls are more stable than single-wythe designs because of the increased weight, wider footprint, and through-wall bonding patterns. Piers may be either flush with the wall or projecting on one or both sides. The coursing of the screen panels must overlap the coursing of the piers to provide adequate structural connection.

Regardless of exact design, the pattern of units in a perforated wall must provide continuous vertical paths for load transfer to the foundation, and the bearing width of the paths or columns should be at least 2 inches.

All freestanding walls require some type of covering over their top. This is usually accomplished with a cap. A cap is usually made from masonry units that cover the top of the wall and is the same thickness. Coping applies to the covering over the top of a wall. These coverings can be of single or multiple units. The tops can also slope in one or both directions.

Figure 25 shows four types of coping on top of different types of garden walls. Copings normally

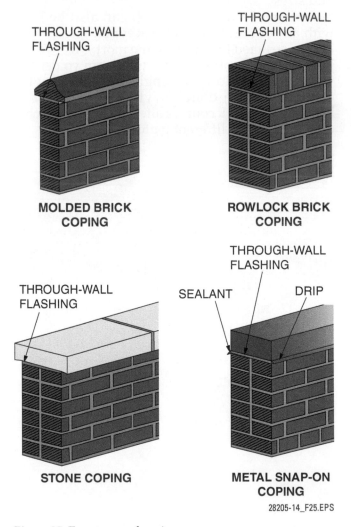

Figure 25 Four types of coping.

do not serve any structural function and do not present any major problems in their construction. Typical materials include brick, precast or cast-in-place concrete, stone, terra-cotta, or metal.

Brick pier and panel walls are composed of a series of thin panels braced intermittently by reinforced masonry piers or pilasters. Details of this type of construction are shown in *Figure 26*. Reinforcing steel is added to each panel at the top and bottom and tied into the piers. Pier construction can be of several different designs. Typical pilaster spacing for a wall is shown in *Table 3*.

Since the panel section is not supported on a continuous footing, it is more like a beam spanning the clear distance between foundation supports. The pier and panels are constructed at the same time course by course. Initially, a 2" × 4" form is constructed under the first course of the panel and kept in place for a minimum of seven days until the mortar has sufficiently cured. Requirements for steel reinforcement will be determined by local codes or by design calculations provided by organizations such as the Brick Industry Association. General requirements are shown in *Table 4*.

A serpentine garden wall does not require vertical support such as pilasters or columns because the shape of the wall gives it stability and resistance to wind pressures. However, each end of the wall needs to be anchored in a pilaster or other intersecting structure, or in a sharp curve. A typical layout for a serpentine wall is shown in *Figure 27*.

While serpentine walls do not require any reinforcing because of the curves, they must be laid out properly as the curve of the wall is related to height, which in turn is related to width. A rule of thumb is that a serpentine wall should not be built higher than 15 times the width of the unit. For example, if a 6-inch-wide block is used, then the wall should not be more than 7½ feet high.

Since the wall depends on its shape for lateral strength, it is important that the degree of curvature be sufficient. Recommendations for block and brick serpentine-wall curvature requirements can be obtained from Brick Industry Association Technical Note 29A, *Brick in Landscape Architecture – Garden Walls*.

Once the design has been established, you must lay out the wall starting with the footing. This is not an easy task, because the footing must be curved in the same shape as the wall and placed so that the wall will always be in the center line. After the footing has been dug and poured, mark the center line of the wall and dry-bond the first course to determine the placement of the units and the joint spacing on both faces.

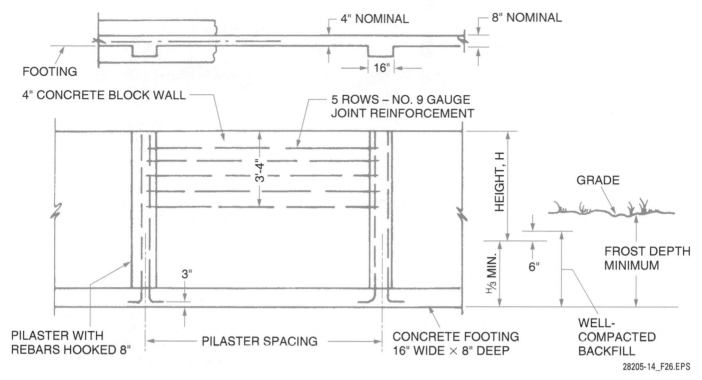

4" NOMINAL **8" NOMINAL**

FOOTING

16"

4" CONCRETE BLOCK WALL

5 ROWS – NO. 9 GAUGE JOINT REINFORCEMENT

3'-4"

3"

HEIGHT, H

H⅓ MIN.

6"

GRADE

FROST DEPTH MINIMUM

PILASTER WITH REBARS HOOKED 8"

PILASTER SPACING

CONCRETE FOOTING 16" WIDE × 8" DEEP

WELL-COMPACTED BACKFILL

28205-14_F26.EPS

Figure 26 Design for pier and panel wall.

Table 3 Pilaster Spacing for Wind Pressure

Wall Height	5 psf	10 psf	15 psf	20 psf
4'-0"	19'-4"	14'-0"	11'-4"	10'-0"
5'-0"	18'-0"	12'-8"	10'-8"	9'-4"
6'-0"	15'-4"	10'-8"	8'-8"	8'-0"

Table 4 Reinforcements for Wind Pressure

Wall Height	5 psf	10 psf	15 psf	20 psf
4'-0"	1–No. 3	1–No. 4	1–No. 5	2–No. 4
5'-0"	1–No. 3	1–No. 5	2–No. 4	2–No. 5
6'-0"	1–No. 4	2–No. 5	2–No. 5	2–No. 5

Because of the curve, this spacing will be different on opposite sides of the wall at all times except for the two joints at each point where the curve changes direction. Sometimes it is helpful to make a plywood form cut to the shape of the curve to use as a guide for dry bonding and as a check after the course has been laid. It takes special skills to plumb and align the masonry units in a serpentine wall. Do not attempt to build this type of wall until you have mastered the other skills of masonry construction.

Serpentine Walls

Serpentine walls have been used successfully for hundreds of years. The serpentine shape provides lateral strength so that it can normally be built with a minimum thickness of 4 inches. Since the shape of the wall provides the strength, it is important that the degree of curvature be sufficient. A rule of thumb states that the radius of curvature of a 4-inch-thick wall should be no more than twice the height of the wall above the grade. The depth of curvature should be no less than half the height.

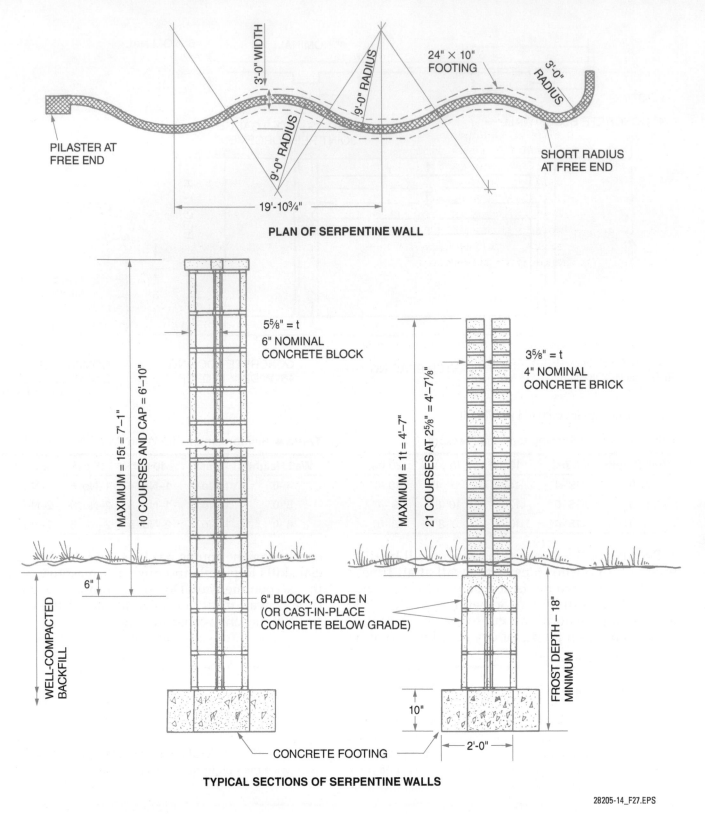

PLAN OF SERPENTINE WALL

TYPICAL SECTIONS OF SERPENTINE WALLS

28205-14_F27.EPS

Figure 27 Layout for a serpentine garden wall.

Additional Resources

ACI 530/ASCE 5/TMS 402, Building Code Requirements for Masonry Structures, Latest Edition. Reston, VA: American Society of Civil Engineers.

ACI 530.1/ASCE 6/TMS 602, Specifications for Masonry Structures, Latest Edition. Reston, VA: American Society of Civil Engineers.

ASTM C90, Standard Specification for Loadbearing Concrete Masonry Units, Latest Edition. West Conshohocken, PA: ASTM International.

ASTM C216, Standard Specification for Facing Brick (Solid Masonry Units Made from Clay or Shale), Latest edition. West Conshohocken, PA: ASTM International.

ASTM C530, Standard Specification for Structural Clay Nonloadbearing Screen Tile, Latest Edition. West Conshohocken, PA: ASTM International.

ASTM C652, Standard Specification for Hollow Brick (Hollow Masonry Units Made From Clay or Shale), Latest edition. West Conshohocken, PA: ASTM International.

Technical Note 29A, Brick in Landscape Architecture – Garden Walls. 1999. Reston, VA: The Brick Industry Association. **www.gobrick.com**

1.0.0 Section Review

1. In 8-inch walls the type of course that can be used for the full width of the wall section is the _____.

   a. rowlock course
   b. soldier course
   c. sailor course
   d. header course

2. Horizontal anchors are usually placed _____.

   a. in alternate courses
   b. in every course
   c. in every third course
   d. in a vertical line

3. In most cavity walls, the vertical load of the structure is carried by _____.

   a. the horizontal joint reinforcement
   b. the middle wythe
   c. the inner wythe
   d. the outer wythe

4. The typical thickness of composite walls that are used as exterior walls is _____.

   a. 2 to 6 inches
   b. 8 to 12 inches
   c. 14 to 18 inches
   d. 20 to 24 inches

5. The minimum recommended width of the open cavity between a veneer and the backing is _____.

   a. 1 inch
   b. 2 inches
   c. 3 inches
   d. 4 inches

6. A wall that is *not* an example of a traditional masonry retaining wall is the _____.

   a. trellis wall
   b. gravity wall
   c. cantilever wall
   d. buttress wall

7. Many codes specify a minimum safe wind-load resistance for fences and walls of _____.

   a. 5 pounds per square foot
   b. 10 pounds per square foot
   c. 15 pounds per square foot
   d. 20 pounds per square foot

2.0.0 CONTROL JOINTS AND EXPANSION JOINTS

Objective

Identify the requirement for and function of control joints and expansion joints.

   a. Identify the effects of temperature and moisture on control joints and expansion joints.
   b. Identify the uses of control joints.
   c. Identify the uses of expansion joints.

Performance Task

Lay out and construct a composite wall with control joints and expansion joints.

Trade Terms

Pencil rod: A type of metallic tie that is similar to the shape of a straight wooden pencil; used for control joints in concrete masonry construction.

The different elements and materials that make up a building are in a constant state of motion. All building materials expand and contract due to changes in temperature and many materials move with changes in moisture content. If this movement is excessive or constrained, the structural capabilities of the building can be reduced. The successful performance of a structure depends upon construction techniques that allow for some degree of movement of the materials.

The expansion and contraction of building materials can result in a number of problems. One problem concerns the amount of movement. Excessive movement will often lead to cracking. For this to occur, the material must be exposed to a wide variation in temperature, and must be a relatively long, continuous structure. This is often the case with exterior masonry walls.

2.1.0 Identifying the Effects of Temperature and Moisture on Control Joints and Expansion Joints

Cracking can be prevented several ways. Either the temperature variation must be controlled, the structure must be strong enough to resist cracking, or the effective length of the structure must

be limited. In masonry walls, the proper placement and construction of joints limits the effective length of the structure and reduces the tendency to crack. Joints used on brick walls are called expansion joints. Joints used on block walls are called control joints.

Another problem can result from surface temperatures of a material. Surface temperatures on a masonry wall facing the sun can reach 140°F. If the temperature on the inside surface of the wall is controlled and limited to a much lower value, there will be a strong tendency for the wall to warp. This warpage will often result in joint cracks.

If building materials are used in combination with one another, they need to be compatible under the expected temperature changes. If one material expands at a much greater rate than that of the other material, problems may be encountered. *Table 5* can be used to determine the compatibility of materials according to their coefficient of linear expansions and the amount of expansion that can be expected. For example, a brick wall 100 feet in length will expand or increase in length approximately 0.43 inch if the temperature increases 100°F.

With the exception of metals, many building materials tend to expand with increases in moisture content and contract with losses of water. For example, brick expands slowly when exposed to water or very humid air. This expansion is very difficult to determine and is unpredictable. Research has been done to determine precisely the expansion of clay masonry products, but investigators have not been able to discover any uniform guidelines. In fact, some structures show signs of distress while others show no ill effects from moisture expansion, although similar materials are used. With this uncertainty, it is best to plan for some expansion or movement.

2.2.0 Identifying the Uses of Control Joints

Cracks that form in a masonry structure are an ever-present problem to the builder. These cracks are caused by various movements in the structure. These movements can cause localized stresses that exceed the breaking point of the masonry and cause a crack to be formed. This does not suggest that the structure is in a weakened condition, only that at a specific location the masonry has become over-stressed to the point of breakage.

You can help limit the size of the cracks or control the location of the cracks by incorporating certain construction techniques, including control joints and expansion joints. In order to do this, you must recognize the reasons for the crack-

Table 5 Thermal Movement of Common Building Materials

Material	Thermal Expansion (in inches per 100 ft) for 100°F Temperature Increase (to closest 1/16 inch)	
Clay Masonry		
Clay or shale brick	0.43	(7/16)
Fireclay brick or tile	0.30	(5/16)
Clay or shale tile	0.40	(3/8)
Concrete Masonry		
Dense aggregate	0.62	(5/8)
Cinder aggregate	0.37	(3/8)
Expanded-shale aggregate	0.52	(1/2)
Expanded-slag aggregate	0.55	(9/16)
Pumice or cinder aggregate	0.49	(1/2)
Stone		
Granite	0.56	(9/16)
Limestone	0.53	(1/2)
Marble	0.88	(7/8)
Concrete		
Gravel aggregate	0.72	(3/4)
Lightweight, structural	0.54	(9/16)
Metal		
Aluminum	1.54	(1 9/16)
Stainless steel	1.15	(1 1/8)
Structural steel	0.80	(13/16)
Wood, Parallel to Fiber		
Fir	0.25	(1/4)
Oak	0.32	(5/16)
Pine	0.43	(7/16)
Wood, Perpendicular to Fiber		
Fir	3.84	(3 13/16)
Oak	3.60	(3 5/8)
Pine	2.28	(2 1/4)
Plaster		
Gypsum aggregate	0.91	(15/16)
Perlite aggregate	0.62	(5/8)
Vermiculite	0.71	(11/16)

ing and properly plan the appropriate preventive measures. Cracks can be caused by the following:

- Expansion and contraction due to temperature changes
- Expansion and contraction due to changes in the material's moisture content
- Movement due to the sudden application of loads such as wind or earthquake loads
- Movement due to the settlement of the structure

Even if these reasons are recognized, cracking is difficult to control because it can be a result of one reason or a number of reasons. For example, if a crack appears after a violent storm, the crack may have been the result of the increased mois-

ture, the sudden drop in temperature, the high winds, or the expansion of the foundation soil as it became wet.

To help control cracks, you can use joint reinforcement, bond beams, reinforced masonry, or flexible joints. You are probably already aware of the value of these in controlling crack formation; however, you must become skilled in laying up expansion joints or control joints to reduce random cracking in the structure. Expansion joints are discussed in the section *Identifying the Uses of Expansion Joints*.

A control joint is a vertical separation built into a block wall at locations where horizontal stresses are likely to cause cracks in the structure. Some shrinkage will occur in long block walls, resulting in a stress trying to pull the block apart. To relieve these stresses and allow foundations to shrink or expand without causing damage to the wall's alignment or strength, control joints are formed.

Each control joint must be able to support lateral loads across the joint (perpendicular to the wall face, such as wind forces) but must allow some movement longitudinally (along the length of the wall). In fact, control joints are designed to allow as much as 1/4-inch movement longitudinally. Control joints may be designed to work with bond beams and joint reinforcement, or may be designed to work independently to increase crack resistance in a structure.

2.2.1 Positioning

The proper positioning of control joints in block walls has evolved over the years with the masonry trade. The positioning of control joints is dependent on local practices, the construction materials in use, and local codes. The following factors should be considered in planning for joint construction:

- The expected movement of the wall
- The resistance of the wall to horizontal tensile stress
- The number and location of wall openings

The consideration of these factors will normally result in the use of a control joint where different materials join, where different wall heights or thicknesses join, or where walls intersect other structural elements. Typical locations for control joints are shown in *Figure 28*. They include the following:

- At changes in wall height or thickness
- At construction joints in foundations, roofs, or floors
- At chases and recesses for piping, columns, fixtures, or wiring
- At the abutment of walls and columns

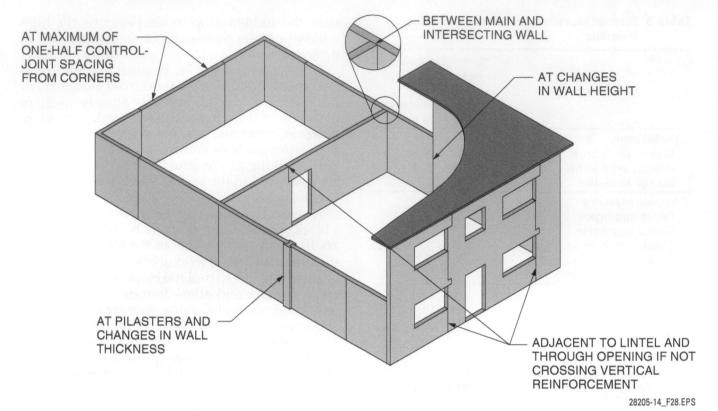

AT MAXIMUM OF ONE-HALF CONTROL-JOINT SPACING FROM CORNERS

BETWEEN MAIN AND INTERSECTING WALL

AT CHANGES IN WALL HEIGHT

AT PILASTERS AND CHANGES IN WALL THICKNESS

ADJACENT TO LINTEL AND THROUGH OPENING IF NOT CROSSING VERTICAL REINFORCEMENT

28205-14_F28.EPS

Figure 28 Locations for control joints.

- At wall intersections in L-, T-, and U-shaped structures
- At wall openings. Generally a control joint is placed at one side of an opening of less than 6 feet in width. When the opening is greater than 6 feet, a control joint is placed at both jambs.

Control joints are also placed at specified intervals in long, continuous block walls. Control joints divide the wall into panels and limit any stresses that have a tendency to build up along the length of the wall. These control joints should be placed every 25 to 30 feet in a continuous block wall. The location of control joints is also dependent on the length-to-height ratio of the wall and the use of joint reinforcement. *Table 6* considers the use of these factors.

For example, to determine the joint spacing for a wall 8 feet high and 100 feet long, the following analysis is necessary:

Step 1 Determine if horizontal joint reinforcement is specified, and if so, at what spacing. If specified at 16 inches on center, a length-to-height ratio (L/H) of 3 should be used.

Step 2 The wall height is 8 feet, so the maximum panel length is:

$$L/H = 3 \text{ and } H = 8$$
$$L/8 = 3$$
$$L = 3 \times 8$$
$$L = 24 \text{ feet}$$

The maximum spacing of control joints for this wall is 24 feet on center.

Not all block walls require the use of control joints. For reinforced walls, the amount of horizontal reinforcing steel is normally more than adequate to resist or control cracking. For concrete masonry used as a backing for other materials, joints may or may not be required. If required, the joint should extend through the facing wythe and the backup wythe if they are masonry bonded or use full collar joints. If the two wythes are bonded only by horizontal anchors, joint construction does not have to be continuous through both wythes.

Table 6 Control-Joint Spacing.

Recommended Spacing of Control Joints	Vertical Spacing of Joint Reinforcement			
	None	24"	16"	8"
Expressed as Ratio of Panel Length to Height (L / H)	2	2½	3	4
With Panel Length (L) Not to Exceed	40'	45'	50'	60'

2.2.2 Types of Control Joints

There are at least five different types of control joints: a control joint using standard block, a control joint using a special tongue-and-groove control-joint block, a premolded gasket control joint, a Z-type unit-tie joint, and a pencil rod control joint. The first three types are commonly used in the masonry trade.

A control joint can be formed using standard block and building paper, as shown in *Figure 29*. To break the mortar bond between two horizontally placed block, you must curl a piece of building paper into the end of the block on one side of the joint. As the block on the other side of the joint is laid, the core is filled with mortar. The mortar fill bonds to one block, but the paper breaks the bond with the other block. After completion of the wall, the head joint is partially raked out and replaced with an elastic sealing compound. Longitudinal movement is allowed, but the mortar fill will resist any lateral loads.

The tongue-and-groove control joint uses a specially designed full- or half-length control-joint block, as shown in *Figure 30*. The tongue of one special unit fits into the groove of another special unit or into the open end of a regular flanged unit. Place the control-joint block in the normal manner, paying particular attention to the alignment of the head joint. The head-joint mortar will later be partially raked out and replaced with a flexible caulk.

A control joint can be formed by inserting a premolded rubber gasket in the joint between sash block (*Figure 31*). The gasket is fairly stiff but will allow some longitudinal movement. The gasket also aids in the prevention of moisture penetration. Again, the head joints are raked and caulked with flexible caulking.

A control joint can be formed using two jamb block and a noncorroding metal Z-type unit tie. This type of assembly is shown in *Figure 32*. The jamb block is placed with mortar only at the face shell of the jamb and not throughout the block thickness. The head joint will be raked and caulked to allow for movement, yet remain weather-tight. Jamb block is not often used for control joints since they upset the modular dimensioning often used for planning. That is, two jamb block used in forming a control joint are less than the standard 32 inches in length.

The pencil-rod joint is a variation of the Z-type unit-tie joint but is made with standard block and a metal pencil rod or Z-type unit tie. The block are tied with a 10- to 12-inch pencil rod across the face shells, as shown in *Figure 33*, or a Z-type unit tie. One leg of the Z-type unit tie or one end of the pencil rod must be greased to prevent a continuous longitudinal bond. This type of control joint does not offer as much lateral support as others.

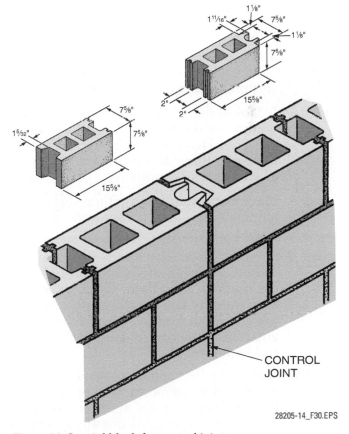

28205-14_F30.EPS

Figure 30 Special block for control joints.

28205-14_F29.EPS

Figure 29 Control joint using standard concrete masonry units.

Figure 31 Premolded gasket control joints.

2.2.3 Placing a Control Joint

You need to follow certain specific steps in placing a control joint in a block wall. If these steps are followed and if the joint is properly positioned, the crack resistance of the structure is greatly increased.

Step 1   Review the construction documents to identify how and where joints should be placed.

Step 2   Mark the location of the joint at the proper distance from the wall end, corner, intersection, or adjacent control joint.

Step 3   Establish the width of the joint by placing a mark for the first unit on either side of the joint. This is typically ⅜ inch, but can be less or more depending on the design as well as field conditions.

Step 4   Place bond marks on either side of the joint, ensuring that proper bond is maintained around the walls or building.

Step 5   Spread the bed joint for the first course of masonry, taking care not to cover the layout marks. Then mortar the cross web on both sides of the control joint. This will prevent grout from seeping from the cells adjacent to the control joint into the control joint itself.

Step 6   Install the first course of masonry at the control joint. No mortar will be applied to the heads of the units at the control joint. Make sure these units are aligned properly with the layout marks. Pay special at-

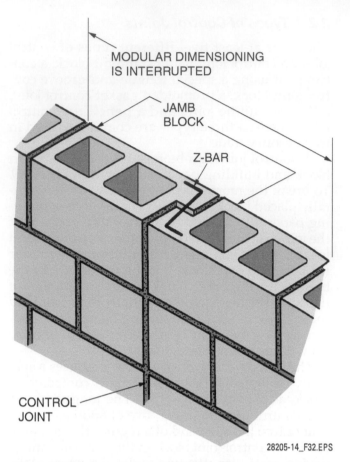

Figure 32 Z-type unit-tie control joint.

tention to the width of the joint; it should be consistent from the top to bottom of the units. If the joint is flared at the top or bottom, adjust the units as needed to correct the width of the joint.

Step 7   Install a second course of masonry at the control joint. Make sure to follow proper bond. Mortar the cross webs against the control joint prior to installing the units. Ensure that the joint is kept plumb in both directions and is of uniform width.

Step 8   Tool the adjacent masonry joints when they are thumbprint-hard. The control joint should be kept free of all mortar and debris. Use a trowel or striking iron to clean out any excess mortar in the control joint.

Step 9   Install the horizontal joint reinforcement at the specified intervals. This is typically every 16 inches vertically, but the spacing can vary by project; refer to the project drawings and specifications for the correct spacing. The joint reinforcement is typically not continuous at the control joints. If this is the case, use small bolt cutters to cut reinforcement at the control joint and make sure the ends of

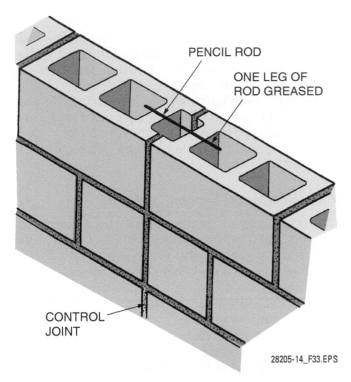

Figure 33 Pencil-rod control joint.

the reinforcement are kept about an inch away from the control joint. Occasionally, the design will require the joint reinforcement to be continuous through the control joint; again, be sure to check the drawings and specifications for this information.

Step 10 Install remaining courses of masonry. Make sure the joint is kept plumb, of a uniform width, and clean as the work progresses.

2.3.0 Identifying the Uses of Expansion Joints

One of the most effective means of preventing damage due to expansion in a brick wall is by using an expansion joint. Although expansion joints can be expensive and difficult to maintain, they provide a method of eliminating cracks that otherwise would form because of movement of the material. An expansion joint is simply a separation of parts.

A brick wall, for example, that will expand to such a degree that cracks occur can be divided into multiple wall segments. The wall segments will have a gap or joint between them that allows room for the expansion without cracking. The joint, of course, must be flexible and weather resistant.

Expansion joints in brick walls can be constructed in a number of ways so long as primary consideration is given to the flexibility and weather resistance of the joint. The joint can be made from a variety of materials. The five most common types of material are the following:

- Backer rod and sealant
- Copper
- Premolded foam pad
- Neoprene
- Extruded plastic

Copper joints have been used for many years. These are made from short pieces of copper sheets, which are lapped to form a continuous weather seal. Copper has the advantages of being flexible, noncorroding (under most conditions), and impervious to moisture. Some of the newer materials used for watertight joints include specially formed rubber and plastic materials. These may be premolded for the correct size and shape, and are extremely flexible and water resistant. Typical construction of expansion joints is shown in *Figure 34*.

The use of any expansion joint requires the use of a sealing compound for the exposed side of the joint. This sealing compound must be an elastic material that will not become brittle with age and must blend with the finish treatment of the wall for an attractive appearance. This compound may be a standard caulking compound, such as silicon caulk, or a specially formulated sealing compound offering elasticity and durability (some of these are classified as polymer materials). Occasionally, a cover plate is used in place of, or in conjunction with, the sealing compound.

2.3.1 Placement of Joints

The placement of expansion joints in brick construction is determined by considering a number of factors. Among these are the loads carried by the wall; the strength of the wall; the maximum and minimum temperature exposure of the wall; the length, height, and thickness of the wall; and the location of any intersecting walls. Due to the factors that must be considered, there are no hard-and-fast rules that can be used for determining the position and spacing of expansion joints. *Figure 35* illustrates building locations where expansion joints are typically located.

2.3.2 Joint Details

Expansion joints should be designed for the specific situation. Each building should be analyzed and separate details developed to meet the requirements for each joint. Many details are very similar in construction, and the diagrams in *Figure 36* are provided as suggested construction details. These are typical of the many possible joints used under a variety of situations.

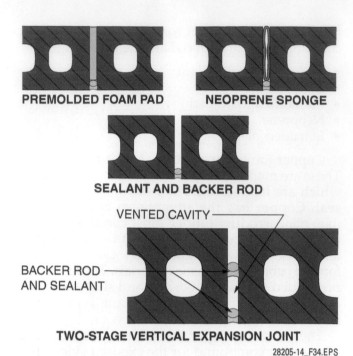

TWO-STAGE VERTICAL EXPANSION JOINT

PREMOLDED FOAM PAD

NEOPRENE SPONGE

SEALANT AND BACKER ROD

VENTED CAVITY

BACKER ROD
AND SEALANT

28205-14_F34.EPS

Figure 34 Expansion joints.

In general, expansion joints, also sometimes referred to as slip planes, should be located at offsets and at junctions of walls in L-shaped, T-shaped, or U-shaped buildings (*Figure 37*). This joint prevents the expansion of one wall from exerting pressures on the other wall.

Expansion joints are frequently placed at or near external corners in both cavity and solid walls. This is particularly true with solid walls that rest on concrete foundations and extend above grade (*Figure 38*).

All expansion joints should be carried the full height of the masonry, even if this includes a parapet. Additional expansion joints are often placed in unreinforced parapets halfway between those running full height.

2.3.3 *Horizontal Expansion Joints*

If possible, brick walls should be supported entirely on the foundation. This is not always possible because of the wall height or a large number of openings. Where this is the case, support points such as shelf angles are required. Horizontal expansion joints may be necessary at these support points. *Figure 39* shows a cross-section of the construction of a wall panel that is supported by a shelf angle but does not fit flush with the exterior of the wall. In this case the edge of the angle is covered by an elastic joint compound that also fills the whole horizontal joint. The sealant should be a color that will closely match the mortar joints.

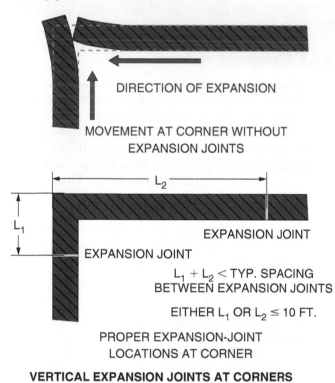

DIRECTION OF EXPANSION

MOVEMENT AT CORNER WITHOUT
EXPANSION JOINTS

L_2

L_1

EXPANSION JOINT

EXPANSION JOINT

$L_1 + L_2 <$ TYP. SPACING
BETWEEN EXPANSION JOINTS

EITHER L_1 OR $L_2 \leq 10$ FT.

PROPER EXPANSION-JOINT
LOCATIONS AT CORNER

VERTICAL EXPANSION JOINTS AT CORNERS

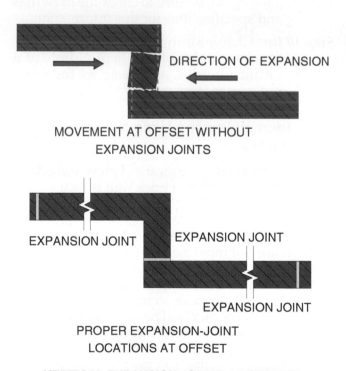

DIRECTION OF EXPANSION

MOVEMENT AT OFFSET WITHOUT
EXPANSION JOINTS

EXPANSION JOINT

EXPANSION JOINT

EXPANSION JOINT

PROPER EXPANSION-JOINT
LOCATIONS AT OFFSET

VERTICAL EXPANSION JOINTS AT OFFSETS

28205-14_F35.EPS

Figure 35 Typical locations of expansion joints.

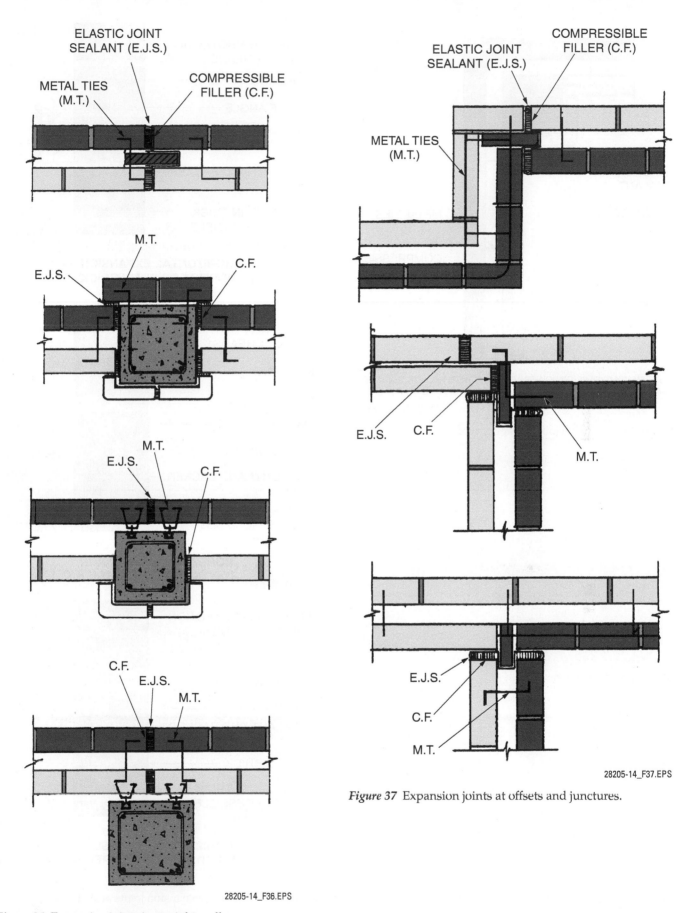

Figure 36 Expansion joints in straight walls.

28205-14_F36.EPS

Figure 37 Expansion joints at offsets and junctures.

28205-14_F37.EPS

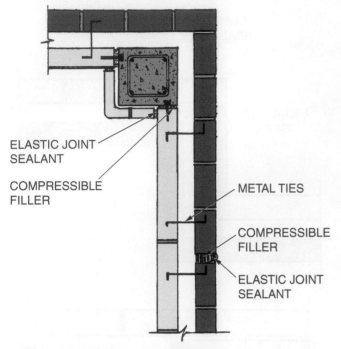

ELASTIC JOINT SEALANT

COMPRESSIBLE FILLER

METAL TIES

COMPRESSIBLE FILLER

ELASTIC JOINT SEALANT

NOTE: WHEN EXPANSION JOINTS ARE EMPLOYED, WALL TIES MUST BE PLACED WITHIN 12 INCHES (MAXIMUM) OF EITHER SIDE OF THE JOINT

28205-14_F38.EPS

Figure 38 Expansion joints near corners.

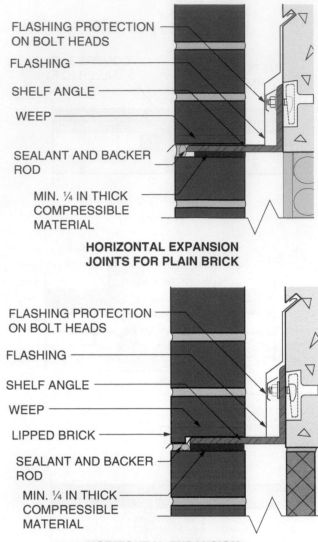

FLASHING PROTECTION ON BOLT HEADS

FLASHING

SHELF ANGLE

WEEP

SEALANT AND BACKER ROD

MIN. ¼ IN THICK COMPRESSIBLE MATERIAL

HORIZONTAL EXPANSION JOINTS FOR PLAIN BRICK

FLASHING PROTECTION ON BOLT HEADS

FLASHING

SHELF ANGLE

WEEP

LIPPED BRICK

SEALANT AND BACKER ROD

MIN. ¼ IN THICK COMPRESSIBLE MATERIAL

HORIZONTAL EXPANSION JOINTS FOR LIPPED BRICK

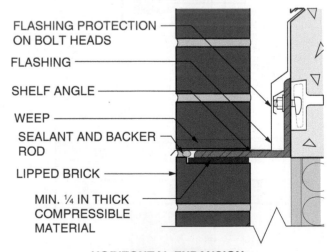

FLASHING PROTECTION ON BOLT HEADS

FLASHING

SHELF ANGLE

WEEP

SEALANT AND BACKER ROD

LIPPED BRICK

MIN. ¼ IN THICK COMPRESSIBLE MATERIAL

HORIZONTAL EXPANSION JOINTS FOR INVERTED LIPPED BRICK

28205-14_F39.EPS

Figure 39 Horizontal expansion joints at shelf angle.

Maintenance Holes

Before the development of precast concrete structures, all shallow, circular, underground structures were made from various types of brick masonry units. These included maintenance holes, catch basins, pump wells, valve vaults, and other structures that were typically connected to wastewater and sewer systems. Catch basins have space at the bottom for the settlement and storage of suspended solids that might otherwise be carried and deposited in the pipeline. If the catch basins are part of a sanitary sewerage system, they should also be provided with solid covers to prevent sewer odors from reaching the street level. Type M mortar is specified for this installation.

Block for underground sewer and wastewater structures should meet the absorption and strength requirements of *ASTM C139, Standard Specifications for Concrete Masonry Units for Construction of Catch Basins and Manholes.* This specification is for segmental masonry units.

Batter block is very useful for cone construction at the top of the barrel of a catch basin. Some unit designs include matching tongue-and-groove ends. These blocks come in standard sizes and are graduated, as shown in the figure, so that they gradually reduce the inside diameter to 2 feet at the top to receive a standard maintenance hole cover. No cutting of units is necessary because block producers have predetermined the exact number and size of batter block required. Use Type N mortar for this type of installation.

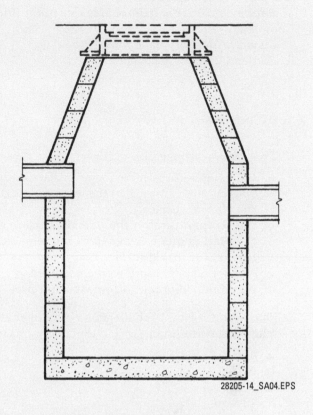

28205-14_SA04.EPS

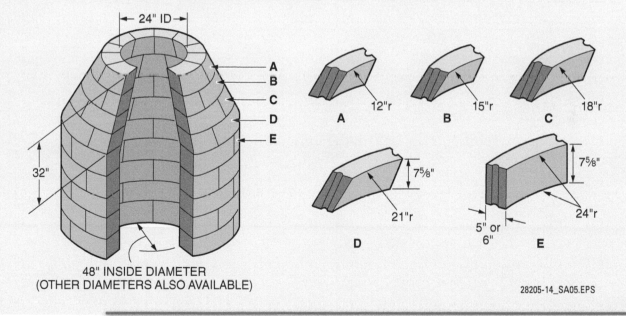

28205-14_SA05.EPS

Additional Resources

Bricklaying: Brick and Block Masonry. 1988. Brick Industry Association. Orlando, FL: Harcourt Brace & Company.

Concrete Masonry Handbook, Fifth edition. W. C. Panerese, S. K. Kosmatka, and F. A. Randall, Jr. Skokie, IL: Portland Cement Association.

2.0.0 Section Review

1. On a wall, warpage may occur if _____.

   a. the temperature on the inside surface of the wall is much lower than the temperature on the outside surface
   b. the temperature on the outside surface of the wall is much lower than the temperature on the inside surface
   c. horizontal joint reinforcement is used
   d. horizontal joint reinforcement is not used

2. Place a control joint at both jambs in an opening when it is wider than _____.

   a. 4 feet
   b. 6 feet
   c. 8 feet
   d. 10 feet

3. On solid walls that rest on concrete foundations and extend above grade, expansion joints are frequently placed at or near _____.

   a. every 400 to 450 feet
   b. doors, windows, and other openings
   c. internal corners
   d. external corners

3.0.0 LAYING OUT AND CONSTRUCTING ANGLED CORNERS AND INTERSECTIONS

Objective

Lay out and construct various corners and intersections.

    a. Lay out and construct toothing.
    b. Lay out and construct corbeling.
    c. Lay out and construct intersecting walls.
    d. Lay out and construct angled corners.

Performance Tasks 2 and 3

Lay out and construct intersections.
Lay out and construct angled corners.

Trade Term

Toothing: Construction of a temporary end of a wall with the end stretcher of every alternate course projecting. The projecting units are called toothers.

Complex masonry structures include more than one wall. Walls can be tied together at corners and at intersections. Masons build corners and intersections that distribute loads evenly to ensure that the walls continue standing even when exposed to external forces like wind and stresses such as the loads caused by the weight of a building. Therefore, it is important to learn how to construct corners and intersections that are structurally sound. In this section, you will learn how to lay out and design a variety of angled corners and intersections that meet the strict standards employed today.

3.1.0 Laying Out and Constructing Toothing

Toothing (*Figure 40*) is the construction of a temporary end of a wall by allowing every other end of stretcher brick to project halfway over the stretcher below. Usually, at the close of the workday you can stop work by stepping the block at the halting point. There are times, however, when you must raise a temporary end over the full height of an area of wall. Toothing is usually reserved for very rare times when normal stepping is not possible. This can be done for both block and brick.

Toothing is not normally permitted because it increases the possibility of poor jointing and has a tendency to result in poor wall alignment. It should only be done with the permission of the contractor and under the foreman's supervision. Toothing is usually done only in the following situations:

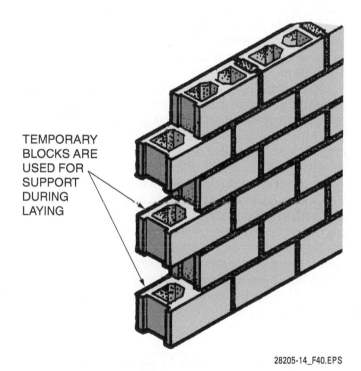

TEMPORARY BLOCKS ARE USED FOR SUPPORT DURING LAYING

28205-14_F40.EPS

Figure 40 Toothing in block.

Use Only When Necessary

Toothing is not recommended for general use. It can result in mortar joints that leak if not very carefully constructed. Toothing is necessary when openings are required for traffic or when windows or doors are not set on time.

    However, since the possibility of poor joints is so great and the chance of knocking the wall out of line increases by removing the toothing bats, architects and builders seldom allow the use of this technique. You should master the skill, but must get the permission of your supervisor before toothing an unfinished corner or opening.

- When a halt must be made in construction for a period of time while waiting for delivery of units
- When construction must be halted for an extended time but is expected to continue at a later date
- When a masonry bond is needed between a wall section made of block and one made of brick
- When very high corner leads must be fully constructed before line block is laid up
- When a walkway for the convenience of moving materials must be provided through the wall being built
- When window or door frames or jamb block have not yet arrived at the work site

Toothing for block is done by laying up the block in alternate courses so that every other end block projects halfway over the block below. This technique is done by laying a block piece in the mortar on the end of the last course laid and then laying a full block directly over the piece. After the block is fairly well set but not completely set, the block pieces are carefully removed. Care must be taken at this time to avoid disturbing the wall's alignment.

Once construction begins again, you must be absolutely sure that the joints around the toothers are full of mortar and that all of the surfaces to be joined are free of dirt and old mortar. Experience has shown that toothed areas are often the areas where the most water damage will occur if you do not take the time to carefully prepare this work area before construction begins again.

Toothing for brick is done by temporarily laying a bat in the mortar on the end of the course just laid and laying the next brick directly over it. After the mortar has set enough to be firm and rigid, the bats are removed very carefully so as not to disturb the mortar bond of the permanent brick. Wooden wedges can be used for temporary support if needed.

3.2.0 Laying Out and Constructing Corbeling

Corbeling is the process used by masons to widen a wall by projecting out masonry units to form a ledge or shelf. To build a corbel, each brick course extends out farther than the one below it. As a general rule, the unit should not project beyond the one below it more than one-third the unit width or one-half the unit height, whichever is less. When corbeling, you should carefully follow good bonding practices and use well-fitted mortar joints to ensure that the projection will remain stable.

In the past, the tops of many buildings were constructed with masonry trim work designed by corbeling brick into intricate patterns. An example is shown in *Figure 41*. While this practice has declined because of cost and design considerations, there is still a need for this type of work in specialty areas and restoration projects. Also, inside walls of chimneys and fireplaces still use corbeling as a standard practice.

Brick should be dry and free from chips and cracks because most edges are exposed to view. A dry brick sets more quickly, which eases the operation. The brick must be level, plumb, and ranged. Well-filled mortar joints give maximum strength and reduce the amount of tooling and jointing required. Building codes generally require the top course to be a full header course. Corbeling should not be done so quickly that the mortar and brick cannot take a firm initial set and remain in position.

To build corbeling for a brick wall (*Figure 42*), follow these steps:

Step 1   Snap a line on the foundation or base where the front edge of the wall is to be constructed.

Step 2   Lay out the first course as a dry course to check spacing. Running bond will be used with a header course every seventh course.

Step 3   Lay the first course with headers on the front wythe and stretchers on the back. Use a full mortar bed. Level, plumb, and range the brick. Check the spacing.

Step 4   Lay the second through sixth courses. Check the height of each course as it is laid, and be sure bed joints are uniform and level.

28205-14_F41.EPS

Figure 41 Corbeling on masonry wall.

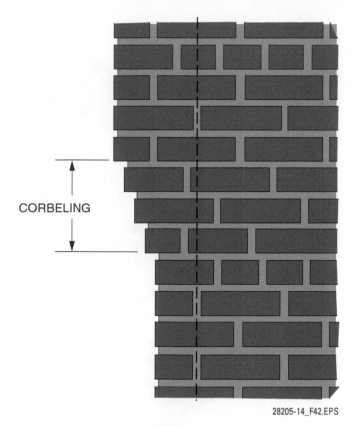

CORBELING

Figure 42 Corbeling in a brick wall.

28205-14_F42.EPS

Step 5 Lay the seventh course as a header course with full mortar joints.

Step 6 Continue to lay the wall until you reach the height where the corbel course is to begin. Check your progress often to be sure the wall is plumb and straight.

Step 7 Begin the corbel course by projecting headers out ⅝ inch beyond the course below. Fill the wide head joint with mortar. If the projection is more than ⅝ inch, you will probably need to use an uncored unit.

Step 8 Lay the second corbel course using stretchers along the front and headers along the back wythe. Fill in the space between the bats. Be sure the course is level and straight.

Step 9 Lay the third corbel course using stretchers on the front and back wythes and ¾ headers between the two wythes. Each corbel should project out ¾ inch beyond the course below.

Step 10 Lay the next course the same as the first course of the wall: headers on the front wythe and stretchers on the back wythe. Continue the wall to the desired height.

Step 11 Finish all joints when the mortar has set to thumbprint-hard.

Step 12 Clean the wall with the trowel and bricklayer's brush.

3.3.0 Laying Out and Constructing Intersecting Walls

When you tie intersecting walls to a main wall, the right-angled wall provides additional support to the main structure. There are a number of different methods that can be used to tie the two walls together. Some walls are connected with a rigid connection, while others are connected in a manner that will allow for some movement. You must consider where lateral support is needed and where joints are required.

The two basic methods of anchoring intersecting walls are true masonry bonding (*Figure 43A*) and flexible connection (*Figure 43B*).

An important requirement in masonry work is that loadbearing walls should be anchored where they meet at corners. In the past, intersections and corners of bearing brick walls were anchored with tie bars that had T-shaped ends, shown in *Figure 44*. Note that this technique is not commonly used anymore.

A variety of tie-bar shapes are available, but all should extend at least 8 inches into the holding wall. If the walls are being carried up together, the anchors are installed when both walls reach bonding height. These anchors are generally provided at least every 36 inches on center. Anchors of this type are designed to resist lateral stress and must be level to obtain maximum strength.

Decorative Corbeling

Corbeling and arches can be used to add decorative accents to brick storefront facades.

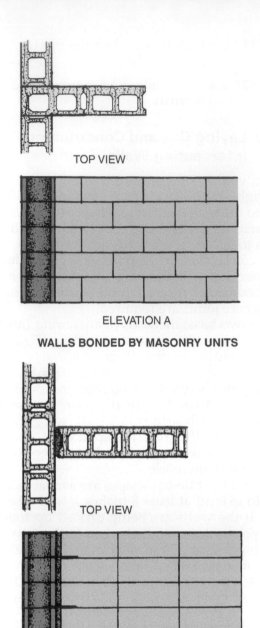

TOP VIEW

ELEVATION A

WALLS BONDED BY MASONRY UNITS

TOP VIEW

ELEVATION B

WALLS BONDED BY FLEXIBLE CONNECTORS

28205-14_F43.EPS

Figure 43 Masonry bonding and flexible connection.

When intersecting block walls are loadbearing and depend directly on each other for support, the bond between the two must resist all movement. For this reason, the bond at such an intersection is usually a rigid, true masonry bond. In fact, such bonding was originally the only means of tying intersecting walls.

This rigid bonding can be created in a number of ways:

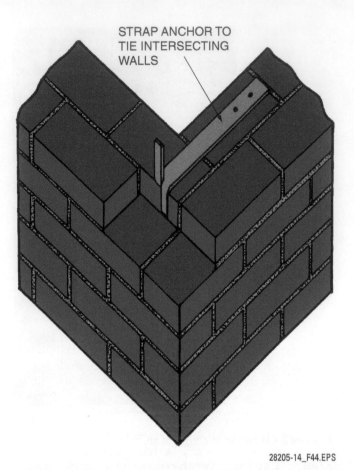

STRAP ANCHOR TO TIE INTERSECTING WALLS

28205-14_F44.EPS

Figure 44 Strap anchor for intersecting brick walls.

- If the courses of intersecting walls are laid up at the same time, a true masonry bond can be established by overlapping masonry units at the wall intersection (refer to *Figure 45*).
- A rigid bond can also be created by butting the intersecting walls together and applying a metal tie or anchor, which creates a physical bond. In such joints, the bond is maintained by grouting the joining cores, by applying the various ties or anchors, and by fully mortaring the vertical joints at the intersection. This is shown in *Figure 46*.
- A true masonry bond can be achieved even if the intersecting walls are not laid up at the same time. The intersecting wall may be tied into the main wall by bonding the masonry units into holes left in the main wall. In such cases, metal bar anchors must be solidly placed in mortared cores to strengthen the bond.

As more-complex masonry structures were designed and built, the need for both rigid and flexible joints was recognized. Newer jointing techniques have been developed that use flexible anchoring and control joints that allow for slight movement but provide lateral support.

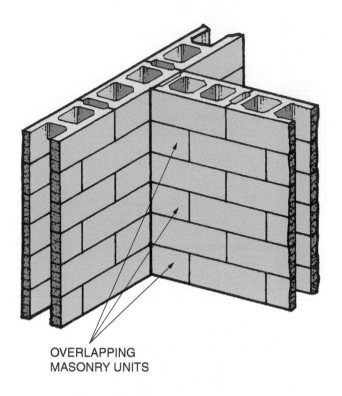

OVERLAPPING
MASONRY UNITS

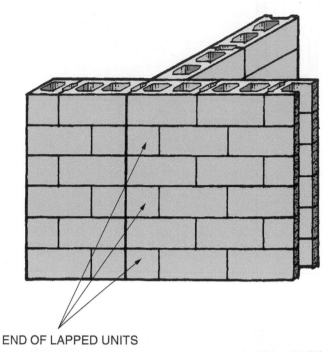

END OF LAPPED UNITS

28205-14_F45.EPS

Figure 45 True masonry bond at intersecting walls.

In one method of creating a flexible joint at intersecting walls, a steel tie bar is used for lateral support (*Figure 47*). These tie bars are usually heavy-gauge steel that is ¼ inch thick, 1¼ inches wide, and 16 to 28 inches long. The bars usually have 2-inch right-angle bends on each end that are locked into the grouted cores. When tie bars are used, they are usually placed at 16 or 24 inches on center, and the cores in which the right-angle bends are placed are filled with mortar or grout. Metal lath should be placed under the grouted core to prevent mortar from seeping out of the core. The joint is completed by placing elastic caulking at the wall intersection to create a control joint.

Other variations of the flexible intersecting wall joint can be made by applying metal lath, Z-type unit ties, or T-shaped wire mesh to provide lateral support, as shown in *Figure 48*. Control joints provide the flexibility needed in these joints. When ¼-inch wire mesh or strips of metal lath are used, they are placed in alternate courses in the wall or at 16 inches on center. The lath should extend at least 8 inches into the intersecting wall. For the sake of safety, the lath should be folded down against the wall until it is tied into the intersecting wall.

3.4.0 Laying Out and Constructing Angled Corners

When masons construct walls, typically they begin with the corners and fill in the rest of the

wall. Corners therefore require special attention because they affect how the rest of the wall is constructed. Follow the floor or foundation plan when building corners. It is important to ensure that modular dimensions are maintained during the construction of corners and walls. Block used in corners is designed with square ends (see *Figure 49*). Corner block may have ends that are beveled or molded with a 45-degree miter. These features are intended to permit the construction of corners with a minimum of cutting while also maintaining modular dimensions. *Figure 50* shows the corner details for 4-inch and 8-inch block walls.

Corners must be designed to withstand stresses caused by structural loads. *ACI 530/ASCE 5/TMS 402, Building Code Requirements for Masonry Structures*, lays out the requirements for the stress requirements of corners in masonry construction. It specifies that stresses may be transferred from one wall to an intersecting wall in one of three ways:

- By using a running bond
- By using steel connectors
- By using bond beams

Running bond provides enough interlock between masonry units to transfer shear (or sliding) stresses from one wall to the interlocking wall. For example, a wall subjected to high winds transfers the load caused by the wind to the walls that intersect it, thereby preventing the wall from

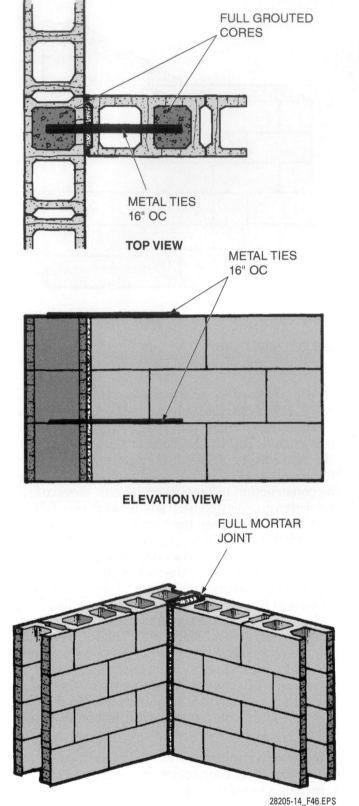

Figure 46 Rigid connections for intersecting walls.

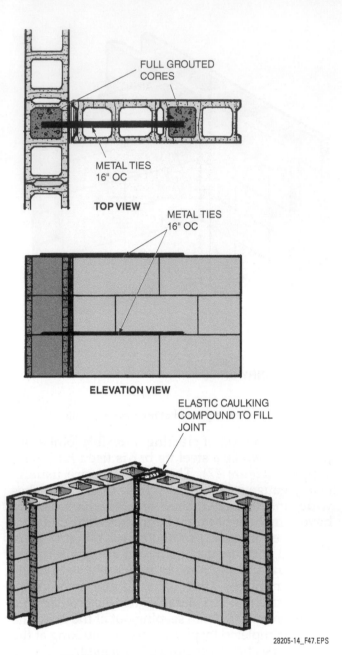

Figure 47 Flexible joints using tie bars.

Anchors vs. Ties

You should be able to recognize the difference in the use of the terms *anchor* and *tie*. Anchors attach a masonry wall to supports such as other walls, floors, beams, columns, or other structural supports. Ties hold the separate units in a masonry wall together. The placement and use of anchors and ties is described in the module titled *Masonry Openings and Metal Work*.

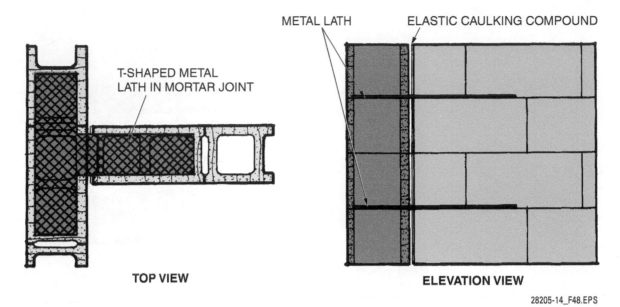

METAL LATH   **ELASTIC CAULKING COMPOUND**

T-SHAPED METAL LATH IN MORTAR JOINT

TOP VIEW   **ELEVATION VIEW**

28205-14_F48.EPS

Figure 48 Flexible joints using metal lath.

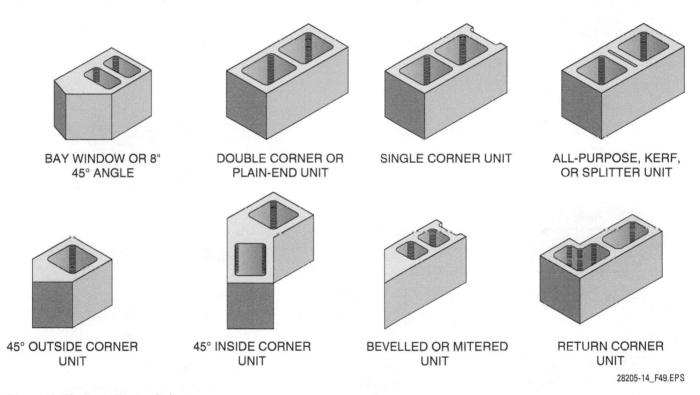

BAY WINDOW OR 8"
45° ANGLE

DOUBLE CORNER OR
PLAIN-END UNIT

SINGLE CORNER UNIT

ALL-PURPOSE, KERF,
OR SPLITTER UNIT

45° OUTSIDE CORNER
UNIT

45° INSIDE CORNER
UNIT

BEVELLED OR MITERED
UNIT

RETURN CORNER
UNIT

28205-14_F49.EPS

Figure 49 Block used in angled corners.

collapsing under the load. Local codes, project specifications, and the latest edition of *Building Code Requirements for Masonry Structures* all spell out the conditions that corners are required to meet to permit the transfer of stresses.

There are two ways to construct masonry corners: by using a corner pole and by using a level. As you learned in *Masonry Level One*, a corner pole (see *Figure 51*) is a post braced into a plumb position so that a line can be fastened to it. Cor-

ner poles allow masons to mark courses, hang a line, and begin laying masonry units to the line at a corner. Corner poles save masons time during construction and result in more accurate results than if the mason had to build a corner lead before completing the balance of the courses.

When using a corner pole to build a corner, begin by placing the corner pole at the location where the plans indicate the corner is to be constructed. If there is no existing structure to which to secure the

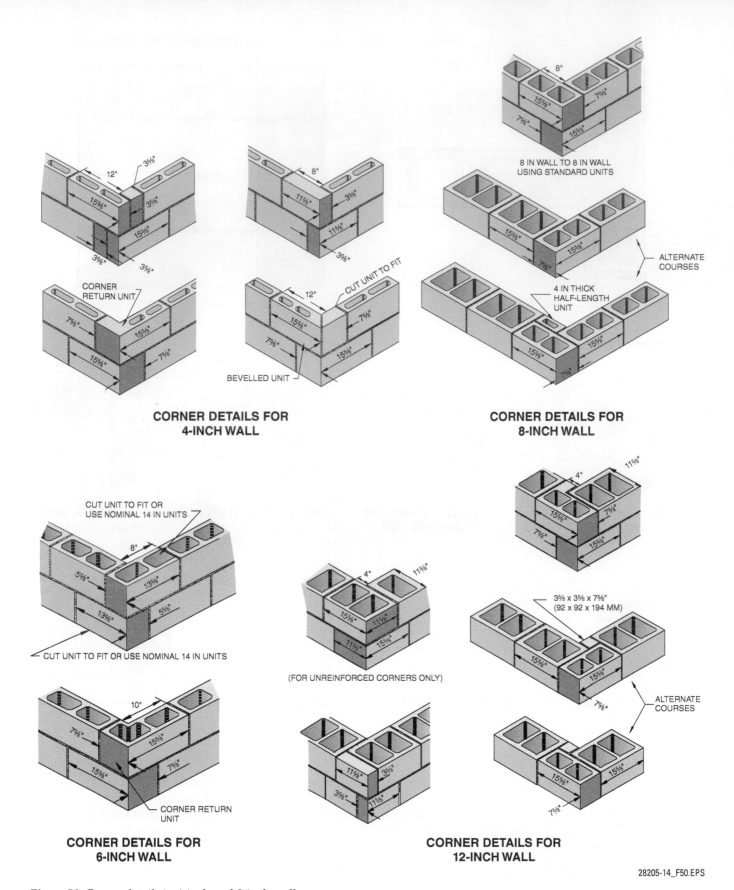

**CORNER DETAILS FOR
4-INCH WALL**

**CORNER DETAILS FOR
8-INCH WALL**

**CORNER DETAILS FOR
6-INCH WALL**

**CORNER DETAILS FOR
12-INCH WALL**

28205-14_F50.EPS

Figure 50 Corner details in 4-inch and 8-inch walls.

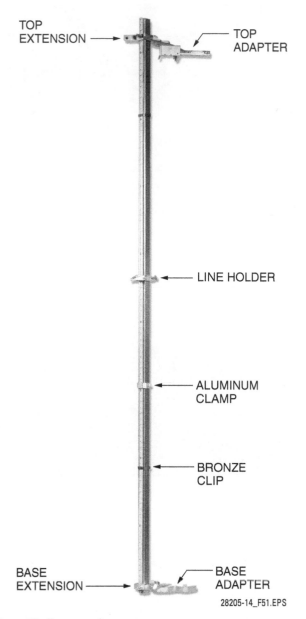

Figure 51 Corner pole.

28205-14_F52.EPS

Figure 52 Corner block attached to a corner pole.

corner pole, use telescopic braces to brace the post. Corner poles used for placing brick-veneer corners are fitted with adjustable brackets that allow you to secure it to the structure. Follow the manufacturer's instructions for securing the pole in position and ensuring that it is plumb. Then use a modular spacing rule to begin marking off course heights on the corner post and begin stringing line using corner blocks (see *Figure 52*). If the corner pole is fitted with line holders, use them to secure the lines to mark the block or brick course.

To use a level to construct a masonry corner, begin by setting up a layout line according to the floor or foundation plan. Next, spread a 16-inch-long mortar bed and place two masonry units in a corner position along the layout line. Measure the height of the brick using a modular spacing rule (*Figure 53A*). Place the level across the brick center line and tap the units level using the handle of a trowel (*Figure 53B*). Then turn the level vertically at the plumb corner and use it as a straightedge to tap the brick into vertical alignment (*Figure 53C*). Repeat this process for additional brick added along the mortar bed.

Lay the second course using the same technique, but reversing the direction to reflect the bond pattern (see *Figure 54*). Tool the joints as you proceed, using the techniques you have already learned for laying block and brick in a wall. Continue this process until reaching the desired number of courses (see *Figure 55*).

A

Figure 54 Laying the second course using a level.

B

Figure 55 A completed corner.

C

Figure 53 Using a level to place corner block.

Additional Resources

ACI 530/ASCE 5/TMS 402, Building Code Requirements for Masonry Structures, Latest edition. Reston, VA: American Society of Civil Engineers.

CM260A, Concrete Masonry Shapes and Sizes Directory. 1997. Herndon, VA: National Concrete Masonry Association.

Reinforced Concrete Masonry Inspector's Handbook, Fourth edition. 2002. Torrance, CA: Masonry Institute of America.

TEK 5-9A, Concrete Masonry Corner Details. 2004. Herndon, VA: National Concrete Masonry Association. **http://www.ncma.org**

3.0.0 Section Review

1. Because it increases the possibility of poor jointing, a method of temporarily ending a wall that is generally *not* permitted is _____.

   a. stepping
   b. cornering
   c. corbeling
   d. toothing

2. When building corbeling for a wall, the course that should be laid as a header course with full mortar joints is the _____.

   a. ninth course
   b. seventh course
   c. fifth course
   d. third course

3. All tie-bar shapes should extend into the holding wall at least _____.

   a. 8 inches
   b. 10 inches
   c. 12 inches
   d. 14 inches

4. A method for transferring stresses from one wall to an intersecting wall that is not specified in *Building Code Requirements for Masonry Structures* is to use _____.

   a. steel connectors
   b. a running bond
   c. bond beams
   d. a stack bond

SUMMARY

Masonry wall systems, control and expansion joints, and corners and intersections are important parts of many structures. Wall systems may be loadbearing or nonbearing. The five basic types of masonry wall systems are solid, hollow, cavity, composite, and veneered. Each type has specific construction techniques that must be followed. Proper bonding and reinforcement must be applied.

Solid masonry walls can be built as single-wythe or multiwythe structures using uncored masonry units. Hollow masonry walls use masonry units that have a void over 25 percent or more of their surface area. These walls may also be laid up as single-wythe or multiwythe walls. The difference between solid- and hollow-wall construction is the type of masonry units used and the reinforcing technique.

Cavity walls have an open space between two wythes. Each wythe is constructed as a separate structure and tied together with metal ties along the inner face of each wall and with a coping at the top. Composite walls are constructed of two different types of masonry units solidly tied together with mortar, grout, and metal ties. Both wythes work together in supporting any applied load. Veneer walls have a masonry exterior surface that is nonbearing. It is tied to an interior wall with metal ties and serves as a protection for the interior wall.

Retaining walls are constructed to hold back earth embankments and water. Various designs include gravity, counterfort, buttress, and cantilever. A cantilever wall must rely on the force of the earth against the footing, as well as its own strength, to resist this pressure. Reinforcement using either steel bars or grout or both is typically used in retaining walls of hollow block or multi-wythe structures. All retaining walls must be tied into a good footing.

Freestanding walls include many designs and patterns. A straight wall can be constructed from material such as brick and concrete masonry units, clay tile, or stone. All freestanding walls must be designed to withstand certain specified minimum wind loads. Other types of freestanding walls include screens, pier and panel, and serpentine.

The proper placement and construction of joints in masonry walls limits the effective length of the structure and reduces the tendency to crack. To help control cracks, masons use joint reinforcement, bond beams, reinforced masonry, and flexible joints. Expansion joints are among the most effective means of preventing damage due to expansion. The joint must be flexible and weather resistant. Expansion joints can be constructed in a number of ways and from a variety of materials.

Corners and intersections are designed and built to distribute loads evenly to ensure that masonry walls are able to resist external stresses and building loads. Because toothing increases the possibility of poor jointing and has a tendency to result in poor wall alignment, it is typically not normally permitted. Corbeling, which is a process for widening a wall by projecting out masonry units to form a ledge or shelf, should not extend beyond the course below it more than one-third the unit width or one-half the unit height, whichever is less.

Corners require special care because they will affect how the rest of the wall is constructed. Corners must be designed to withstand stresses caused by structural loads. A running bond provides sufficient interlock between masonry units to transfer shear stresses to interlocking walls.

1. To completely bond two wythes of masonry in a multiwythe wall, fill the collar joint with _____.
   a. foamed-in-place insulation
   b. reinforcing bars
   c. mortar
   d. rubble

2. The typical horizontal spacing between steel anchors used to bond wythes of masonry is typically _____.
   a. 36 inches
   b. 32 inches
   c. 28 inches
   d. 24 inches

3. Improved insulating value, reduced rain penetration, and prevention of condensation on interior walls are the main advantages of the _____.
   a. cavity wall
   b. solid masonry
   c. hollow masonry wall
   d. partition wall

4. Flashing is installed at the base of a wall cavity to _____.
   a. direct moisture to weepholes
   b. bond base courses together
   c. improve air circulation
   d. prevent entry by insect pests

5. Weepholes should be provided on center in head joints immediately above flashing sheets every _____.
   a. 5 to 7 inches
   b. 7 to 14 inches
   c. 16 to 32 inches
   d. 24 to 48 inches

6. Composite wall systems are effective in preventing rain penetration because they are built with a(n) _____.
   a. continuous masonry header
   b. integral moisture barrier
   c. layer of flashing beneath the top course
   d. continuous collar joint

7. Continuous joint reinforcement is manufactured in _____.
   a. 50-foot coils
   b. 10-foot lengths
   c. 5-foot lengths
   d. 3-foot sections

8. A height of 30 feet is the maximum allowed by code prior to a horizontal expansion joint for an anchored veneer attached to a backing of _____.
   a. hollow masonry
   b. cold-formed steel framing
   c. wood framing
   d. solid masonry

9. A retaining wall that depends primarily upon its own weight to hold back the soil pressure is a _____.
   a. gravity wall
   b. counterfort wall
   c. cantilever wall
   d. buttress wall

10. Seepage of water through a retaining wall can cause an aesthetic condition known as _____.
    a. blooming
    b. efflorescence
    c. inflorescence
    d. scale

11. The rule of thumb for building a serpentine wall is that the height should be no greater than the width of the structural unit times _____.
    a. 10
    b. 12
    c. 15
    d. 18

12. Limiting the size or locations of cracks in block masonry structures is the reason for using _____.
    a. continuous reinforcement
    b. collar joints
    c. adjustable wall ties
    d. control joints

13. Control joints are designed to allow longitudinal movement of up to _____.
    a. ⅛ inch
    b. ³⁄₁₆ inch
    c. ¼ inch
    d. ⁵⁄₁₆ inch

14. When installing a typical control joint, partially rake the head joint and replace it with _____.
    a. a silicone rod
    b. elastic sealing compound
    c. a Z-shaped metal filler
    d. rubber gasket

15. The type of expansion-joint material made from short pieces of material lapped to form a continuous weather seal is _____.
    a. copper
    b. extruded plastic
    c. premolded plastic
    d. neoprene

16. Toothing should only be used in special circumstances, since it can result in _____.
    a. additional expense
    b. wall collapse
    c. poor wall alignment
    d. wasted material

17. A common projection for each course of corbeling with standard brick is _____.
    a. ½ inch
    b. ⅝ inch
    b. ⅞ inch
    d. 1⅛ inch

18. A flexible joint must allow slight movement while providing _____.
    a. longitudinal support
    b. compressive strength
    c. torsional stability
    d. lateral support

19. When using a corner pole to position masonry units in a corner position, the length of the initial mortar bed should be _____.
    a. 8 inches
    b. 16 inches
    c. 24 inches
    d. 32 inches

20. When exposed to water or very humid air, brick _____.
    a. contracts slowly
    b. expands slowly
    c. contracts quickly
    d. expands quickly

Trade Terms Quiz

Fill in the blank with the correct term that you learned from your study of this module.

1. The construction of a temporary end of a wall with the end stretcher of every alternate course projecting is called _____.

2. The materials or masonry units used to form a cap or finish on top of a wall, pier, chimney, or pilaster to protect the masonry below from water penetration are called _____.

3. A(n) _____ is a masonry unit laid on top of a finished wall.

4. A wall made of segmental block stacked on top of each other without mortar bonding is called a(n) _____.

5. A design based on the application of physical limitations learned from experience or observations gained through experience, without structural analysis, is said to be _____.

6. A section of masonry that is adjusted gradually in order to correct alignment issues is said to be _____.

7. A(n) _____ is a type of metallic tie that is similar to the shape of a straight wooden pencil.

8. Starting a course with a cut so as to center the header unit over the head joints in the course below is called _____.

Trade Terms

Breaking the bond
Cap
Coping
Empirically designed

Humored
Pencil rod
Segmental retaining wall (SRW)
Toothing

Trade Terms Introduced in This Module

Breaking the bond: Starting a course with a cut so as to center the header unit over the head joints in the course below.

Cap: Masonry units laid on top of a finished wall.

Coping: The materials or masonry units used to form a cap or finish on top of a wall, pier, chimney, or pilaster to protect the masonry below from water penetration. Coping is usually projected from both sides of the wall to provide a protective covering as well as an ornamental design.

Empirically designed: Design based on the application of physical limitations learned from experience or on observations gained through experience, but not based on structural analysis.

Humored: A slang term for gradually adjusting a section of masonry in order to correct alignment issues.

Pencil rod: A type of metallic tie that is similar to the shape of a straight wooden pencil; used for control joints in concrete masonry construction.

Segmental retaining wall (SRW): A wall made of segmental block stacked on top of each other without mortar bonding.

Toothing: Construction of a temporary end of a wall with the end stretcher of every alternate course projecting. The projecting units are called toothers.

Additional Resources

This module presents thorough resources for task training. The following resource material is suggested for further study.

ACI 530/ASCE 5/TMS 402, Building Code Requirements for Masonry Structures, Latest Edition. Reston, VA: American Society of Civil Engineers.

ACI 530.1/ASCE 6/TMS 602, Specifications for Masonry Structures, Latest Edition. Reston, VA: American Society of Civil Engineers.

ASTM C90, Standard Specification for Loadbearing Concrete Masonry Units, Latest Edition. West Conshohocken, PA: ASTM International.

ASTM C216, Standard Specification for Facing Brick (Solid Masonry Units Made from Clay or Shale), Latest edition. West Conshohocken, PA: ASTM International.

ASTM C530, Standard Specification for Structural Clay Nonloadbearing Screen Tile, Latest Edition. West Conshohocken, PA: ASTM International.

ASTM C652, Standard Specification for Hollow Brick (Hollow Masonry Units Made From Clay or Shale), Latest Edition. West Conshohocken, PA: ASTM International.

Bricklaying: Brick and Block Masonry. 1988. Brick Industry Association. Orlando, FL: Harcourt Brace & Company.

CM260A, Concrete Masonry Shapes and Sizes Directory. 1997. Herndon, VA: National Concrete Masonry Association.

Concrete Masonry Handbook, Fifth edition. W. C. Panerese, S. K. Kosmatka, and F. A. Randall, Jr. Skokie, IL: Portland Cement Association.

Reinforced Concrete Masonry Inspector's Handbook, Fourth edition. 2002. Torrance, CA: Masonry Institute of America.

Technical Note 29A, *Brick in Landscape Architecture – Garden Walls*. 1999. Reston, VA: The Brick Industry Association. **www.gobrick.com**

TEK 5-9A, Concrete Masonry Corner Details. 2004. Herndon, VA: National Concrete Masonry Association.

Figure Credits

Section Review Answers

Answer	Section Reference	Objective
Section One		
1. d	1.1.0	1a
2. a	1.2.0	1b
3. c	1.3.0	1c
4. b	1.4.0	1d
5. b	1.5.1	1e
6. a	1.6.0	1f
7. d	1.7.0	1g
Section Two		
1. a	2.1.0	2a
2. b	2.2.1	2b
3. d	2.3.2	2c
Section Three		
1. d	3.1.0	3a
2. b	3.2.0	3b
3. a	3.3.0	3c
4. d	3.4.0	3d

NCCER CURRICULA — USER UPDATE

NCCER makes every effort to keep its textbooks up-to-date and free of technical errors. We appreciate your help in this process. If you find an error, a typographical mistake, or an inaccuracy in NCCER's curricula, please fill out this form (or a photocopy), or complete the online form at **www.nccer.org/olf**. Be sure to include the exact module ID number, page number, a detailed description, and your recommended correction. Your input will be brought to the attention of the Authoring Team. Thank you for your assistance.

Instructors – If you have an idea for improving this textbook, or have found that additional materials were necessary to teach this module effectively, please let us know so that we may present your suggestions to the Authoring Team.

NCCER Product Development and Revision

13614 Progress Blvd., Alachua, FL 32615

Email: curriculum@nccer.org
Online: www.nccer.org/olf

❏ Trainee Guide   ❏ Lesson Plans   ❏ Exam   ❏ PowerPoints   Other _____

Craft / Level: _____   Copyright Date: _____

Module ID Number / Title: _____

Section Number(s): _____

Description: _____

_____

_____

_____

_____

Recommended Correction: _____

_____

_____

_____

_____

Your Name: _____

Address: _____

_____

Email: _____   Phone: _____

28206-14

Effects of Climate on Masonry

There is more to masonry than just laying block and brick to create horizontal and vertical surfaces. Masons must understand the ways that heat transfer and moisture absorption can affect masonry elements and the soundness of masonry structures, and must also take into account the effects of weather. This module describes techniques used to apply insulation, to eliminate problems caused by moisture, and to protect masonry construction, materials, and equipment during hot and cold weather. These skills are just as essential to good masonry as knowing how to lay block and brick.

Module Six

Trainees with successful module completions may be eligible for credentialing through NCCER's National Registry. To learn more, go to **www.nccer.org** or contact us at **1.888.622.3720**. Our website has information on the latest product releases and training, as well as online versions of our *Cornerstone* magazine and Pearson's product catalog.

Your feedback is welcome. You may email your comments to **curriculum@nccer.org**, send general comments and inquiries to **info@nccer.org**, or fill in the User Update form at the back of this module.

This information is general in nature and intended for training purposes only. Actual performance of activities described in this manual requires compliance with all applicable operating, service, maintenance, and safety procedures under the direction of qualified personnel. References in this manual to patented or proprietary devices do not constitute a recommendation of their use.

Objectives

When you have completed this module, you will be able to do the following:

1. Identify the various types of insulation used in conjunction with masonry construction, and explain installation techniques.
   a. Explain the concept of heat transfer.
   b. Explain the purpose of and installation procedures for internal insulation.
   c. Explain the purpose of and installation procedures for external insulation.
2. Identify the need for moisture control in various types of masonry construction, and describe the techniques used to eliminate moisture problems.
   a. Explain the purpose of and installation procedures for flashing.
   b. Explain the purpose of and installation procedures for weep vents.
   c. Explain the purpose of and installation procedures for waterproofing.
3. Explain the various techniques used to provide adequate protection during hot- and cold-weather masonry construction.
   a. Explain the role played by weather data and information in masonry construction.
   b. Explain the various techniques used to provide adequate protection during hot-weather masonry construction.
   c. Explain the various techniques used to provide adequate protection during cold-weather masonry construction.

Performance Task

Under the supervision of your instructor, you should be able to do the following:

1. Install a 4-foot section of base flashing.

Trade Terms

Chase	Thermal mass
Membrane	Waterproofing
Plasticity	Wick
Reveal	

Industry-Recognized Credentials

If you're training through an NCCER-accredited sponsor, you may be eligible for credentials from NCCER's Registry. The ID number for this module is 28206-14. Note that this module may have been used in other NCCER curricula and may apply to other level completions. Contact NCCER's Registry at 888.622.3720 or go to **www.nccer.org** for more information.

Code Note

Codes vary among jurisdictions. Because of the variations in code, consult the applicable code whenever regulations are in question. Referring to an incorrect set of codes can cause as much trouble as failing to reference codes altogether. Obtain, review, and familiarize yourself with your local adopted code.

Contents

Topics to be presented in this module include:

Figures and Tables

SECTION ONE

1.0.0 INSULATION IN MASONRY CONSTRUCTION

Objective

Identify the various types of insulation used in conjunction with masonry construction, and explain installation techniques.

a. Explain the concept of heat transfer.
b. Explain the purpose of and installation procedures for internal insulation.
c. Explain the purpose of and installation procedures for external insulation.

Trade Terms

Thermal mass: The ability of a building wall to alternately absorb and release heat energy in response to temperature changes.

For the purpose of masonry construction, the term *insulation* can be defined in two ways. Insulation is any material that reflects heat or simply provides a barrier that retards heat penetration. Insulation is also defined as any material that resists the flow of heat because of the physical properties of that material.

Generally, denser materials will be less effective in resisting heat transfer. For instance, lightweight block has a greater resistance to heat movement than normal block. Those insulation materials that best resist heat transfer have the physical property of containing millions of tiny air cells or spaces that resist the flow of heat by trapping a great deal of the heat as it moves through the material.

Insulation has become a major consideration in the design and construction of all buildings since the cost of energy began increasing dramatically in the early 1970s. Because of this energy awareness, the general public now insists that the construction industry provide insulation in almost every type of construction in all areas of the country. Federal and state government agencies have also developed recommendations and regulations concerning insulation.

While most masons will find that the building plans provide for the type of insulation to be used, you should recognize the basic physical properties of heat transfer and be able to recommend an insulation type if asked.

1.1.0 Understanding Heat Transfer

In both summer and winter, the comfort of any structure depends on a number of physical features of the building and the natural laws of heat transfer. Heat transfer is the movement of heat energy from a warmer to a colder area by conduction, radiation, or convection.

Heat moves through different building materials at different rates. *Table 1* shows several types of materials and their resistance to heat. Thermal resistance (or R-value) is a measure of the rate of heat transfer through a material. The higher the R-value, the better a material's insulating ability.

It is important to remember that during the course of a day, the indoor and outdoor temperatures fluctuate and can even reverse. This is especially true for locales that experience extreme temperature changes between daytime to nighttime, such as in the desert. As the temperature difference changes, so does the direction of the heat flow. Depending on what materials are used in a wall, the flow of heat from the outside of a building to the inside could be greater than the heat flow from inside to outside, or vice versa. This is commonly referred to as thermal mass. As a result, the wall's thermal performance will differ from the steady-state R-value assigned to the materials used in the wall.

Table 1 Heat Transfer Coefficients

Material or Product	Resistance per Inch of Thickness	Resistance per Thickness Shown
Concrete (per 1" thickness)	0.08	0.08
Face brick (per 1" thickness)	0.11	0.11
Hollow concrete block, 8"	—	1.11
Stucco (per 1")	0.20	0.20
Metal lath and plaster, ¾"	0.17	0.13
Gypsum board, ¾"	0.90	0.45
Plywood, ½"	1.32	0.62
Pine, fir, other softwoods (per 1")	1.25	1.25
Oak, maple, other hardwoods (per 1")	0.91	0.91
Asphalt shingles	—	0.44
Built-up roofing, ⅜"	—	0.33
Wood shingles	—	0.94

1.1.1 Conduction

Conduction is the passage of heat or electricity through a medium. Some materials transmit heat more quickly than others and are said to have a higher conductivity. The degree to which any building material resists the passage of heat is called its resistance to conductivity or conductance.

Heat travels or flows from warm areas to colder areas. In the winter, when outside temperatures are lower than indoor temperatures, the heat from the room flows through the walls toward the colder outside areas (*Figure 1*). The wall temperature itself is lower than the room air temperature. According to the law of conductivity, the heated air moves first through the wall and on to mix with the outside air. In the summer, the process and the direction of heat flow reverse.

Generally, dense building materials conduct heat more quickly and are called conductors. For example, you can easily feel how quickly metals transfer heat. On the other hand, less dense materials, such as concrete masonry units, wood, and plastics, are often called insulators since they prevent or slow the flow of heat. A wooden match, for example, does not become too hot to hold after it is lit.

1.1.2 Radiation

The surface of any object, building material, or human body transfers heat to any nearby colder surface. Heat transfer by radiation is by invisible light and heat waves. For example, the sun heats the earth by radiation. A radiator heats a room by radiant heat and not by blowing warm air. The warmer wall loses heat to the colder outside air by the same law of radiation.

1.1.3 Convection

Heat moves from one area to another by a process called convection. Hot air expands and rises as cool air falls, creating convection loops, or constant flows of air, that heat and cool an area as shown in *Figure 2*. In fact, most heating and cooling systems directly depend on convection.

Heat transfer is typically a combination of conduction, radiation, and/or convection. For example, the air that comes into contact with a heat source is heated by conduction and rises along a wall. As heat from this air is lost to the walls, ceiling, and window areas by conduction and radiation, it becomes cooler and heavier. When the cooler air falls, it again comes in contact with the heat source and repeats the convection loop.

The walls of a structure are directly responsible for the loss of heat in the winter and for the loss of cooler air in the summer. You need to be aware of construction materials that will best resist the flow of heat through walls and use additional material as insulation to impede the flow of heat through the walls.

The type of insulation material used should be determined by a number of factors:

- The resistance or R-value of the various insulation types

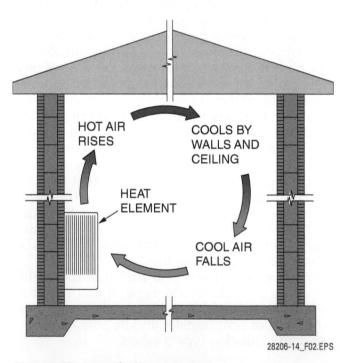

28206-14_F01.EPS

Figure 1 Heat transfer by conduction.

28206-14_F02.EPS

Figure 2 Heat transfer by convection.

- The cost of materials, application, and maintenance
- The point during construction when the insulation will be installed
- The geographical location of the structure, and the changing climates and temperatures
- Code specifications

There are six basic categories of insulation and several types of insulation within each category. Again, the category and type of insulation used should be determined by cost, R-value, and time or position of application. These are the basic types of insulation:

- Loose fill
- Rigid, or slab
- Reflective
- Foamed-in-place
- Insulation inserts
- Flexible blanket or batt

Most building materials suppliers can also advise you on the proper R-value for the region in which the structure is built. You should be aware that the R-value differs according to the type and thickness of the insulation being used. *Table 2* gives values for several different types of insulation.

For masonry work, these general types of insulation are classified by their position in a structure. When insulation is placed in individual cores or in cavities between wythes, it is often referred to as internal insulation. When the insulation material is applied to a wall surface, the insulation is often called external insulation.

1.2.0 Installing Internal Insulation

Insulation placed between the interior and exterior faces of the masonry work is known as internal insulation. This may be loose-fill insulation placed in block cores or cavity walls, insulation inserts in the cores of concrete masonry units, or rigid insulation placed between the masonry wythes in a cavity wall.

1.2.1 Loose-Fill Insulation

Loose-fill insulation is manufactured as fibers, granules, or chips and is delivered by the bag or in bulk. These loose fibers can be poured, placed by hand, or blown into cavities or cores. In choosing a loose-fill insulation, make certain that the desired properties of the material are met.

In most situations, loose-fill insulation must meet at least five specific criteria:

- It must not transmit moisture from the exterior wythe to the interior wythe.

Table 2 R-values for Different Insulation Types

Product	R per 1" Thickness
LOOSE FILL	
Mineral fiber (rock, slag, or glass)	2.20–3.00
Cellulose	3.70
Perlite	2.70
Vermiculite	2.13
FLEXIBLE	
Mineral wool batts	3.10–3.70
RIGID	
Cellular glass	2.63
Expanded polystyrene (extruded)	5.00
Expanded polystyrene (molded)	3.57
Expanded polyurethane	6.25
Mineral fiberboard	3.45
Polyisocyanurate	7.20
REFLECTIVE	
Aluminum foil	3.48
FOAMED-IN-PLACE	
Urea formaldehyde	4.20
Polyurethane	6.25

- It must not absorb water or prevent water from escaping the cavity.
- It must be strong enough to support its own weight. In other words, it should not settle or pack down over a period of time. Such settling eliminates the dead airspaces between particles that give the material much of its insulating qualities.
- Loose-fill insulation must be inorganic or treated organic material. Organic material should be treated to resist fire, decay, insects, and rodents.
- It must be easy enough to handle so that it can be poured into wall cavities in lifts of about 4 feet high.

Two of the more popular loose-fill insulation materials are water-repellent vermiculite and silicone-treated perlite. Both vermiculite and perlite meet all five of the necessary properties for loose-fill insulations. Vermiculite is a form of mica that is expanded and is resistant to fire, decay, moisture, and vermin. Perlite is a volcanic glass that is also expanded by a heating process to create a lightweight insulation with all the required properties. Both of these loose-fill insulations resist packing when they are applied in 4-foot lifts.

Loose-fill insulation can readily be placed in block cores, as shown in *Figure 3*, and in the cavity between masonry wythes. The insulating value will depend on the amount of insulating material that is used, and this can be substantial. For example, a normal-weight 8-inch block has an R-value of about 1.11; a lightweight 8-inch block has an R-value of about 2.18; and an 8-inch lightweight block with loose-fill insulation in the cores has an R-value of about 5.03. The trade-off is between the initial cost of the block and the amount of savings that can be realized due to lower heating/cooling costs.

1.2.2 Rigid or Slab Insulation

Most masons prefer to use preformed rigid nonstructural insulation board rather than loose-fill insulation when they build cavity walls. These forms of insulation meet the requirements of loose-fill insulation, but they also allow you to install the necessary flashing and weep vents at the base of the cavity wall.

Rigid-board insulation is manufactured from inorganic or synthetic materials. It is made in various sheet sizes, normally from ½- to 1-inch thick. Slab insulation, like the rigid-board insulation, is also a synthetic product and is found in 1- to 3-inch thicknesses. *Figure 4* shows rigid foam board.

The most common types of rigid, nonstructural insulation include the following:

- Rigid urethane foam
- Expanded, molded polystyrene foam
- Polyisocyanurate foam
- Expanded perlite
- Expanded polyurethane
- Cellular glass fiberboard
- Preformed fiberglass

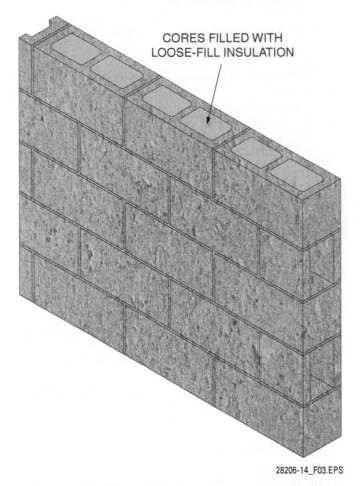

CORES FILLED WITH LOOSE-FILL INSULATION

28206-14_F03.EPS

Figure 3 Loose-fill insulation in concrete masonry walls.

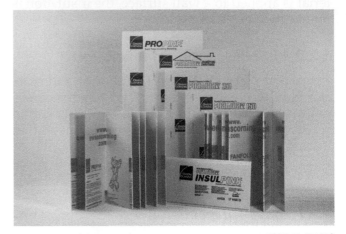

28206-14_F04.EPS

Figure 4 Rigid foam board.

R-value

The actual R-value of an insulation type is a number calculated from laboratory experiments. It is a measure of the overall thermal resistance of the material. The R-value is usually supplied by the insulation manufacturer. It represents the amount of resistance to heat flow between the air on the warm side and the air on the cold side of the building section.

When these rigid boards are used as an internal insulation, they are fastened to the outside face of the inner wythe by applying mastic to the block or by using metal ties to clip the board in place. In this way, the cavity still has an airspace left between the outer wythe and the insulation. This airspace ventilates the cavity. *Figure 5* shows rigid board installed on an interior block wall.

Urethane foam is a plastic noted for its resistance to heat transfer as well as its durability. However, urethane will expand when damp for long periods of time and will burn if exposed to flames. Polyisocyanurate will also burn but has the highest R-value of any of the rigid-board insulations. Perlite board has one of the lowest R-values and must be kept dry because it absorbs water. By far, the most popular rigid-board materials used in cavity wall applications are the polystyrene foam boards, the cellular glass fiberboard, and the preformed fiberglass boards.

Polystyrene boards are formed by molding or by extrusion. The molded type is often called beadboard because it is made from small, plastic beads forced or molded together into a rigid board. Extruded polystyrene is more rigid than the molded boards and has a higher R-value. Both types are resistant to moisture but give off carbon monoxide gas when they burn. Cellular glass fiberboard and fiberglass board insulation are even more resistant to water vapor and moisture absorption. Cellular glass fiberboard is available in thicknesses between 1½ and 4 inches. Fiberglass boards are usually delivered in 2 foot × 4 foot × 1 inch boards that can be cut to size at the job.

1.2.3 Reflective Insulation

Reflective insulation is often used on the inside of heated spaces to reduce the amount of heat moving through the masonry walls. This type of insulation is made from one or more layers of metal foil. The degree of insulation that reflective foil gives depends on the length of time it remains shiny and on the proper use of an airspace. To be effective, the foil must face an airspace in which the heat can be reflected. Reflective insulations include sheets of paper-backed aluminum foil in blankets and foil-backed gypsum wallboard.

1.2.4 Foamed-in-Place Insulation

Foamed-in-place insulation can be used in the same situations where batts are used, but is most often used for reinsulating existing buildings. These insulating products are made in a liquid foam or in expanding pellets. When they are poured or foamed-in-place, they expand to fill a cavity. After curing, they form a solid mass. When masonry work is involved, these foams are usually placed into the core of the block on the interior face of the wall (*Figure 6*). Urethane foam is particularly good where a high R-value is needed in a tight spot, but it will expand after it has been injected. Once the extruded foam has set, clean off the excess and fill in the holes.

1.2.5 Insulated Block

As you learned in *Masonry Level One*, insulated block uses foam insulation inserts to provide better energy efficiency and soundproofing than standard block (*Figure 7*). The block is designed to provide a continuous and uninterrupted thermal barrier that absorbs and stores energy, allowing buildings to stay cooler in warm weather and warmer in cool weather. This allows designers to install smaller HVAC (heating, ventilating, and

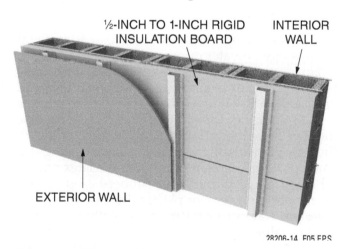

½-INCH TO 1-INCH RIGID INSULATION BOARD

INTERIOR WALL

EXTERIOR WALL

28206-14_F05.EPS

Figure 5 Installing rigid insulation board.

Select the Best Type of Insulation

Two kinds of loose-fill insulation are available but should not be used in masonry construction. These types fail to fulfill all of the desired physical properties of insulation. Mineral-fiber and cellulose-fiber insulations are very good insulators, but should not be used in confined areas susceptible to moisture.

Mineral-fiber loose fill made from rock, slag, or glass wool easily compresses and often clumps on mortar joints or burrs. Cellulose-fiber loose fill made from newsprint, wood chips, or other organic materials also has a tendency to compress, and is not as resistant to moisture and fire as vermiculite and perlite.

Figure 6 Foamed-in-place insulation.

air conditioning) systems, which cost less to operate over time. It is available in standard dimensions in a variety of colors and finishes, and can be installed like conventional block.

Because additional wall insulation is not required, the use of insulated block can reduce construction time. The foam inserts are nontoxic and moisture resistant. Insulated block offers the advantage of allowing the insulation to be placed at the same time as the masonry unit along with flashing or joint reinforcement. Care must be taken to avoid getting mortar droppings in the block cells.

1.3.0 Installing External Insulation

External insulation is any type of insulation that is placed on the external face of the masonry units. The application of external insulation is often not the task of the mason. However, you need to understand the basic application process and the value of these insulation types. The basic types of external insulation include the following:

- The addition of a coating on the exterior face of the masonry
- The addition of rigid-board insulation on the exterior surface of the masonry
- The use of blanket or batt insulation between furring strips

1.3.1 Finishes as Insulating Agents

The insulation created by surface finishes is quite low, but many of these coatings are used as a protective coating for insulation materials. These surface finishes are primarily used to change the color or texture of the project, to enhance the

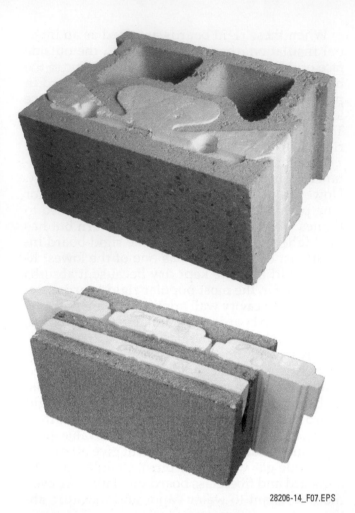

Figure 7 Common types of insulated block.

sound-insulating ability of a wall, or to prevent the passage of water or moisture.

However, the heat resistance of such a covering is quite low. For example, an exterior finish will only add an R-value of about 0.17 to 0.68, while a ½-inch-thick rigid insulation board will increase the R-value of the wall by 1.31 to 3.6, depending on the type of insulating material. Surface coatings include portland cement paints, acrylic latex paints, oil or oil alkyd paints, and rubber-based paints; stains; portland cement plaster finishes; clear coatings; and waterproof bituminous coatings.

The value of paint as an insulation material directly relates to the amount of moisture the paint will allow to pass through a wall. A type of paint that does not allow the passage of water vapor should be applied to the side of a wall where moisture enters. The rubber-based paints are the most moisture resistant and are often used as primers for the less resistant paints. Clear coatings and stains are primarily used to increase the coloration of the wall or to increase the water and dirt resistance. Portland cement mixed as stucco plaster is the best coating application for its insulating value. Portland cement plaster forms a hard, durable, and decorative finish.

Plaster application can be done in two or three coats, by hand or by machine. Two coats are usually enough when applied to masonry units. These two coats usually consist of a ⅜-inch, rough-floated base coat and a ¼-inch finish coat, as shown in *Figure 8*. Three coats are required if the masonry work is not very smooth or if metal lath is used as a base for the plaster. In both cases, the finish should be damp-cured.

1.3.2 Rigid-Board Insulation

Rigid panel or board insulation may be applied directly to the exterior face of a masonry wall using mastic. It may also be applied over furring strips attached to the interior face. Insulation boards are normally used as a base for plaster, bituminous coatings, or gypsum board, but manufacturers also make decorative rigid insulation board that can be used as a finished surface. The most popular of the rigid-board insulations are rigid glass fiberboard, polystyrene panels, and polyurethane boards or sheets.

1.3.3 Blanket or Batt Insulation

Blanket and batt insulation (*Figures 9* and *10*) are considered flexible insulation. It is usually applied between furring strips anchored directly to the interior of a masonry wall or between framing members on the interior of the wall. These flexible insulation types are usually made of mineral wool or cellulose fibers that resist fire, moisture, rodents, and insects. Batts are usually 48 inches long and either 16 or 24 inches wide to fit most standard framing patterns. Blanket insulation is usually made in longer rolls that can be cut to size.

Flexible insulations come in three specific varieties:

- *Wrapped* – A tape binding on one or both edges and a vapor barrier on one or both sides
- *Faced* – A vapor barrier on only one side
- *Friction fit* – Without a covering, but designed to fit tightly in the frame or furring space

Faced and wrapped insulations usually have a stapling or tacking flange for installation. Builders should make certain the vapor barrier is placed toward the living space and that the taped edge of wrapped insulation overlaps on the face of the furring strips. If the insulation does not have a vapor barrier, the builder will often place continuous polyethylene sheeting over the inside stud faces.

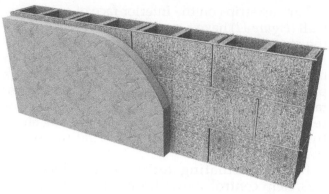

28206-14_F08.EPS

Figure 8 Block coatings.

28206-14_F09.EPS

Figure 9 Fiberglass blanket insulation.

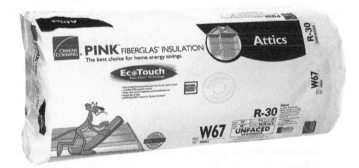

28206-14_F10.EPS

Figure 10 Fiberglass batt insulation.

When placing the furring strips, the safety precautions necessary for attaching materials to masonry structures must be followed. The furring strips should be placed on 16 or 24 inches on-center spacing. They can be attached by anchor bolts using a masonry drill or by specially hardened nails using a nail gun or sledge.

Furring strips on the interior face of a masonry wall (*Figure 11*) have a number of advantages from the standpoint of insulating efficiency. First, the furring strip adds an airspace of at least ¾ inch between the masonry and the wall finish. This provides improved heat and sound transfer resistance. Rigid insulating boards, batt insulation, or reflective films placed between the furring strips further increase the insulating qualities of the wall.

Proper insulating techniques reduce heat loss, help control sound transmission from area to area, and help to maintain energy efficiency. However, the mason and builder must consider cost savings as well. The amount of savings in energy is not directly proportional to the thickness of the insulation or the kinds of insulation used. For instance, the use of two types of insulation or two layers reduces heat transfer, but the savings may not be doubled. Normally, the use of lightweight block and any one of the more popular insulation types will be the best insulating solution. You need to recognize the most efficient means of creating a well-insulated structure and be ready to recommend the appropriate applications.

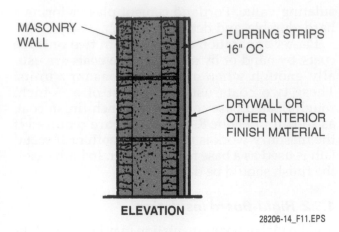

Figure 11 Use of furring strips.

Test and Prepare Masonry before Plastering

The insulating value of plaster is due largely to the bond achieved between the plaster coating and the masonry. The surface preparation of the block is very important in achieving the proper bond. For new construction, the wall face must be free from oil and dirt, and must be dampened before the plaster is applied. Older concrete masonry surfaces should be tested before the plaster is applied.

By spraying the surface with water, you can see how well the water is absorbed. If the water is readily absorbed, the masonry can be dampened and an adequate bond will be achieved. If the water beads up or runs off, sandblasting or other surface preparations may be required before plastering.

Fiberglass Insulation Safety

Flexible fiberglass is probably the first thing that comes to mind when the average person thinks of insulation. Most of us don't realize that this common material must be handled carefully. The tiny strands of glass in fiberglass insulation can irritate the skin, injure eyes, and cause a variety of respiratory problems. Always wear proper personal protective equipment when handling fiberglass insulation.

However, fiberglass safety is more than just personal protection during installation. If debris from the installation is not properly removed or if existing insulation is disturbed, fiberglass particles could spread through the building. Fiberglass that enters the HVAC system will be carried to all parts of the building. Always use care when handling fiberglass insulation. This protects you and the building's current and future occupants.

Additional Resources

Alex Wilson. "Thermal Mass and R-value: Making Sense of a Confusing Issue." *Environmental Building News*, April 1998. **www.buildinggreen.com**

TEK 6-11A, Insulating Concrete Masonry Walls. 2010. Herndon, VA: National Concrete Masonry Association. **www.ncma.org**

1.0.0 Section Review

1. The passage of heat or electricity through a medium is called _____.

   a. convection
   b. conduction
   c. radiation
   d. resonance

2. When insulating cavity walls, masons typically prefer to use _____.

   a. reflective insulation
   b. loose-fill insulation
   c. rigid-board insulation
   d. foamed-in-place insulation

3. An example of a type of insulation that is *not* classified as a flexible insulation is _____.

   a. loose fill
   b. wrapped
   c. faced
   d. friction fit

2.0.0 MOISTURE CONTROL IN MASONRY CONSTRUCTION

Objective

Identify the need for moisture control in various types of masonry construction, and describe the techniques used to eliminate moisture problems.

 a. Explain the purpose of and installation procedures for flashing.
 b. Explain the purpose of and installation procedures for weep vents.
 c. Explain the purpose of and installation procedures for waterproofing.

Performance Task

Install a 4-foot section of base flashing.

Trade Terms

Chase: A continuous recess built into a wall to receive pipes, wires, or heating ducts.

Membrane: A layer of thin, pliable material used to waterproof masonry.

Reveal: The visible portion of a masonry jamb between the face of the wall and the frame.

Waterproofing: The process of treating masonry with a material that will retard penetration of moisture.

Wick: A bundle of fibers that are loosely twisted or braided and woven together to form a cord that can carry water away from an area by capillary action, which occurs as long as the drip end is lower than the absorption end.

Moisture can have an adverse effect on masonry construction. It can cause cracks in mortar and masonry units due to repeated expansion and contraction as water is absorbed and then evaporates. Moisture can also corrode exposed metal elements such as rebar, bolts, and ties and anchors. Moisture that is rich in minerals can cause chemical staining of mortar and masonry units as it evaporates.

Weather is perhaps the most common means by which masonry elements encounter moisture. Throughout their life cycles, masonry structures are exposed to years of rain and snow, mist and fog, standing and running water, and freeze/thaw cycles. Masons design structures to withstand the damaging effects of humidity by ensuring that the proper design techniques and materials are used. In this section, you will learn how to use flashing, weep vents, and waterproofing to protect masonry structures against damage caused by moisture.

2.1.0 Installing Flashing

In masonry, flashing is the application of any material that prevents water penetration or provides water drainage. It is difficult to completely prevent rainwater from entering a wall at places like roof intersections, parapets, sills, or other projections. Flashing provides a degree of control over moisture movement and penetration within masonry walls. It is installed at these vulnerable locations to collect water that may enter the wall and then to divert it through weep vents to the exterior.

2.1.1 Flashing Materials

Suitable flashing materials must possess at least four important properties:

- Resistance to moisture penetration
- Resistance to corrosion caused either by exposure to the atmosphere or by the caustic alkalies present in mortars
- Toughness to resist puncture, abrasion, or other damage during installation
- Flexibility so that the flashing can be worked to the correct shape and retain this shape after installation

Flashings are usually formed from sheet metals, rubberized asphalt, plastics, or a combination of these materials. The selection of material is determined by cost and suitability, but suitability should never be compromised because of cost. Replacement costs always exceed original costs, so it is wise to select a permanent flashing material for the original installation rather than have to replace it.

The following are the most commonly used materials for flashing applications:

- *Copper* – Though more costly than most other flashing materials, copper is durable, available in special preformed shapes, and an excellent moisture barrier. It is not affected by the caustic alkalies present in masonry mortars, so it can be safely embedded in fresh mortar and will not deteriorate unless excessive chlorides are present. When using copper flashing, the use of chloride-based additives in mortar (such as calcium chloride to speed the set of mortar in cold weather) should be prohibited. One objection to copper flashing is that when exposed to weather, the copper wash may stain or discolor light-colored masonry surfaces. Where copper staining would be undesirable, coated copper can be used.

- *Galvanized steel coatings* – These materials can corrode in fresh mortar. When placed in contact with mortar, it is usually necessary to protect them with a rubberized asphalt coating. Galvanized steel is widely used in residential building and generally performs adequately. However, exposed galvanized steel flashing requires painting maintenance.

- *Stainless steel* – Stainless steel flashings are available in several gauges and finishes. They are durable, highly resistant to corrosion, excellent moisture barriers, and surprisingly workable. Various types of stainless steel are available, the varieties depending on the proportions of chromium and nickel in the steel. These varieties have differing degrees of resistance to corrosion and are priced accordingly. Stainless steel should be specified by number and not by the generic term alone. For instance, stainless steel flashing exposed to weather should be 28 gauge, while that completely concealed should be 30 gauge. In addition to other favorable properties, stainless steel resists rough handling; can be soldered, welded, or brazed; and will not stain adjacent areas. The main drawback of stainless steel is its higher cost.

- *Thermal plastics* – As a group, plastics are becoming one of the most widely used flashing materials. The better plastic flashings are tough, resilient materials that are highly resistant to corrosion. However, because the chemical compositions of plastics differ widely, it is impossible to lump all plastic flashings into one general group. Some plastics will not withstand the corrosive effects of masonry mortars. Rely on performance records of the material, the reputation of its manufacturer, and where possible, test data to ensure satisfactory performance of any one specific type of plastic.

- *Rubberized asphalt* – Rubberized asphalt is commonly used in residential and commercial construction for many different flashing applications. It is constructed using modified asphalt adhesives that nearly completely eliminate the "drool" of the bituminous fabrics that masons used in the past. Primers are typically recommended for use with rubberized asphalt, along with drip edges and termination bars.

- *Combination flashings* – Combination flashings consist of materials combined to gain the best properties of each, such as sheet metal coated with plastic or rubberized asphalt materials. The combination of materials can provide a lower-cost flashing by reducing the thickness of metal required, or it may permit the use of corrodible metals that might otherwise prove unsatisfactory unless coated.

Other materials can also be used as flashing. These include thermoplastic polyolefins (TPOs), ethylene-propylene-diene monomer (EPDM) synthetic rubber, liquid rubberized materials, and air barriers. Refer to the local applicable code and manufacturer's specifications to identify materials that are suitable for the particular application.

2.1.2 Flashing Placement

When the construction calls for cavity walls or for veneering over existing framework, the application of flashing and weep vents is a must. Since it is impossible to keep all moisture out of an open cavity, and since all entering moisture eventually travels downward, you must provide a way for the moisture to be channeled outside the wall. Cavity walls are far higher in weather resistance than are solid masonry walls if the proper techniques are applied. When the proper flashing is placed in the cavity, the cavity wall is highly water-resistant.

When the flashing is properly installed, it should provide a channel at the bottom of the cavity where moisture can collect and flow to a weep vent and out of the cavity. To properly bed the flashing, most masons apply a thin mortar coat at the bed joint, place the flashing, and finish the bed joint, making sure that the joint size remains the same for all joints. In this way, when the mortar sets, the flashing becomes a permanent part of the structure, and no open joint is left where water or moisture may drip past the flashing. The flashing is placed the full length of the cavity.

Exposure to rain varies greatly throughout the United States. For instance, exposure is severe along the Gulf Coast, the Atlantic Seaboard, and the Great Lakes; moderate through the Midwest; and slight in the far Southwest. The following recommendations apply to areas of severe or moderate exposure. Where exposure is slight, internal flashing, such as lintel or sill flashing, may be eliminated or reduced to a minimum.

Where flashing extends to the interior, place its end between furring and the interior finish and turn it up at least 1 inch to collect moisture that may penetrate the wall.

Any moisture that enters a wall gradually travels downward. Place flashing above grade at the wall base to divert this water to the exterior (*Figure 12*). In areas where first-floor wood joists require protection from termites, use metal flashing that projects at least 2 inches past the inside face of the wall and is bent down at an angle.

Place flashing under and behind all sills. The ends of sill flashing should extend beyond both sides of the opening and turn up at least 1 inch into the wall. In addition, sills should slope and project away from the wall to drain water away from the building. When the underside of a sill is not sloped, it should have a drip notch or the flashing should be extended and bent down to form a drip (*Figure 13*).

Install flashing over all openings. At steel lintels, the flashing should be placed under and

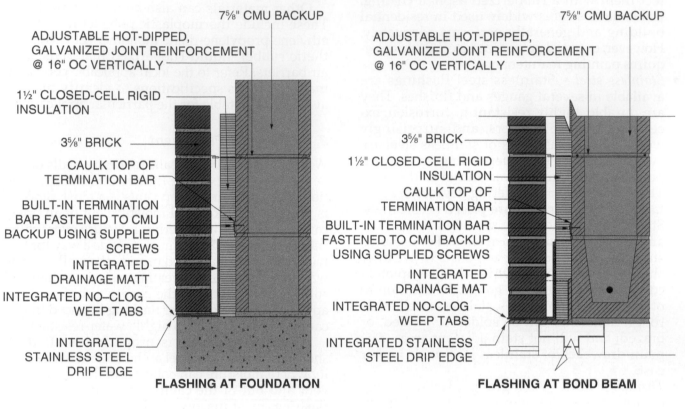

FLASHING AT FOUNDATION

FLASHING AT BOND BEAM

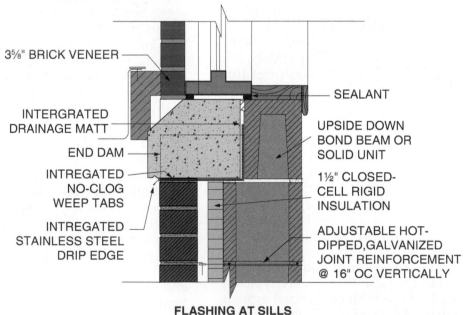

FLASHING AT SILLS

28206-14_F12.EPS

Figure 12 Flashing in a cavity wall.

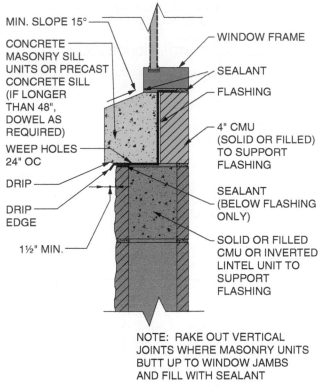

MIN. SLOPE 15°

CONCRETE
MASONRY SILL
UNITS OR PRECAST
CONCRETE SILL
(IF LONGER
THAN 48",
DOWEL AS
REQUIRED)

WEEP HOLES
24" OC

DRIP

DRIP
EDGE

1½" MIN.

WINDOW FRAME

SEALANT

FLASHING

4" CMU
(SOLID OR FILLED)
TO SUPPORT
FLASHING

SEALANT
(BELOW FLASHING
ONLY)

SOLID OR FILLED
CMU OR INVERTED
LINTEL UNIT TO
SUPPORT
FLASHING

NOTE: RAKE OUT VERTICAL
JOINTS WHERE MASONRY UNITS
BUTT UP TO WINDOW JAMBS
AND FILL WITH SEALANT

28206-14_F13.EPS

Figure 13 Flashing in a single-wythe windowsill.

behind the facing material with its outer edge bent down over the lintel to form a drip. *Figure 14* shows flashing installed in a composite wall over a window. Stepped flashing should be turned up on the inside and folded to form an end dam to protect the window from moisture.

Projections and recesses tend to hold rainwater and snow. They should have a top slope for drainage, and if possible, a drip to keep the water away from the wall surface below. Place the flashing over the top of projections and recesses, with the outer edge bent down to form a drip and the back edge turned up at the inner face of the wall.

Face both sides of parapet walls with the same durable masonry materials. Using an inferior material on the side not exposed to view is a frequent cause of trouble. Do not paint or coat the backs of parapet walls; they must be left free to dry rapidly.

If required, place flashing in the mortar bed beneath the coping. Where the coping provides an adequate drip, the flashing may stop at the wall surface. However, where the copings are flush with the surface of the wall, extend the flashing at least ¼ inch on both sides and turn it down to provide a drip (*Figure 15*).

Because perimeter flashing occurs at vulnerable points, it must be designed and installed with great care. Base-flashing design depends upon the type of roofing material. Where base flash-

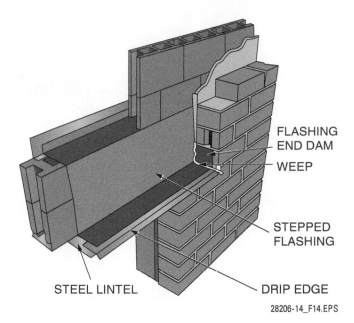

FLASHING
END DAM

WEEP

STEPPED
FLASHING

STEEL LINTEL

DRIP EDGE

28206-14_F14.EPS

Figure 14 Flashing at a window.

ing is metal, the counter flashing should also be metal, extending into the wall and overlapping the base flashing (*Figure 16*).

Flashing at chimney walls consists of base flashing placed under the roofing and turned up against the masonry. Counter flashing overlaps the base flashing and extends into the mortar joints, where it is securely embedded (*Figure 17*).

Flashing built on the back of a chimney will prevent the buildup of melting snow and ice. The flashing is held in place by a triangular wood structure called a cricket or a saddle. The triangular shape of the flashing allows water to flow around the chimney without pooling. The height of the flashing is typically half the width of the chimney, although local codes vary. The ridge is level and extends back to the roof slope.

The cricket consists of a horizontal ridge piece and a vertical piece at the back of the chimney, as shown in *Figure 18*. The support can be covered with heavy-gauge metal, which is nailed to the roof deck and sealed (refer to *Figure 17*). The support can also be covered with plywood and shingled.

2.1.3 Installation of Flashing

Masonry must be relatively smooth and free of projections that might puncture the flashing and destroy its effectiveness. Through-wall flashing should be placed on a thin bed of mortar with another thin bed laid on top to bond the next masonry course. Seams in the flashing must be completely bonded to prevent water penetration. Although most sheet-metal flashings can be soldered, lock-slip joints are required at intervals to permit thermal expansion and contraction. Many

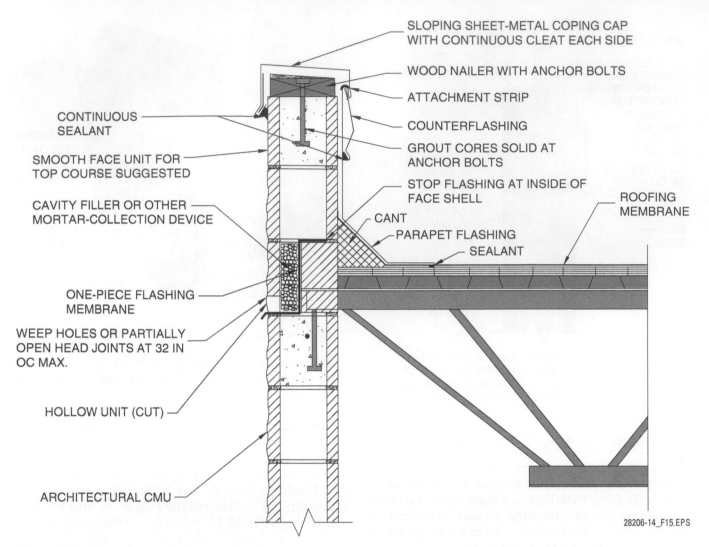

SLOPING SHEET-METAL COPING CAP
WITH CONTINUOUS CLEAT EACH SIDE

WOOD NAILER WITH ANCHOR BOLTS

ATTACHMENT STRIP

COUNTERFLASHING

GROUT CORES SOLID AT
ANCHOR BOLTS

STOP FLASHING AT INSIDE OF
FACE SHELL

ROOFING
MEMBRANE

CANT

PARAPET FLASHING

SEALANT

CONTINUOUS
SEALANT

SMOOTH FACE UNIT FOR
TOP COURSE SUGGESTED

CAVITY FILLER OR OTHER
MORTAR-COLLECTION DEVICE

ONE-PIECE FLASHING
MEMBRANE

WEEP HOLES OR PARTIALLY
OPEN HEAD JOINTS AT 32 IN
OC MAX.

HOLLOW UNIT (CUT)

ARCHITECTURAL CMU

28206-14_F15.EPS

Figure 15 Flashing of coping and parapet wall.

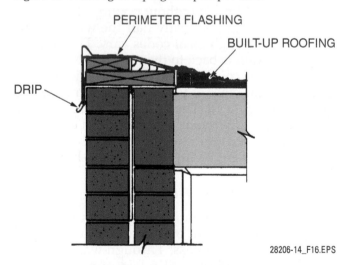

PERIMETER FLASHING

BUILT-UP ROOFING

DRIP

28206-14_F16.EPS

Figure 16 Perimeter flashing on a parapet wall.

plastics can be permanently and effectively joined by heat or an appropriate adhesive. The elasticity of plastic flashings eliminates the need for expansion joints.

The effect of flashing on wall strength depends on mortar placement and bond to both flashing and masonry. Through-wall flashing does not affect a wall's compressive strength, but it will reduce bending and shearing strengths. If flashing is placed directly on a masonry unit with a mortar bed only on one side of the flashing, the bending strength of the wall at that point is zero. It is therefore very important that the flashing be properly bedded in mortar to preserve the strength of the wall.

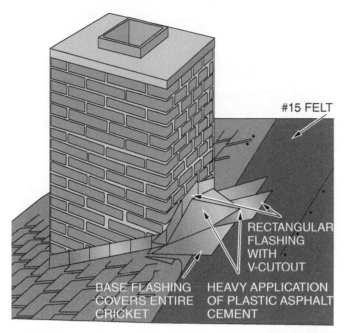

Figure 17 Chimney flashing.

28206-14_F17.EPS

28206-14_F19.EPS

Figure 19 Unitized flashing.

2.1.4 Unitized Flashing

Unitized flashings (see *Figure 19*) are factory fabricated and combine all the important components of a flashing system. Made from high-quality products under factory conditions to ensure consistency, unitized flashing is delivered to the project ready to use directly out of the box. Unitized flashings can also be produced in the field using components supplied by the manufacturer and can be assembled to meet the project specifications of any project. Unitized flashings are typically precut for masonry openings to ensure better overall performance.

2.1.5 Single-Wythe Concrete Masonry Unit Through-Wall Flashing

Ensuring proper drainage in a single-wythe block wall using traditional construction methods typically requires additional steps and specially shaped materials in order to create a functional wall-flashing system. Traditional through-wall flashing for single-wythe walls can present a bond break at the flashing course. An alternative option to the traditional single-wythe flashing method is to use a BlockFlash® pan, set on the bed of the single-wythe wall, that vents every cell of the wall with a pitched pan and attached weep vent (see *Figure 20*). The BlockFlash® pan does not require additional materials or special shapes and will not create a bond break at the flashing course.

2.2.0 Installing Weep Vents

Flashing should drain to the outside; however, base flashing concealed in tooled mortar joints is not self-draining. Weep vents, also called weeps or weep holes (see *Figure 21*), allow for the escape of collected moisture through the outer wythe

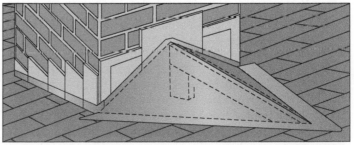

28206-14_F18.EPS

Figure 18 Chimney cricket.

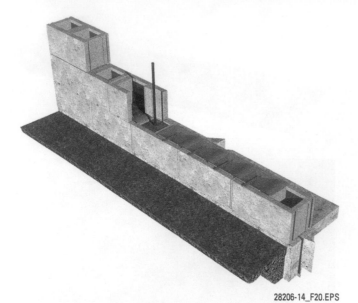

Figure 20 BlockFlash ® pan set in a single-wythe wall.

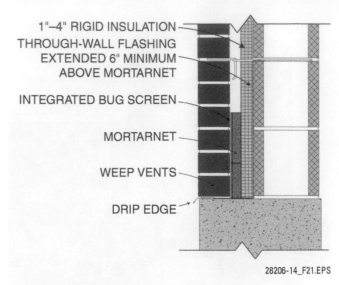

1"–4" RIGID INSULATION
THROUGH-WALL FLASHING EXTENDED 6" MINIMUM ABOVE MORTARNET
INTEGRATED BUG SCREEN
MORTARNET
WEEP VENTS
DRIP EDGE

28206-14_F21.EPS

Figure 21 Weep vents with MortarNet® in a multiwythe wall foundation.

of a cavity wall. Placed directly on the flashing at the base of a wall, shelf angle, sill, or masonry opening, weep vents allow air and water to pass from the cavity to the exterior of the wall. Some of the performance characteristics of a properly functioning weep include:

- Prevents insects from entering the cavity
- Allows for a dryer cavity, reducing presence of mold and mildew
- Helps prevent moisture condensation on structural components
- Helps prevent efflorescence
- Reduces the presence of saturated masonry and reinforcement corrosion

Weep vents are typically provided every 24 inches immediately above flashing sheets to prevent the buildup of water (see *Figure 22*). Weep vents are also placed wherever there is a risk of water penetration. For instance, weep vents may be placed over heads of windows, under window-sills, over steel lintels, or at every floor height on a concrete structure that is veneered with masonry work. Remember that weep vents should never be placed where they will be covered with fill earth after the construction is complete. Most codes require weep vents to be spaced no more than 33 inches apart. Weeps in foundation bases should be installed at least 6 inches above grade.

Weep vents can be made in several different ways. The recommended way to form a weep vent is to simply leave the head joints open. Open-joint weep vents should be a minimum of 2 inches high, with a minimum diameter of ³⁄₁₆ inch. Install screens or mesh made of noncorrosive metal or plastic in open head weep joints if called for in

the project specifications. Open head weep joints should be spaced a maximum of 24 inches apart.

Several manufacturers make plastic tubes that can be inserted into the mortar to serve as small weep vents (see *Figure 23*). These plastic tubes can also be fitted with **wicks** to help encourage moisture drainage. Weep tubes should be spaced no more than 16 inches apart. This is because they do not drain moisture as quickly as open-joint weep vents. Position wicks so that they are flush with the face of the wall, or else allow them to extend ½ inch from the face of the wall to encourage moisture evaporation. Place wicks along the back of the brick and ensure that they extend at least 16 inches into the airspace.

Cell vents (*Figure 24*) are metal or plastic devices designed to be the width of a mortar joint. They are manufactured with a cellular design consisting of a series of parallel channels and are manufactured to match mortar colors. The channels allow moisture to drain and air to enter to encourage evaporation. Cell vents are installed in mortar joints in the bottom course of a brick wall. They are useful for keeping insects from entering the cavity.

2.3.0 Installing Waterproofing

In masonry construction, the term *waterproofing* is used to describe the treatment of a wall with material that will retard the penetration of moisture. The term is also used in reference to protection against seepage that may or may not involve hydrostatic pressure. Masons can choose from a wide variety of protective coatings for practically any type of block or brick wall.

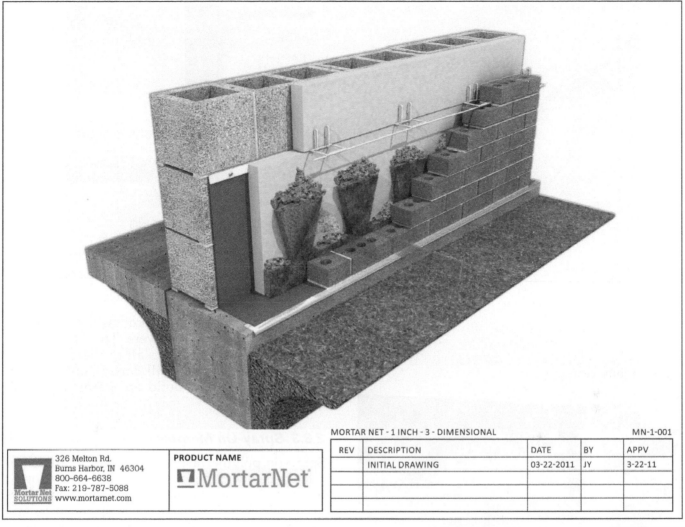

| MORTAR NET - 1 INCH - 3 - DIMENSIONAL | | | | MN-1-001 |

326 Melton Rd. Burns Harbor, IN 46304 800-664-6638 Fax: 219-787-5088 www.mortarnet.com	**PRODUCT NAME** **MortarNet**

REV	DESCRIPTION	DATE	BY	APPV
	INITIAL DRAWING	03-22-2011	JY	3-22-11

28206-14_F22.EPS

Figure 22 Weep vents at a wall foundation.

Below-grade foundation walls are the most critical areas requiring waterproofing. They require protection against rising water tables, hydrostatic heads, structural movement, and groundwater. A liquid waterproofing system applied by spraying ensures the high buildup of film thickness needed to cope with these problems. For all below-grade applications of waterproofing, be sure to fill all cracks, crevices, and grooves. Ensure that the coating is continuous and free from breaks or pinholes. Asphalt coatings and membrane systems, which involve the use of thin, pliable waterproofing materials, are two widely used methods for waterproofing foundation walls.

Waterproof coatings should be carried over the exposed tops and outside edges of the footing to form a cove at the junction of the wall and footing (*Figure 25*). Spread the coating around all joints, grooves, and slots and into all chases, corners, reveals, and soffits. Bring coating up to the finished grade. Apply waterproofing material on the outside face of backup block when plastering is eliminated. Apply waterproofing material to the inside face of the block once the wall has been plastered. Commercial admixtures are also available. These admixtures improve the waterproofing properties of mortar, stucco, and concrete.

2.3.1 Asphalt Coating

When treating a foundation with asphalt coating, carefully prepare the wall by removing all dust, dirt, or mortar drippings that may prevent the tar preparation from sticking to the block surfaces. Apply the waterproof coating with a broom, brush, or mop. Be very careful and wear appropriate personal protective equipment when working with hot asphalt. For mild climates that have little ground moisture, a single coating will likely be sufficient. Refer to your local applicable code for the standards that apply in your area.

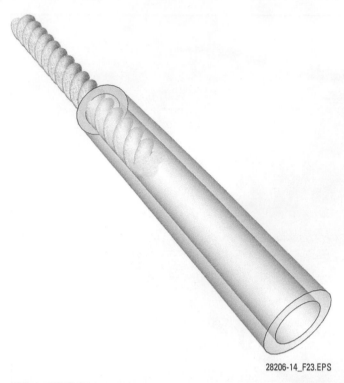

28206-14_F23.EPS

Figure 23 Tube vents.

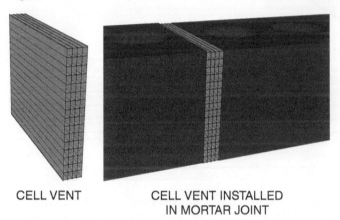

CELL VENT   CELL VENT INSTALLED IN MORTAR JOINT

28206-14_F24.EPS

Figure 24 Cell vent.

Hot tar preparations will dry to a fairly hard surface. However, many types of cold coating materials are designed to remain flexible. This allows them to resist shrinkage and cracking, and to fill smaller mortar cracks on the surface of the masonry structure.

2.3.2 Membrane Systems

To install a membrane system, after the masonry has been cleared of dirt and mortar droppings, a coat of coal tar, wood tar, or pitch primer is applied with a mop. While the primer is still sticky, apply a layer of geotextile fabric. The project spec-

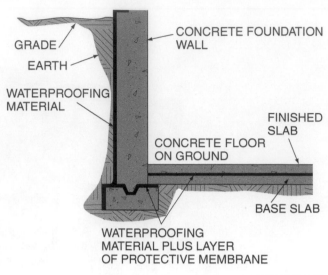

28206-14_F25.EPS

Figure 25 Below-grade waterproofing application.

ifications may call for this process to be repeated for a total of two to six layers. The final layer is a coat of primer that seals all of the layers. The number of layers used will depend on the water pressure expected to build up against the wall's surface.

2.3.3 Spray-On Membrane Systems

Spray-on polymer/asphalt emulsions can be applied to below-grade foundation walls and allowed to dry to form a membrane. Spray-on membranes are applied with special spray applicators, and they require adequate ventilation and the use of appropriate personal protective equipment. Watchdog Waterproofing® by Tremco Barrier Solutions is a popular type of spray-on membrane system that can be applied to block walls. Spray-on membranes are designed to span cracks of up to $\frac{1}{16}$ inch and remain flexible at low temperatures. Spray-on membrane systems are designed to withstand hydrostatic pressure more effectively than other types of waterproofing.

Ensure that the foundation wall has been cleaned in accordance with the manufacturer's instructions. Cover brick and other elements to protect against coming into contact with the membrane material. The membrane temperature when applied should be between 110°F and 130°F. The membrane should be applied to the required thickness, keeping in mind that the thickness will decrease as the membrane dries. Ensure that the membrane is applied evenly to provide a smooth coat.

Surface Preparation Is Important

The success or failure of any masonry coating is dependent on the adequacy of surface preparation. The masonry must be free of dirt, dust, oil, grease, and efflorescence. Efflorescent deposits may originate within the masonry material or mortar, and are a problem only when water can enter the masonry wall. Efflorescence can be removed by washing with a proprietary cleaner followed by thoroughly rinsing with water. Grease and oil can be removed with solvents or with a lye solution followed by rinsing.

Don't Break the Seal

Do not place backfill for 24 to 48 hours after application of waterproofing. Where possible, place backfill within seven days to avoid unnecessary damage to the waterproofing due to construction activities. Place the backfill carefully, in a manner that will not rupture or damage the finish and break the seal.

Additional Resources

Below Grade Waterproofing Manual, Latest Edition. Kansas City, MO: Sealant, Waterproofing & Restoration Institute.

A Practical Guide to Waterproofing Exterior Walls, Latest Edition. Kansas City, MO: Sealant, Waterproofing & Restoration Institute.

Technical Note 7, *Water Penetration Resistance – Design and Detailing*. 2005. Reston, VA: The Brick Industry Association. **www.gobrick.com**

Technical Note 7A, *Water Penetration Resistance – Materials*. 2005. Reston, VA: The Brick Industry Association. **www.gobrick.com**

Technical Note 7B, *Water Penetration Resistance – Construction and Workmanship*. 2005. Reston, VA: The Brick Industry Association. **www.gobrick.com**

TEK 3-13, *Construction of Low-Rise Concrete Masonry Buildings*. 2005. Herndon, VA: National Concrete Masonry Association. **www.ncma.org**

TEK 19-4A, *Flashing Strategies for Concrete Masonry Walls*. 2008. Herndon, VA: National Concrete Masonry Association. **www.ncma.org**

TEK 19-5A, *Flashing Details for Concrete Masonry Walls*. 2008. Herndon, VA: National Concrete Masonry Association. **www.ncma.org**

2.0.0 Section Review

1. A flashing material that can corrode in fresh mortar and that requires painting maintenance if installed exposed is _____.

   a. PVC
   b. galvanized steel
   c. copper
   d. stainless steel

2. The minimum height of an open-joint weep vent should be _____.

   a. 3½ inches
   b. 3 inches
   c. 2½ inches
   d. 2 inches

3. Apply waterproofing material to the inside face of block once the wall has been _____.

   a. grouted
   b. plastered
   c. painted
   d. tied

3.0.0 EXTREME WEATHER IN MASONRY CONSTRUCTION

Objective

Explain the various techniques used to provide adequate protection during hot- and cold-weather masonry construction.

a. Explain the role played by weather data and information in masonry construction.
b. Explain the various techniques used to provide adequate protection during hot-weather masonry construction.
c. Explain the various techniques used to provide adequate protection during cold-weather masonry construction.

Trade Terms

Plasticity: A material's ability to be easily molded into various shapes.

In every phase of masonry construction, adequate planning can save time and money. Weather conditions can seriously affect the mason's work. In fact, the weather can seriously damage the desired properties of the completed structure. Always take precautions to avoid damage from bad weather. This includes any type of extreme weather: hot, cold, wet, or extremely dry.

Weather affects mortar in several ways. Hot weather can cause mortar joints to set too rapidly. This increases the difficulty of tooling joints. High wind velocities, high relative humidity, and too much direct sunlight lower the water-retaining quality of the mortar. The mortar will lose its workability. Its bonding ability is decreased and it will be more difficult to create weather-tight joints.

Cold weather, ice, sleet, and snow can cause failure in the structural strength and durability of mortar. For instance, if the temperature falls below 40°F, hydration is impaired. The mortar will not develop much strength. If the masonry units are cold when they are laid up, they will drain the heat from the mortar. This will impair the strength development, initial set, and bonding ability.

Rainy weather will directly affect joint strength and the appearance of the completed structure. Rain can dilute the mortar. If the chemical mix-ture is affected, its strength may be reduced. Partially completed work must be protected from rain or it may become saturated. The mortar can be washed from the joints. This may cause efflorescence when the area dries.

Rapid drying of the mortar must be avoided during extremely dry weather. In addition to evaporation, dry masonry units will absorb water from the mortar. Mortar that dries too rapidly has poor workability.

All-weather construction has become more desirable. There is greater demand for low-cost, durable masonry structures. Hot-weather problems have often occurred in construction, but they have only been fully recognized in recent years. Recognition of these problems improves both the economy of the structure and the quality of the workmanship. Cold-weather construction techniques have enabled the construction industry to extend its season into the winter months.

Take the following additional steps to produce effective masonry structures under all weather conditions:

- Understand and review masonry performance at varying temperatures under different weather conditions.
- Gather the appropriate weather data.
- Know the typical values for wind and rain in your area.
- Modify the selection and storage of materials according to the weather data.
- Modify the construction procedures and post-construction protection features appropriately.

Admixtures are sometimes added to the mix to chemically change the properties of the mortar. Admixtures are often used in difficult construction situations, such as hot- or cold-weather construction, because they can speed up or slow down the setting time. This will partially protect mortar against the effects of extreme temperature.

Different admixtures produce different results, so it is important to use the correct one. Common admixtures and their effects are listed in *Table 3*. Choose one that is compatible with the mortar, the construction practices, and the job specifications. For the best result, all admixtures should meet the requirements established in *ASTM* (American Society for Testing and Materials) *C494/C494M, Standard Specification for Chemical Admixtures for Concrete*, and should be used throughout the entire project. Retarders, accelerators, and water reducers are commonly used.

Admixtures may change the strength, color, and speed of setting for mortar. Adding them must be cleared with the project engineer. In

Table 3 Admixtures and Their Effects

Admixture	Work-Ability	Strength	Weather Resistance
Air-entraining agents	Increase	Decrease slightly	Increase
Bonding agents	–	Increase	–
Plasticizers	Increase	–	–
Set accelerators	Decrease	Increase	–
Set retarders	Increase	–	–
Water reducers	Decrease	Increase	Increase
Water repellants	–	–	Increase
Pozzolanic agents	Increase	Increase	–

some cases, the project specifications will list the acceptable kinds and amounts of admixtures that can be used under different circumstances.

3.1.0 Using Weather Data and Information

Accurate weather forecasting is an important part of planning all-weather construction. Important weather factors include temperature, wind, rain, snow, ice, and humidity. Although *ideal*, *cold*, and *hot* are relative terms for masonry construction, ideal is generally considered to be between 40°F and 90°F. Many codes specify a minimum working temperature and direct how to protect the work when the temperature falls below that mark. However, they do not address procedures for high temperatures.

Some problems may occur even with ideal temperatures. For example, wind and humidity can cause problems that are normally associated with higher temperatures.

Construction schedules are based on long-term weather expectations, local weather information, and the experience of the contractor. Long-term weather expectations are called climatology data. Local weather information is called meteorological data.

Meteorological information can be obtained from the National Weather Service. Climatological information can be obtained from the National Climatic Data Center. They usually provide climatic information in the form of maps (*Figure 26*). These maps contain daily, monthly, and annual data for a region. They may be obtained for a nominal fee by contacting the Center.

3.2.0 Protecting Masonry during Hot-Weather Construction

Hot temperatures, particularly when combined with high winds and direct sunlight, have a greater effect on mortar than on masonry units. However, a number of changes can occur in the physical properties of both.

3.2.1 Mortar and Masonry Unit Performance

Hot weather is defined as temperatures above 90°F. However, temperature, wind speed, relative humidity, and solar radiation all influence masonry materials. These factors affect the absorption of masonry units, the rate of set, and the drying rate of mortar. The primary concern in hot weather is evaporation of water from the mortar. The bond between the brick and mortar diminishes if there is not enough water present.

Low temperatures and low humidity affect masonry more than high temperatures and high humidity. Hot, humid weather can increase the rate of hydration of the cement. If sufficient water is present, these favorable curing conditions can help develop masonry strength.

The masonry units themselves are the least affected by hot weather. However, the interaction

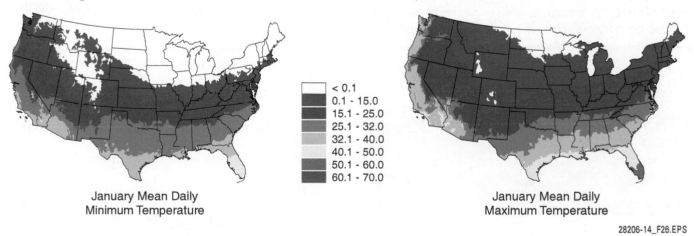

January Mean Daily Minimum Temperature	January Mean Daily Maximum Temperature

< 0.1
0.1 - 15.0
15.1 - 25.0
25.1 - 32.0
32.1 - 40.0
40.1 - 50.0
50.1 - 60.0
60.1 - 70.0

28206-14_F26.EPS

Figure 26 Examples of climatology data.

between the masonry units and the mortar or grout is critical. Warmer units absorb more water from the mortar. If there is not enough water present in the mortar when the units are laid, the mortar will have lower bond strength.

Mortar will tend to lose its plasticity, or ability to be molded, rapidly in hot weather. This is due to the evaporation of the water from the mix and the increased rate of hydration of the cement. The use of admixtures to increase plasticity is not recommended unless their full effect on the mortar is known. Instead of an additive, use mortar with a high lime content and high water retention. Keep mortar at the proper consistency by tempering (*Figure 27*). This may be required at more frequent intervals during hot weather. To avoid overtempering, mix smaller batches so the mortar will not lose water before it is applied.

Grout reacts to hot weather in a manner similar to mortar. Water evaporates more easily in hot weather and reduces the water-to-cement ratio. Grout requires a high slump for placement. Maintain a high water-to-cement ratio by reducing evaporation and initially mixing grout with sufficient water.

3.2.2 *Materials Selection and Storage*

The primary objective in hot-weather masonry construction is to avoid moisture loss. The mason must maintain cooler temperatures in all equipment, mortar ingredients, and masonry units. If possible, when preparing the job site, arrange for materials to be delivered to and stored in the shade.

Sandpiles should be covered or shaded with tarpaulins. This will lessen the evaporation of the natural moisture of the sand. Typically, sand has a free moisture content of about 8 percent. If the sand's free moisture content drops, it can be cooled by lightly sprinkling water on the sand-

Figure 27 Tempering mortar.

pile. Do not wet the sand, however. A good rule to remember is that 1 gallon of water will cool 1 cubic yard of sand by 20°F. Stockpiled masonry units may also be sprinkled with water. However, do not soak them. Never lay up wet units.

Consider the comfort of workers and the protection of materials throughout the construction process. If necessary, set up sunscreens to protect the materials, the mixing area, and the construction area. Make use of windbreaks or fogger-sprayers if the job site is hot, windy, and dry. Fogger-sprayers spray a fine mist of water.

3.2.3 *Construction Practices*

During the actual construction process, there are several effective steps that can be taken to maintain lower temperatures and therefore lower the rate of evaporation. *Table 4* lists steps that should be followed for hot-weather masonry construction.

In addition to the practices listed in *Table 4*, the following procedures are suggested:

- Use cool water or ice in the mortar mixture (*Figure 28*). The ice should be melted by the time the mixture leaves the mixer.
- Keep equipment cool by flushing with cool water before use. Both wooden and metal tools absorb heat. If possible spray any tools, particularly those that contact mortar. Keep items such as levels in the shade.
- Cover each section of wall as it is completed, to slow the rate of moisture loss. When a structure must meet high standards for tensile strength, it should be damp-cured by covering with wet burlap during the entire curing period.

To ensure that you have considered everything relative to hot-weather masonry construction, use the following checklists as a guide to your work on the next hot-weather project.

During the preliminary planning phase, do the following:

- Check if there are any provisions for hot-weather construction in the contract specifications.
- Lay out the site to accommodate material deliveries during high temperatures.
- Provide covered storage at the site for masonry materials.
- Review available weather data for the project area:
  - Expected daytime and nighttime temperatures
  - Expected wind speed
  - Expected relative humidity

Table 4 Brick Masonry Construction in Hot Weather

Temperature[1]	Preparation Requirements (Prior to Work)	Construction Requirements (Work in Progress)	Protection Requirements (After Masonry Is Placed)
Above 115°F or 105°F with a wind velocity over 8 mph.	Shade materials and mixing equipment from direct sunlight. Comply with hot-weather requirements below.	Use cool mixing water for mortar and grout. Ice must be melted or removed before water is added to other mortar or grout materials. Comply with hot-weather requirements below.	Comply with hot-weather requirements below.
Above 100°F or 90°F with 8 mph wind	Provide necessary conditions and equipment to produce mortar having a temperature below 120°F. Maintain sandpiles in a damp, loose condition.	Maintain mortar and grout at a temperature below 120°F. Flush mixer, mortar transport container, and mortarboards with cool water before they come into contact with mortar ingredients or mortar. Maintain mortar consistency by retempering with cool water. Use mortar within 2 hours of initial mixing.	Fog-spray newly constructed masonry until damp, at least three times a day until the masonry is three days old.

1. Preparation and construction requirements are based on ambient temperatures. Protection requirements, after masonry is placed, are based on mean daily temperatures.

- Select the type of required protection:
  - Cover or shade for mortar preparation
  - Cover or shade, windscreen, and fogger-sprayer for masons on the wall
  - Protection for completed walls
- Arrange for short-term, daily weather forecast service.
- Keep a thermometer at the job site and check it often.

During the execution phase, do the following:

- Anticipate the temperature and windward humidity.
- Provide a shaded area for the storage of mortar materials and masonry units, and for mortar preparation.

28206-14_F28.EPS

Figure 28 Mix mortar with cold water.

National Weather Service

The National Weather Service is a branch of the National Oceanic and Atmospheric Administration (NOAA). The National Weather Service has information centers located at major airports in cities throughout the country. These centers provide current weather information and regularly scheduled weather forecasts for the local region.

- Set the work area for the masons:
  - Use sunscreens when direct sunlight is excessive and the temperature is high
  - Use windscreens and fogger-sprayers when winds and low humidity are expected
- Review the procedures for cooling materials and equipment:

  - Sprinkle sand and masonry units with water if moisture loss is too great
  - Use cool mix water
  - Consider increasing the amount of cementitious materials
  - Rinse all equipment with cool water before use
  - Protect completed wall sections from early set by covering
  - Avoid rapid set by damp-curing

During the protection phase, do the following:

- At any temperature, protect a completed wall by covering the top 2 feet with plastic or canvas to prevent the entry of water.
- Maintain windscreens if continued winds are expected.
- Maintain damp-curing by wetting burlap at close of work schedule.

3.3.0 Protecting Masonry during Cold-Weather Construction

Cold weather during masonry construction can affect material and craftsmanship. Consider the effects of cold weather in planning, scheduling, and setup of masonry work. In addition to anticipating specific weather conditions, you must determine what the possible effects of the weather will be on the materials and the work effort. Plan how to protect materials and workers, and how to store the materials. Determine the procedures to use to meet the contract requirements.

In the United States, all model building codes have requirements relating to the construction of masonry during cold weather. While not identical, each of the codes has similar general requirements. These address materials protection, materials heating, use of frozen materials, and protection of completed work.

3.3.1 Mortar and Masonry Unit Performance

The major concern during cold-weather construction is to keep all equipment, materials, and supplies at a temperature well above the freezing point. As in hot-weather construction, the temperature will affect mortar hydration and bond strength development in several ways:

- Mortars mixed at lower temperatures have lower water contents, longer setting and hardening times, higher air contents, and lower early strength than those mixed at normal temperatures.
- As the temperature falls below 40°F, the heat of hydration decreases or stops altogether. This results in a decrease in the strength of the mortar.
- Although cold concrete masonry units may appear much like normal units in every way, they will be slightly smaller in volume than normal. When the temperature returns to normal, the blocks expand and joints may crack.
- Wet or frozen masonry units will have a reduced water absorption rate. This will increase the possibility of early freezing, mortar expansion, and joint cracking. It will also create bonding problems.

Keep Materials Close

Find the balance between temperature control and logistics. It is better to have materials close at hand and under a tarpaulin, even if they will be somewhat warmer. This sandpile is close to the work area. Tarp the sandpile if it gets too hot, to prevent extensive evaporation.

28206-14_SA01.EPS

These problems can be prevented. Select mortar ingredients to maintain hydration. Reduce the water content of mortar to below 8 percent. Heat mortar materials. Select the correct masonry units. Protect materials during storage and construction.

3.3.2 Materials Selection and Storage

Although the materials used on the job are specified by contract and code requirements, you may have several options for dealing with cold-weather problems. For example, high-early-strength (Type III) portland cement substituted for normal (Type I) cement will give increased protection for the mortar because it increases the rate of strength development. The physical properties of mortar made with high-early-strength cement instead of normal cement will not be changed.

When you understand cold-weather effects, it is easier to make decisions about the use of particular masonry units and available mortar admixtures. An absorptive masonry unit will suck water from the mortar and lessen the problem of initial freezing and mortar-joint expansion. On the other hand, a masonry unit with a very low absorption rate will not collect enough water to prevent mortar-joint expansion as ice forms. Considerations should include the unit's absorptive rate, volume, density, and temperature when used in cold-weather construction.

Carefully consider the use of admixtures in cold-weather construction. Most admixtures reduce the strength of the mortar. In cold weather, there is even more strength reduction. Only two types of admixtures, accelerators and air-entraining agents, are recommended during cold weather. Accelerators offer protection against cold weather by increasing the high-early-strength development of the cement in the mortar (*Figure 29*).

The most frequently used accelerator is calcium chloride. Soluble carbonates, silicates and fluosilicates, calcium aluminate, and some aluminous cements have also been used as accelerators. Using calcium chloride may increase shrinkage, efflo-

28206-14_F29.EPS

Figure 29 Set accelerator.

rescence, and corrosion of metal ties, joint reinforcement, flashing, and anchors.

Calcium chloride is not recommended in reinforced masonry construction. If it is used, it should be strictly regulated to 2 percent by weight of portland cement or 1 percent by weight of masonry cement. Add the calcium chloride to the mix water. To prepare the proper solution, dissolve 100 pounds of flaked calcium chloride in 25 gallons of water. Do not use more than 1 quart of solution with each 94-pound bag of masonry cement.

Air-entraining agents are sometimes used in cold climates to increase workability and freeze/thaw durability. Air-entraining admixtures cause millions of tiny air bubbles to form in the mortar. These air bubbles provide room for expansion during freezing cycles. This results in a more durable mortar. Avoid corrosion inhibitors and color pigment with dispersing agents. Experience has shown that these admixtures often slow hydration.

Regardless of the materials selected, they must be protected to prevent freezing. Build a temporary shelter to store mortar materials. If a shelter cannot be built, cover all materials when the temperature is below 40°F. Completely cover bagged

Beat the Heat

Hot weather can cause heat exhaustion, cramps, and even heat stroke. To prevent these from slowing you down, take the following precautions when working in extremely hot weather:

- Drink plenty of water
- Do not overexert yourself
- Wear lightweight clothing
- Keep your head and face shaded
- Take frequent short breaks

materials with canvas or plastic sheeting (*Figure 30*). Store them on raised platforms so they cannot absorb cold and moisture from the ground. Also cover the mortar sand and masonry units to protect them from water, snow, and ice.

3.3.3 Construction Practices

There are several important factors in cold-weather masonry construction. First, maintain safe and comfortable working conditions. You must also maintain the physical properties of the mortar and masonry units. And finally, protect the finished portions of the construction. Heat is the most important ingredient in solving these problems. Specific procedures for providing heat and protection during the construction process are provided in *Table 5*.

Consider using windbreaks or temporary heated enclosures when low temperatures are expected. There are several areas that need to be protected. Protect the materials, the mortar mixing area, the work areas, and the finished areas of the masonry construction. In fact, US federal regulations now call for heated enclosures on all government projects.

Figure 31 shows typical scaffold enclosures. They are erected over an exterior wall using the tubular scaffold as part of the system. There are several factors to consider when creating an en-

28206-14_F30.EPS

Figure 30 Cover bagged materials.

closure. Think about the following: material strength, durability, transparency, fire resistance, flexibility, ease of installation, cost, severity of the weather, the type of masonry structure being built, and Occupational Safety and Health Administration (OSHA) safety requirements.

Often, an unheated structure or covering will be sufficient. It will reduce the windchill factor and prevent the masonry from freezing. Enclosures are typically built using lumber, steel, canvas, building paper, fiberglass, and clear plastic sheets.

A widely used technique is to make polyethylene panels. These can be fastened and braced to the scaffolding around the work area. The panels are made of lumber framing with plastic film 6 to 12 mm thick, reinforced with fiber or wire mesh. They are particularly effective because they resist the wind but allow the light and heat of the sun to pass.

Other techniques for covering construction areas include independent framework covering the entire building or portable framework that can be raised with swinging scaffolds as the height of the structure increases. The enclosure should be braced to provide a safe area for the masons to work. It should be draft-proof enough to protect the structure from early freezing.

If the enclosure is heated, ensure that the heat does not cause the mortar to dry out too quickly. When planning a structure, consider wind force, loading, counterweights, and bracing, and their effect on the scaffolding. Keep the plastic from blowing or flapping, using frames or ties. The flapping plastic creates a safety hazard.

There are several additional points to be considered in regard to heat: the most economical and efficient source of heat, the area to be heated, the volume of air to be heated, and the degree of safety of the heating device. It may be possible to use the stationary heating plant intended for the finished structure with the permission of the heating contractor. More often, the source of heat will be portable vented heaters. Typically they use electricity, steam, fuel oil, or bottled propane. *Figure 32* shows a heating unit that operates on propane. These heaters are often called salamanders. OSHA regulations require that all temporary

It's Cooler at Night

In areas with extremely high temperatures and dry winds, consider avoiding hot midday temperatures. Work schedules can be altered to early-morning or late-evening hours. This may create difficulties with the other construction trades. However, it benefits worker comfort, safety, and efficiency. The ability to maintain the desired physical properties of the structure may outweigh the inconvenience.

Table 5 Brick Masonry Construction in Cold Weather

Temperature[1]	Preparation Requirements (Prior to Work)	Construction Requirements (Work in Progress)	Protection Requirements (After Masonry Is Placed)
40°F to 32°F	Do not lay masonry units either having a temperature below 20°F or containing frozen moisture, visible ice, or snow on their surface. Remove visible ice and snow from the top surface of existing foundations and masonry to receive new construction. Heat these surfaces above freezing, using methods that do not result in damage.	Heat mixing water or sand to produce mortar between 40°F and 120°F. Do not heat water or aggregates used in mortar or grout above 140°F. Heat grout materials when their temperature is below 32°F.	Completely cover newly constructed masonry with a weather-resistive membrane for 24 hours after construction.
32°F to 25°F	Comply with cold-weather requirements above.	Comply with cold-weather requirements above. Maintain mortar temperature above freezing until used in masonry. Heat grout materials so grout is at a temperature between 70°F and 120°F during mixing and placed at a temperature above 70°F.	Comply with cold-weather requirements above.
25°F to 20°F	Comply with cold-weather requirements above.	Comply with cold-weather requirements above. Heat masonry surfaces under construction to 40°F and use wind breaks or enclosures when the wind velocity exceeds 15 mph. Heat masonry to a minimum of 40°F prior to grouting.	Completely cover newly constructed masonry with weather-resistive insulating blankets or equal protection for 24 hours after completion of work. Extend time period to 48 hours for grouted masonry, unless the only cement in the grout is Type III portland cement.
20°F and Below	Comply with cold weather requirements above.	Comply with cold-weather requirements above. Provide enclosure and heat to maintain air temperatures above 32°F within the enclosure.	Maintain newly constructed masonry temperature above 32°F for at least 24 hours after being completed, by using heated enclosures, electric heating blankets, infrared lamps, or other acceptable methods. Extend time period to 48 hours for grouted masonry, unless the only cement in the grout is Type III portland cement.

1. Preparation and construction requirements are based on ambient temperatures. Protection requirements, after masonry is placed, are based on mean daily temperatures.

Icy Scaffolding Can Be Deadly

A laborer was working on the third level of a tubular welded-frame scaffold that was covered with ice and snow. Planking on the scaffold was weak and a guardrail had not been set up. The worker slipped and fell headfirst approximately 20 feet to the pavement below.

The bottom line: Don't work on a wet or icy scaffold. Make sure that all scaffolding is sturdy and includes proper guardrails.

Figure 31 Scaffold enclosures.

enclosures heated with combustible fuel have separate venting systems for the heater, to avoid carbon monoxide buildup.

Properly vented oil or gas heaters with blowers are recommended for projects where complete enclosures are provided and where heat is not available from a stationary heating plant. Ideally, locate the heater outside the enclosure and blow hot air in. This way the heater can blow warm air across both sides of the wall being constructed. Workers can be ensured of fresh air rather than the polluted air created by burning fuel.

> **WARNING!**
>
> Carbon monoxide is a gas produced by combustion. Breathing fumes can cause serious illness and death.

Job-site preparation also includes setting up heat sources for the heating of mortar materials and masonry units, specialized tools for cold-weather masonry, and postconstruction protection for the structure. The final steps in preparing the work area should be to provide walkways over icy or hazardous areas and to thaw the base where the masonry structure is to be laid up. Masonry work should never be laid up on frozen surfaces or on a snow-covered base. Frozen surfaces may shift or move as they thaw. Ice-covered surfaces destroy the bonding property of mortar. Both will result in cracks in the masonry wall.

Figure 32 Portable heater.

3.3.4 Heating the Materials

When the site has been prepared and the construction begins, the immediate objective is to ensure proper hydration of the mortar. This may require heating the various materials at the time of construction.

The mixing water is the most logical mortar material to be heated. It is the easiest material to heat. Also, it can store the most heat. Do not heat the water to a temperature above 160°F. Heat water in a 55-gallon drum. Build a fire under or around it. You can also use an immersion heater, flame guns, or steam probes to heat the water. Always be careful not to pollute the water.

Heat the mixing water to produce mortar temperatures between 70°F and 120°F. Maintain that temperature in later batches. If the mix water is too hot, the mortar may flash-set when the portland cement is added. Prevent this by combining sand and water in the mixer before adding the cement. This will lower the temperature and prevent flash setting.

Masonry sand contains moisture, which will turn to ice at temperatures below 32°F even if it is covered. When the temperature of the sand is below freezing, it should be heated slowly and evenly. Take care to avoid scorching the sand by excessive heating. A temperature of 45°F to 50°F is

Antifreeze

Do not use an antifreeze agent as an admixture. Antifreeze agents, such as alcohol, are sometimes used to lower the freezing point of mortar. Experience has shown, however, that so much of the antifreeze admixture is needed to lower the freezing point that the mortar's strength is reduced.

acceptable. Sand may be heated as high as 180°F without scorching. A practical and easy method of checking the sand temperature is simply to touch it. If the sand can be held in your hand, the temperature is not too high.

The most commonly used method of heating sand is to pile it over a heated metal pipe (*Figure 33*). Culvert pipe, a steel drum, or a smokestack are typically used. Other methods serve equally well. A heater box with an ordinary hot-water heater may be used. Extend a pipe from the box under the sandpile. Circulate the water continually through the pipe and back to the box. This method provides both heated water and heated sand at the same time. Other methods of heating sand include steam boilers with steam coils placed under the sandpile, or commercial oil-fired heating devices that use thermostats to automatically control the temperature.

Whatever the method used, avoid contaminating the sand with foreign substances. Mix the sand periodically to produce uniform heating and to avoid scorching.

When dry masonry units reach 20°F, they should be heated to 40°F. This increased temperature prevents the sudden cooling of mortar when it is laid up with the units. When the weather is damp and cold, heating the units will increase your efficiency. Thaw wet-frozen masonry units but do not overheat them.

The most effective method of heating masonry units is to direct the blower from an oil, gas, or electric heater directly against the pile of units. If heated enclosures are used for construction, a supply of masonry units can be thawed on the scaffold. Another method of heating masonry

28206-14_F33.EPS

Figure 33 Sandpile warmed by heated pipe.

units is to simply stack them around gas- or oil-burning heaters.

3.3.5 Mortar Mixing and Tooling

During cold-weather construction, mix mortar in smaller quantities than usual. Limit the mortar batch size to an amount that can be used in a one-hour period. Once the mortar is mixed, carry it immediately to the mason. Do not leave it to cool in a mortar box or wheelbarrow. If the mortar must be placed in a mortar box, use a metal box that has been raised off the ground and leveled. If necessary, the mortar box can be heated to maintain the required temperature.

Keep the following points in mind when preparing cold-weather mortar:

- Maintain the mortar temperature between 70°F and 90°F.
- When mixing the mortar, do not add heated water that will raise the temperature of the heated mix above 105°F. This prevents the danger of flash, or false, set when the water and cement are mixed.
- Never use frozen sand in the mix; the sand must be thawed.
- Never use sand that has been scorched as a result of overheating.
- Never use antifreeze admixtures to lower the freezing point of mortar.
- Follow the *International Building Code*® requirements for protection against freezing. Masonry must be protected against freezing for at least 24 hours after being laid. Grouted masonry must be protected for up to 48 hours, unless the only cement in the grout is Type III portland cement.

Tooling joints in cold-weather masonry construction creates little difficulty. In fact, you may be able to lay more units before tooling. However, allow the same set time for each section of completed masonry, to ensure the uniform coloring of joints.

3.3.6 Protecting Completed Structures

Protect each section of the masonry wall from freezing as it is completed. This allows it to cure properly without being damaged. When the temperature is above 25°F, cover the wall with a tarpaulin. This will allow the mortar to set properly if it has been previously heated. When the temperature falls below 25°F, use an insulated blanket or heater to raise the temperature. In all cold-weather construction, use tarpaulins or plastic coverings to protect the top, or last-laid, units from rain or snow. Extend these covers at least 2 feet down the sides of the construction.

Use the following checklists to review the job-site conditions and specifications for cold-weather masonry construction.

During the preliminary planning phase, do the following:

- Check provisions for winter construction in contract specifications.
- Lay out the site to accommodate material deliveries during poor weather conditions.
- Review code requirements for cold-weather protection of masonry.
- Plan dry storage at the site for masonry materials.
- Review available weather data for the project area:
  - Expected daytime and nighttime temperatures
  - Expected windchill factor and wind speeds
- Select the type of required protection:
  - Enclosure for mortar preparation
  - Enclosure for masons on the wall if heat is required
  - Protection of completed walls
- Arrange for short-term, daily weather forecast service during winter months.

During the execution phase, do the following:

- Anticipate temperature, wind, and rain.
- Do not lay masonry on frozen surfaces.
- Provide an enclosed area for storage of mortar materials and mortar preparation.

- Set the work area for the masons:
  - Employ windbreaks when the windchill factor approaches zero
  - Employ complete enclosure when the daytime temperature is in the low teens
  - Heat the enclosed work area when required for protection of completed work
- Review the procedures for heating of materials:
  - Heat the mortarboard and mortar to a temperature between 70°F and 90°F
  - Use heated mix water (70°F to 105°F) when the air temperature is below 40°F
  - Do not use frozen sand
  - Heat and completely thaw sand when temperature is below 32°F
  - Heat units as required so that the temperature of units when laid is not less than 20°F
- Protect the wall from early freezing, if not enclosed.
- Avoid rapid dry-out of mortar due to excessive heating.

During the protection phase, do the following:

- Provide minimum protection by covering the top 2 feet of completed wall with plastic or canvas, to prevent the entry of water.
- When the night temperature is expected to be below freezing, cover the newly built wall completely.
- If the expected temperature is in the low teens, or if the expected windchill is zero or less, cover the completed masonry and maintain it above freezing for 48 hours or more with insulated blankets or auxiliary heat.

Cold-Weather Clothing Tips

Use the following tips to prevent injury due to cold weather:

- Dress in layers.
- Wear thermal underwear.
- Wear outer clothing that will resist wind and moisture.
- Wear a face mask and head and ear coverings.
- Carry an extra pair of dry socks when working in snowy or wet conditions.

Additional Resources

ASTM C494/C494M, Standard Specification for Chemical Admixtures for Concrete. 2013. West Consohocken, PA: ASTM International.

International Building Code, latest edition. Falls Church, VA: International Code Council.

Technical Note 1, *Cold and Hot Weather Construction.* 2006. Reston, VA: The Brick Industry Association. **www.gobrick.com**

TEK 3-1C, All-Weather Concrete Masonry Construction. 2002. Herndon, VA: National Concrete Masonry Association. **www.ncma.org**

3.0.0 Section Review

1. When developing a construction schedule, obtain meteorological information from _____.

   a. the US Coast Guard
   b. the National Weather Service
   c. the Weather Channel
   d. the National Climatic Data Center

2. In hot-weather masonry construction, the primary objective is to _____.

   a. avoid moisture loss
   b. facilitate moisture loss
   c. increase the use of admixtures in mortar
   d. decrease the use of admixtures in mortar

3. A set accelerator that may increase shrinkage and efflorescence is _____.

   a. calcium carbonate
   b. calcium silicate
   c. calcium chloride
   d. calcium aluminate

Summary

The installation of insulation in and on masonry walls is an important requirement in today's energy-conscious society. Heat is transferred through material and space by conduction, radiation, and convection. Various types of insulation will reduce the loss of heat or protect from overheating when properly applied to masonry structures. All materials are rated for their effectiveness in resisting the transfer of heat by an index called the R-value.

The main categories of insulation include loose fill, rigid, slab, reflective, foamed-in-place, insulation inserts for block, and flexible blankets or batts. Internal insulation is placed between the interior and exterior faces of masonry work. External insulation is placed on the face of the masonry units, and includes coatings, rigid-board insulation, and blanket or batt insulation.

Effective control of moisture is very important in masonry construction. Damage that results from the presence of water in a wall is a problem you cannot afford to ignore. Replacement costs far exceed initial construction costs. Even if the wall can be repaired, your reputation may not mend as easily. Be sure that the workmanship is professional and the materials are adequate.

In every type of outdoor masonry construction, adequate planning must take the weather into account. Both hot weather and cold weather can have a serious effect on the work. Hot weather can cause the mortar joints to set too rapidly and increase the difficulty of tooling joints. Cold weather, ice, sleet, and snow can cause failure in the strength and durability of mortar. Accurate weather information is needed to aid in planning for all-weather construction. Important weather factors include temperature, wind, rain, snow, ice, and humidity.

1. The length of time it takes a given amount of heat to move through a material is a measure of its thermal _____.
   a. density
   b. resistance
   c. conductivity
   d. inductance

2. Basic insulation can be classified into _____.
   a. two categories
   b. four categories
   c. six categories
   d. eight categories

3. When rigid boards are used as internal insulation they are applied to _____.
   a. the outside face of the inner wythe
   b. the inside face of the outer wythe
   c. the inside face of the inner wythe
   d. both sides of the cavity

4. The type of material that is most often used when reinsulating existing buildings is _____.
   a. slab insulation
   b. expanded polystyrene foam
   c. reflective insulation
   d. foamed-in-place insulation

5. Flexible insulation with a vapor barrier only on one side is described as _____.
   a. reflective
   b. friction-fit
   c. faced
   d. wrapped

6. Unless the structure has been designed to withstand the damaging effects of humidity, repeated absorption and evaporation of moisture in masonry structures can cause _____.
   a. cracking
   b. erosion
   c. fungus growth
   d. warping

7. Stainless steel flashing that is exposed to weather should be _____.
   a. 24 gauge
   b. 26 gauge
   c. 28 gauge
   d. 30 gauge

8. Flashing and weep vents are vital to controlling moisture in _____.
   a. foundation walls
   b. cavity walls
   c. garden walls
   d. retaining walls

9. Weep vents are typically placed immediately above the flashing, spaced every _____.
   a. 12 inches
   b. 16 inches
   c. 24 inches
   d. 33 inches

10. A membrane waterproofing system involves applying one or more layers of _____.
    a. geotextile fabric
    b. heavy-gauge plastic sheeting
    c. roofing paper
    d. fiberglass membrane

11. Fresh mortar washed from joints by rain may dry on masonry surfaces and show _____.
    a. fluorescence
    b. yellow staining
    c. efflorescence
    d. mildew deposits

12. Ideal weather for masonry construction is generally considered to be a temperature range between 40°F and _____.
    a. 60°F
    b. 70°F
    c. 80°F
    d. 90°F

13. During hot weather, the primary concern is _____.
    a. overheating of exposed surfaces
    b. evaporation of water from the mortar
    c. excessive mixing of mortar
    d. expansion of masonry materials

14. Stored sand should be covered or shaded in hot weather to maintain a free moisture content of about _____.

　　a. 5 percent
　　b. 7 percent
　　c. 8 percent
　　d. 11 percent

15. OSHA requires that all temporary enclosures heated with combustible fuel _____.

　　a. be inspected daily
　　b. include at least two exits
　　c. be supplied with fire extinguishers
　　d. have a separate venting system

Trade Terms Quiz

Fill in the blank with the correct term that you learned from your study of this module.

1. A layer of thin, pliable material used to waterproof masonry is called a(n) _____.

2. A(n) _____ is a continuous recess built into a wall to receive pipes, wires, or heating ducts.

3. The visible portion of a masonry jamb between the face of the wall and the frame is called the _____.

4. A(n) _____ is a bundle of fibers that are loosely twisted or braided and woven together to form a cord that can carry water away from an area by capillary action.

5. The process of treating masonry with a material that will retard penetration of moisture is called _____.

6. _____ is a material's ability to be easily molded into various shapes.

7. The ability of a building wall to alternately absorb and release heat energy in response to temperature changes is called _____.

Trade Terms

Chase
Membrane
Plasticity
Reveal

Thermal mass
Waterproofing
Wick

Trade Terms Introduced in This Module

Chase: A continuous recess built into a wall to receive pipes, wires, or heating ducts.

Membrane: A layer of thin, pliable material used to waterproof masonry.

Plasticity: A material's ability to be easily molded into various shapes.

Reveal: The visible portion of a masonry jamb between the face of the wall and the frame.

Thermal mass: The ability of a building wall to alternately absorb and release heat energy in response to temperature changes.

Waterproofing: The process of treating masonry with a material that will retard penetration of moisture.

Wick: A bundle of fibers that are loosely twisted or braided and woven together to form a cord that can carry water away from an area by capillary action, which occurs as long as the drip end is lower than the absorption end.

Additional Resources

This module presents thorough resources for task training. The following resource material is suggested for further study.

Alex Wilson. "Thermal Mass and R-value: Making Sense of a Confusing Issue." *Environmental Building News*, April 1998. **www.buildinggreen.com**

ASTM C494/C494M, Standard Specification for Chemical Admixtures for Concrete. 2013. West Consohocken, PA: ASTM International.

Below Grade Waterproofing Manual, Latest Edition. Kansas City, MO: Sealant, Waterproofing & Restoration Institute.

International Building Code, Latest Edition. Falls Church, VA: International Code Council.

A Practical Guide to Waterproofing Exterior Walls, Latest Edition. Kansas City, MO: Sealant, Waterproofing & Restoration Institute.

Technical Note 1, *Cold and Hot Weather Construction*. 2006. Reston, VA: The Brick Industry Association. **www.gobrick.com**

Technical Note 7, *Water Penetration Resistance – Design and Detailing*. 2005. Reston, VA: The Brick Industry Association. **www.gobrick.com**

Technical Note 7A, *Water Penetration Resistance – Materials*. 2005. Reston, VA: The Brick Industry Association. **www.gobrick.com**

Technical Note 7B, *Water Penetration Resistance – Construction and Workmanship*. 2005. Reston, VA: The Brick Industry Association. **www.gobrick.com**

TEK 3-1C, All-Weather Concrete Masonry Construction. 2002. Herndon, VA: National Concrete Masonry Association. **www.ncma.org**

TEK 3-13, Construction of Low-Rise Concrete Masonry Buildings. 2005. Herndon, VA: National Concrete Masonry Association. **www.ncma.org**

TEK 6-11A, Insulating Concrete Masonry Walls. 2010. Herndon, VA: National Concrete Masonry Association. **www.ncma.org/etek/Pages**

TEK 19-4A, Flashing Strategies for Concrete Masonry Walls. 2008. Herndon, VA: National Concrete Masonry Association. **www.ncma.org**

TEK 19-5A, Flashing Details for Concrete Masonry Walls. 2008. Herndon, VA: National Concrete Masonry Association. **www.ncma.org**

Figure Credits

Courtesy of Dennis Neal, FMA & EF, Mod Opener, Figure 6, Figure 28

Courtesy of National Concrete Masonry Association, Figure 3, Figure 5, Figure 8, Figure 13, Figure 15

Photo Courtesy of Owens Corning. THE PINK PANTHER™ and © 1964–2013 Metro-Goldwyn-Mayer Studies, Inc. All Rights Reserved. The color PINK is a registered trademark of Owens Corning, © 2013 Owens Corning, Figure 4, Figures 9–10

Courtesy of NRG Insulated Block, Figure 7A

Courtesy Of Rae Paravia, Las Vegas, NV, Figure 7B

Courtesy of Associated Builders & Contractors, Figure 11, Figure 16

Mortar Net Solutions, Figure 12, Figures 19–22

Courtesy: Heckmann Building Products, Inc., Figures 23–24

Courtesy of RLT Construction, SA01

Courtesy of the Brick Industry Association, Table 4–Table 5, Figure 31, Figure 33

Quikrete, Figure 29

Dennis Neal, Figure 30

Mr. Heater, Figure 32

Answer	Section Reference	Objective
Section One		
1. b	1.1.1	1a
2. c	1.2.2	1b
3. a	1.3.3	1c
Section Two		
1. b	2.1.1	2a
2. d	2.2.0	2b
3. b	2.3.0	2c
Section Three		
1. b	3.1.0	3a
2. a	3.2.2	3b
3. c	3.3.2	3c

NCCER CURRICULA — USER UPDATE

NCCER makes every effort to keep its textbooks up-to-date and free of technical errors. We appreciate your help in this process. If you find an error, a typographical mistake, or an inaccuracy in NCCER's curricula, please fill out this form (or a photocopy), or complete the online form at **www.nccer.org/olf**. Be sure to include the exact module ID number, page number, a detailed description, and your recommended correction. Your input will be brought to the attention of the Authoring Team. Thank you for your assistance.

Instructors – If you have an idea for improving this textbook, or have found that additional materials were necessary to teach this module effectively, please let us know so that we may present your suggestions to the Authoring Team.

NCCER Product Development and Revision
13614 Progress Blvd., Alachua, FL 32615

Email: curriculum@nccer.org
Online: www.nccer.org/olf

❏ Trainee Guide   ❏ Lesson Plans   ❏ Exam   ❏ PowerPoints   Other _____

Craft / Level: _____ Copyright Date: _____

Module ID Number / Title: _____

Section Number(s): _____

Description: _____

_____

_____

_____

_____

Recommended Correction: _____

_____

_____

_____

_____

Your Name: _____

Address: _____

_____

Email: _____ Phone: _____

28207-14

Construction Inspection and Quality Control

Quality inspection and testing are important for several reasons. Building owners want to be sure they get what they have paid for in terms of building quality. Architects and engineers want to be sure they get what they have specified in terms of loadbearing, performance, and appearance. Contractors want to be sure they get what they have paid for in terms of materials and craftworker skills. Masons want to be sure the materials they are using are the right materials, and the techniques they are using will give the right results for that job. This module introduces the concepts and tasks related to quality inspection and testing.

Module Seven

Trainees with successful module completions may be eligible for credentialing through NCCER's National Registry. To learn more, go to **www.nccer.org** or contact us at **1.888.622.3720**. Our website has information on the latest product releases and training, as well as online versions of our *Cornerstone* magazine and Pearson's product catalog.

Your feedback is welcome. You may email your comments to **curriculum@nccer.org**, send general comments and inquiries to **info@nccer.org**, or fill in the User Update form at the back of this module.

This information is general in nature and intended for training purposes only. Actual performance of activities described in this manual requires compliance with all applicable operating, service, maintenance, and safety procedures under the direction of qualified personnel. References in this manual to patented or proprietary devices do not constitute a recommendation of their use.

Objectives

When you have completed this module, you will be able to do the following:

1. Describe how standards and specifications are used to ensure quality control throughout the masonry industry.
   a. Describe the standards and specifications that apply to masonry units, mortar, grout, and accessories.
   b. Describe the standards that apply to laboratory and field testing of masonry construction.
2. Describe how masonry sample panels and prisms are built and tested to ensure quality control on a project.
   a. Describe how to build sample panels.
   b. Describe how to build hollow masonry prisms.
   c. Describe how to build grouted masonry prisms.
   d. Describe how to prepare and test mortar and grout prisms.
   e. Describe how to conduct masonry tests.
3. Describe how mortar is tested to ensure quality control on a project.
   a. Describe how to perform sand tests.
   b. Describe how to perform mortar consistency tests.
   c. Describe how to perform brick absorption tests.
   d. Describe how to perform laboratory tests.
4. Describe how field inspections and observations are used to ensure quality control on a project.
   a. Describe why and how standards and codes inspections are performed.
   b. Describe why and how materials inspections are performed.
   c. Describe the types of observations that are undertaken during construction.
   d. Describe why and how construction tolerances are monitored.

Performance Tasks

Under the supervision of your instructor, you should be able to do the following:

1. Build a prism for mortar testing.
2. Perform a slump test.

Trade Terms

Flexural	Quality assurance	Slump
Prism	Quality control	Tensile
Quality	Silt	

Industry-Recognized Credentials

If you're training through an NCCER-accredited sponsor, you may be eligible for credentials from NCCER's Registry. The ID number for this module is 28207-14. Note that this module may have been used in other NCCER curricula and may apply to other level completions. Contact NCCER's Registry at 888.622.3720 or go to **www.nccer.org** for more information.

Code Note

Codes vary among jurisdictions. Because of the variations in code, consult the applicable code whenever regulations are in question. Referring to an incorrect set of codes can cause as much trouble as failing to reference codes altogether. Obtain, review, and familiarize yourself with your local adopted code.

Contents

Topics to be presented in this module include:

Figures and Tables

1.0.0 STANDARDS AND SPECIFICATIONS IN MASONRY CONSTRUCTION

Objective

Describe how standards and specifications are used to ensure quality control throughout the masonry industry.

a. Describe the standards and specifications that apply to masonry units, mortar, grout, and accessories.

b. Describe the standards that apply to laboratory and field testing of masonry construction.

Trade Terms

Prism: A sample prepared under controlled conditions specifically for testing that uses the same materials, techniques, and craftsmanship as those used on the job site.

Quality: The totality of features and characteristics of a product or service that bear on its ability to satisfy stated or implied needs.

Quality assurance: All planned and systematic actions necessary to provide adequate confidence that an item or a facility will perform satisfactorily in service.

Quality control: Procedures for testing, inspecting, checking, and verifying to ensure that the work meets the required standard of quality established by the contract documents.

Quality is defined by the International Organization for Standardization as the totality of features and characteristics of a product or service that bear on its ability to satisfy stated or implied needs. On most construction jobs, the standards of quality are set in the contract documents and specifica-

tions. These usually incorporate specifications set in national and local building codes, and American Society for Testing and Materials (ASTM) materials specifications. ASTM also publishes procedures for testing materials to make sure they meet the specifications.

Architects and project engineers set up procedures for checking materials and work-in-progress against project specifications. This is known as quality assurance, which is defined as actions planned to provide adequate confidence that an item or facility will perform satisfactorily in service. The architect or engineer can try to ensure quality, but assurance is not the same as control.

One way architects and engineers ensure quality is to request a list of submittals. These are documents, samples, test results, and reports listed in the contract to be delivered over the course of the work. They can include shop drawings, product data, and samples. A typical submittal would include the following:

- *Product data* – Submit manufacturer's catalogs, test reports, and data for all products proposed for use, including installation, cleaning, and maintenance recommendations.
- *Samples* – Submit three units showing colors and textures.
- *Accessories* – Submit small samples of anchors, insulation, adhesive, vapor barrier, and other accessories for approval by the architect or engineer.

Quality control is usually defined in specifications as verification that eliminates the possibility of error. Quality control tasks are the procedures for testing, inspecting, checking, and verifying to make sure that the work meets the required standard of quality specified in the contract. Since the contractor has control over when and how the work is done, the contractor has the responsibility for quality control. Quality control and quality assurance both include field inspections and testing.

Testing requirements can be specified in the contract submittals. Typically submittals include samples, test reports, and certificates. Contractors

---

Death for Poor Workmanship

The first building code, *Hammurabi's Code of Laws*, 1780 BC, set fairly strict standards to ensure buildings were built safely:

"If a builder builds a house for someone, and does not construct it properly, and the house which he built falls in and kills its owner, then that builder shall be put to death."

"If it kills the son of the owner, then the son of the builder shall be put to death."

Laws 229-233

---

must provide samples of masonry units showing extreme variations in color and texture. The contractor must certify that the materials comply with the applicable specifications for grade, type, or class.

Test reports must be completed for each type of building and facing brick. The testing and reporting must be done by an independent laboratory. Typical tests include the following:

- Compressive strength
- Twenty-four-hour cold-water absorption
- Five-hour boil absorption
- Saturation coefficient
- Initial rate of absorption

1.1.0 Identifying the Standards and Specifications for Masonry Units, Mortar, Grout, and Accessories

Most of the masonry specifications in construction contracts reference building codes or standards. A standard usually has a single subject. The standard is based on studies, research, and advances in materials and construction techniques. Building codes are enforceable standards.

Model codes are developed by national or international organizations. These model codes are then adopted by states and municipalities. Once adopted they become enforceable laws. The *International Building Code®* (IBC) and the National Fire Protection Association's *NFPA 5000, Building Construction and Safety Code®*, are the primary codes used in the United States.

> **NOTE**
> Always check to make sure what building codes are used in your area. Codes are updated periodically; make sure you are using the latest edition.

Standards are technical specifications that establish minimum requirements for all aspects of masonry units and masonry work. ASTM standards are the most commonly used. The following organizations publish consensus standards:

- American Society for Testing and Materials (ASTM)
- The American Society of Civil Engineers (ASCE)
- The American Concrete Institute (ACI)
- American Institute of Steel Construction (AISC)

Trade associations also publish technical notes and advisories. While these are not standards, they are considered best practices. Some contracts require the use of best practices. These technical notes are available online at the organizations' websites. The following trade associations publish technical notes:

- The Brick Industry Association (BIA), **www.bia.org**
- The National Concrete Masonry Association (NCMA), **www.ncma.org**

General contract specifications usually include clauses such as "materials and equipment must be of good quality and new, and work must be free of defects and conform to specifications." A standard of quality can be set in different ways, depending on the method used to describe it. Quality standards can be set in one of the following ways:

- Descriptive specifications name exact properties of materials and methods of installation without using brand names.
- Proprietary specifications list specific products by brand name, model number, etc.
- Performance specifications describe the results of the work, the measures or criteria by which the results will be judged, and the methods by which the evaluation will be done.
- Reference standard specifications name and incorporate established industry quality and performance standards for products and processes.

There are more than 100 ASTM standards for masonry and masonry-related products, and others are currently in development. They are listed in the *Appendix*. Some of these standards apply to specialty products such as sewer block and firebrick, while other standards are designed for

---

Building Codes Converge

Prior to 2000 there were several organizations that issued model building codes. The most widely used were the *Southern Standard Building Code*, the *National Building Code*, and the *Uniform Building Code*. In 2000, these organizations merged into the International Code Council and issued the *International Building Code®* (IBC).

The National Fire Protection Association also issued a model code called *NFPA 5000*. It is used in a few communities and by the State of California.

---

research rather than construction work. Nevertheless, there are many standards that do apply to masonry work on the typical job. The most intensive use of reference standards is on large commercial projects.

The following sections list the most commonly referenced detailed standards for clay and concrete masonry units, mortar and grout, accessories, and product testing.

1.1.1 Clay Masonry Units

Most clay masonry units fall under *ASTM C216, Standard Specification for Facing Brick (Solid Masonry Units Made from Clay or Shale)*, and *ASTM C652, Standard Specification for Hollow Brick (Hollow Masonry Units Made from Clay or Shale)*. *ASTM C216* lists the following three types of facing brick:

- Facing brick standard (FBS), the industry standard
- Extra select brick (FBX), with tighter size tolerances
- Architectural brick (FBA), with greater size irregularities

If no type of brick is specified in a contract, the default is type FBS.

Brick is also graded according to its resistance to damage from freezing. The grades are: SW for severe weathering, MW for moderate weathering, and NW for no weathering. Grade SW is specified when brick is likely to freeze when it is saturated with water, and grade MW is specified when brick is likely to freeze when it is damp but not saturated. The grade is tested in accordance with *ASTM C67, Standard Test Methods for Sampling and Testing Brick and Structural Clay Tile*. *ASTM C216* also sets compressive strength standards for face brick, and requires efflorescence testing according to procedures in *ASTM C67*. If the contractor does not specify the grade, the default is SW.

Common or standard brick shapes, tolerances, and performance are set under *ASTM C62, Standard Specification for Building Brick (Solid Masonry Units Made from Clay or Shale)*. Grade requirements for SW and MW are summarized in *Table 1*. The last grade is used only for interior work with no weather exposure. The compressive-strength

requirements for face brick listed under *ASTM C216* also apply to common brick.

ASTM C652, Standard Specification for Hollow Brick (Hollow Masonry Units Made from Clay or Shale), covers hollow brick in several classes and grades. *ASTM C126, Standard Specification for Ceramic Glazed Structural Clay Facing Tile, Facing Brick, and Solid Masonry Units*, covers glazed brick and structural clay tile.

1.1.2 Concrete Masonry Units

The majority of CMUs fall under *ASTM C90, Standard Specification for Loadbearing Concrete Masonry Units*. There are two types of units: Type I

Table 1 Grade Requirements for Brick

Exposure	Weathering Index	
	Less than 50	50 and greater
In vertical surfaces: In contact with earth	SW or MW	SW
Not in contact with earth	SW or MW	SW
In other than vertical surfaces: In contact with earth	SW or MW	SW
Not in contact with earth	MW	SW

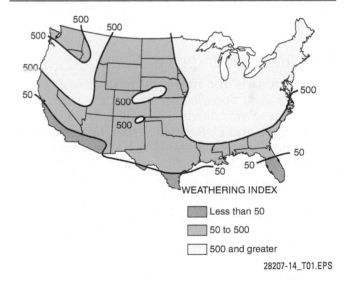

WEATHERING INDEX

■ Less than 50
▨ 50 to 500
□ 500 and greater

28207-14_T01.EPS

Masonry Building Code

The ACI and ASCE collaborated with other organizations and issued the *Building Code Requirements for Masonry Structures* and the *Specification for Masonry Structures*, through the Masonry Standards Joint Committee. They are commonly referred to as the *MSJC Code* and *MSJC Specification*, respectively. Both the IBC and NFPA 5000 reference the *MSJC Code*.

---

and Type II. Type I units have limits on shrinkage and moisture content, and must be kept dry at the plant, in transit, and at the job site. Type II units have limits on shrinkage but not on moisture content.

Weight classifications include light, medium, and normal. These weight classifications affect water and sound absorption, and fire resistance. Compressive strength requirements are the same for all types and weights: 1,900 pounds per square inch (psi). Compliance with *ASTM C90* is verified by testing in accordance with *ASTM C140, Standard Test Methods for Sampling and Testing Concrete Masonry Units and Related Units*.

ASTM C129, Standard Specification for Nonloadbearing Concrete Masonry Units, specifies unit types and weight categories for nonbearing CMUs. These have the same Type I and II classes and three weight categories as for loadbearing block. The major difference is in the loadbearing requirement, which is only 600 psi for these units.

ASTM C55, Standard Specification for Concrete Building Brick, specifies standards for loadbearing and nonbearing concrete brick. Grading is based on strength and weather resistance, with Grade N giving high strength and freeze resistance, and Grade S giving moderate strength and freeze resistance. Each grade can be either Type I, moisture controlled, or Type II, not moisture controlled. Minimum compressive strength is 3,500 psi for Grade N, and 2,500 psi for Grade S. Sampling and testing are done according to *ASTM C140*.

1.1.3 Masonry Mortar and Grout

The standards for masonry mortar are *ASTM C270, Standard Specification for Mortar for Unit Masonry*. These standards cover four types of mortar: M, S, N, and O. Reference specifications cover the cementitious agents, which can be portland cement (*ASTM C150, Standard Specification for Portland Cement*), masonry cement (*ASTM C91, Standard Specification for Masonry Cement*), hydraulic and slag cement (*ASTM C595, Standard Specification for Blended Hydraulic Cements*), quicklime (*ASTM C5, Standard Specification for Quicklime for Structural Purposes*), hydrated masonry lime (*ASTM C207, Standard Specification for Hydrated Lime for Masonry Purposes*), and mortar cement (*ASTM C1329, Standard Specification for Mortar Cement*).

ASTM C270 specifies both proportion and strength requirements for each of the four types of mortar. The proportion requirements are used by default in most contracts, unless the strength requirements are specifically included. Strength-specified mortar requires preconstruction laboratory testing according to *ASTM C270*. If field testing will also be done, both preconstruction and construction testing are done according to *ASTM C780, Standard Test Method for Preconstruction and Construction Evaluation of Mortars for Plain and Reinforced Unit Masonry*. This is necessary because laboratory-mixed mortar has less water in it than job-mixed mortar. The difference allows for the absorption of the masonry units. If both laboratory and field testing is to be done, the water content must be the same. This is covered under the specifications in *ASTM C780*.

Grout mixtures are specified by proportion or compressive strength in *ASTM C476, Standard Specification for Grout for Masonry*. Two types of grout, fine and coarse, are specified by proportions of ingredients. *ASTM C476* allows pumping admixtures for high-lift grouting. Strength testing for grout is done in accordance with *ASTM C1019, Standard Test Method for Sampling and Testing Grout*.

1.1.4 Masonry Accessories

Several different specifications cover masonry accessories. In most cases, the manufacturer certifies that the materials conform to the standards.

ASTM A36, Standard Specification for Carbon Structural Steel, covers size and strength requirements for the steel used for angle lintels and shelf angles in masonry construction. It also applies to bent bar or strap anchors. Coatings must be specified separately.

ASTM A82, Standard Specification for Steel Wire, Plain, for Concrete Reinforcement, covers steel wire used in joint reinforcement, some anchors, and ties. It includes strength and size requirements, but not coatings.

ASTM A951, Standard Specification for Steel Wire for Masonry Joint Reinforcement, covers materials, fabrication, tests and tolerances, and some coatings for prefabricated wire-joint reinforcement.

ASTM A153, Standard Specification for Zinc Coating (Hot-Dip) on Iron and Steel Hardware, covers galvanized coatings for steel-joint reinforcement, anchors, and ties. It also includes various masonry accessories.

ASTM A167, Standard Specification for Stainless and Heat-Resisting Chromium-Nickel Steel Plate, Sheet, and Strip, covers the steel most commonly used in anchors, ties, and flashing. It lists more than 20 types of stainless steel according to chemical composition.

ASTM A615, Standard Specification for Deformed and Plain Carbon-Steel Bars for Concrete Reinforcement, covers the deformed steel reinforcing bars used for vertical reinforcement and bond beams.

1.2.0 Identifying the Standards for Laboratory and Field Testing of Masonry Construction

Laboratory and field testing are mostly concerned with the compressive strength of mortar, grout, and masonry performing as a unit, and mortar or grout performing alone. On large commercial projects, these materials are tested before and during construction. Test samples are built before and after the actual laying has started.

ASTM C780 is a complex standard. It covers methods of sampling and testing mortar for plastic and hardened properties. This standard calls for a high water-to-cement mix design in laboratory testing to simulate actual construction conditions.

The tests performed under *ASTM C270* use a low water-to-cement ratio for the test mortars. These tests cannot be used to compare results with tests on mortars mixed on the job. Job-mixed mortars use a much higher water-to-cement ratio to overcome the absorption of the masonry units. The formulas in *ASTM C780* calculate a wetter mix design for laboratory testing to simulate job-mix conditions. These tests can then be used to accept or reject job-mixed mortars during construction.

ASTM C1019 covers both field and laboratory sampling for grout. This standard is referenced when compressive strength is to be tested by unit strength or by the prism method.

ASTM C1314, Standard Test Method for Compressive Strength of Masonry Prisms, is another complex standard. Project engineers specify the compressive strength of masonry for a particular job. The contractor verifies this in one of two ways: by unit strength or by prism test.

Verifying by unit strength calls for the contractor to submit the manufacturer's certification on the masonry units, and to specify that the mortar is mixed according to *ASTM C270* proportions. These documents allow the engineer to check compressive strength in lookup tables in *ASTM C1314,* without any further testing.

Verifying by prism test calls for a mason to build a test panel, or prism, using the selected units and mortar. The prism is then tested according to *ASTM C1314* procedures. This test can be used for preconstruction and construction evaluation of the masonry strength.

ASTM C67 sets procedures for sampling and testing individual clay or shale brick and structural clay tile. This standard covers procedures for sampling and testing brick for modulus of rupture, compressive strength, absorption, saturation coefficient, effect of freezing and thawing, initial rate of absorption (suction), and efflorescence. *ASTM C67* also includes procedures for measuring size and warpage. *ASTM C140* defines the procedures for testing CMUs. These and other test specifications are designed to measure the quality of the ingredients used in masonry construction.

Additional Resources

ACI 530/ASCE 5/TMS 402, Building Code Requirements for Masonry Structures, Latest Edition. Reston, VA: American Society of Civil Engineers.

ASTM A36, Standard Specification for Carbon Structural Steel, Latest edition. West Conshohocken, PA: ASTM International.

ASTM A82, Standard Specification for Steel Wire, Plain, for Concrete Reinforcement, Latest Edition. West Conshohocken, PA: ASTM International.

ASTM A153, Standard Specification for Zinc Coating (Hot-Dip) on Iron and Steel Hardware, Latest Edition. West Conshohocken, PA: ASTM International.

ASTM A167, Standard Specification for Stainless and Heat-Resisting Chromium-Nickel Steel Plate, Sheet, and Strip, Latest Edition. West Conshohocken, PA: ASTM International.

ASTM A615, Standard Specification for Deformed and Plain Carbon-Steel Bars for Concrete Reinforcement, Latest Edition. West Conshohocken, PA: ASTM International.

ASTM A951, Standard Specification for Steel Wire for Masonry Joint Reinforcement, Latest Edition. West Conshohocken, PA: ASTM International.

ASTM C5, Standard Specification for Quicklime for Structural Purposes, Latest Edition. West Conshohocken, PA: ASTM International.

ASTM C55, Standard Specification for Concrete Building Brick, Latest Edition. West Conshohocken, PA: ASTM International.

ASTM C62, Standard Specification for Building Brick (Solid Masonry Units Made from Clay or Shale), Latest Edition. West Conshohocken, PA: ASTM International.

ASTM C67, Standard Test Methods for Sampling and Testing Brick and Structural Clay Tile, Latest Edition. West Conshohocken, PA: ASTM International.

ASTM C90, Standard Specification for Loadbearing Concrete Masonry Units, Latest Edition. West Conshohocken, PA: ASTM International.

ASTM C91, Standard Specification for Masonry Cement, Latest Edition. West Conshohocken, PA: ASTM International.

ASTM C126, Standard Specification for Ceramic Glazed Structural Clay Facing Tile, Facing Brick, and Solid Masonry Units, Latest Edition. West Conshohocken, PA: ASTM International.

ASTM C129, Standard Specification for Nonloadbearing Concrete Masonry Units, Latest Edition. West Conshohocken, PA: ASTM International.

ASTM C140, Standard Test Methods for Sampling and Testing Concrete Masonry Units and Related Units, Latest Edition. West Conshohocken, PA: ASTM International.

ASTM C150, Standard Specification for Portland Cement, Latest Edition. West Conshohocken, PA: ASTM International.

ASTM C207, Standard Specification for Hydrated Lime for Masonry Purposes, Latest Edition. West Conshohocken, PA: ASTM International.

ASTM C216, Standard Specification for Facing Brick (Solid Masonry Units Made from Clay or Shale), Latest Edition. West Conshohocken, PA: ASTM International.

ASTM C270, Standard Specification for Mortar for Unit Masonry, Latest Edition. West Conshohocken, PA: ASTM International.

ASTM C476, Standard Specification for Grout for Masonry, Latest edition. West Conshohocken, PA: ASTM International.

ASTM C595, Standard Specification for Blended Hydraulic Cements, Latest Edition. West Conshohocken, PA: ASTM International.

ASTM C652, Standard Specification for Hollow Brick (Hollow Masonry Units Made from Clay or Shale), Latest Edition. West Conshohocken, PA: ASTM International.

ASTM C780, Standard Test Method for Preconstruction and Construction Evaluation of Mortars for Plain and Reinforced Unit Masonry, Latest Edition. West Conshohocken, PA: ASTM International.

ASTM C1019, Standard Test Method for Sampling and Testing Grout, Latest Edition. West Conshohocken, PA: ASTM International.

ASTM C1314, Standard Test Method for Compressive Strength of Masonry Prisms, Latest Edition. West Conshohocken, PA: ASTM International.

ASTM C1329, Standard Specification for Mortar Cement, Latest Edition. West Conshohocken, PA: ASTM International.

International Building Code®, Latest edition. Falls Church, VA: International Code Council.

NFPA 5000, Building Construction and Safety Code®, Latest edition. Quincy, MA: National Fire Protection Association.

1.0.0 Section Review

1. Technical notes published by trade associations are considered _____.

   a. specifications
   b. model codes
   c. standards
   d. best practices

2. Compared to the water-to-cement ratio used in mortar on the job, the water-to-cement ratio used for test mortar in tests performed under *ASTM C270* is _____.

   a. high
   b. low
   c. identical
   d. proportional

2.0.0 BUILDING AND TESTING MASONRY SAMPLE PANELS AND PRISMS

Objective

Describe how masonry sample panels and prisms are built and tested to ensure quality control on a project.

   a. Describe how to build sample panels.
   b. Describe how to build hollow masonry prisms.
   c. Describe how to build grouted masonry prisms.
   d. Describe how to prepare and test mortar and grout prisms.
   e. Describe how to conduct masonry tests.

Performance Task

Build a prism for mortar testing.

Trade Terms

Flexural: Capable of turning, folding, or bending.

Tensile: Capable of being stretched or extended, ductile.

The testing of masonry requires masons to perform several tasks. They build sample panels, which may be submittals under the contract documents; they make masonry, grout, and mortar prisms for laboratory testing; and they perform field tests and field observations. The next sections describe these tasks.

2.1.0 Building Sample Panels

On many jobs, a sample panel, or mock-up, must be built by the masonry contractor on the construction site before the masonry work begins. The sample panel serves two purposes:

- It can be observed throughout the construction period for any change or damage as a result of changing weather conditions or poor craftwork.
- It demonstrates the appearance and quality of the materials and the masonry skills that will be used on the project.

The size of the sample panel can vary with the complexity of the project. A typical sample panel is shown in *Figure 1*. The sample panel is made from the same materials that will be used in the structure; therefore, it will include all structural elements that are visible as well as those that are behind the wall (refer to *Figure 1*).

Because the panel is an example of what the finished structure will look like, the contractor typically sets the requirements for constructing the panel. In general, the panel should:

- Demonstrate the average skill of the masons working on the project
- Use the mortar type, ingredients, and admixtures specified for the project
- Use masonry units specified for the project and selected from the production run intended for shipment to the job
- Have the same coursing, bonding, and wall thickness as that specified for the project
- Have the same joint tooling as that specified for the project
- Have the same finishing operations or coatings as those specified for the project

The masonry units used in the sample should be randomly selected from those delivered to the construction site. Some units should be selected from each cube delivered so that any variations in color, texture, or strength can be determined before construction begins.

Since the sample panel will show the effects of weathering on the finished structure, it should be placed where it will be subject to the same weather conditions as the finished construction. No special curing or protective steps or skills should be applied that will not be used in the finished structure. It is the responsibility of the quality assurance representative, the architect, or the owner to approve the sample panel. You can also use the sample panel to check your own work and to fore-

---

Test Panels

The test panels, both sample panels and prisms, should be the same quality as the rest of the project. Masons building the panels should not build a test panel with more or less care or better materials than will be used on the project. The owner has a right to expect the work on the project to match that on the test panel.

FRONT OF SAMPLE PANEL

DETAIL OF SAMPLE PANEL
SHOWING PINTLE

DETAIL OF SAMPLE PANEL
SHOWING FLASHING AND
MORTARNET

28207-14_F01.EPS

Figure 1 Sample panel.

see any problems that may arise. For instance, if efflorescence begins to appear in the sample, the masonry supervisor may need to ask the contractor about changes in the mortar mixture.

2.2.0 Building Hollow Masonry Prisms

Masonry prisms are samples used for testing in a laboratory. The tests will determine the flexural, tensile, and compressive strength of the combined masonry and mortar. Just as with sample panels, prisms should be built using the same materials, techniques, and craftsmanship as the final structure. You may be asked to make as many as 10 test prisms for laboratory testing before construction begins. Prisms can be made in different sizes, as shown in *Figure 2*.

Test prisms are also made during construction for quality control of laying procedures and field operations. Typically, a minimum of three test prisms are made for every 5,000 square feet of wall area, or every story built. Two sets of three or five prisms each are sometimes required. These will be tested at 3, 7, and 28 days after building. The 7-day test is often used to approximate the 28-day strength; tests have shown that the 7-day strength usually equals 90 percent of the 28-day strength.

$$S_7 = 0.9 \times S_{28} \; or \; S_7 \div 0.9 = S_{28}$$

For example, if the seven-day test results in strength of 180 psi, the 28-day test can be calculated at 200 psi (180 ÷ 0.9 = 200).

The size may be specified for any prism. The thickness of the prism should be the same as the thickness of the masonry wall in the structure. The length of the prism should be equal to or greater than the thickness.

The tests more nearly reflect the performance of the wall if the prism is five times as high as it is thick, as shown in *Figure 3*. If there are no specifications, select the prism height using the parameters above; however, check with the laboratory to make sure the testing equipment is large enough to take a prism with the 5:1 dimension.

The mortar bedding, the thickness and tooling of joints, the moisture content of the masonry units at the time of laying, and the bonding should be as close as possible to that used in the structure. The prism should be laid up in stack bond unless specified otherwise. All masonry in the prism should be selected at random from the construction stockpile. Block should be laid with the thicker end of the face shell up. Full mortar bedding should be used in the prism construction

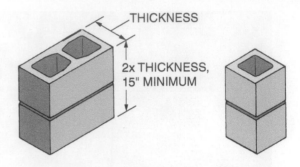

HOLLOW CONCRETE BLOCK PRISMS

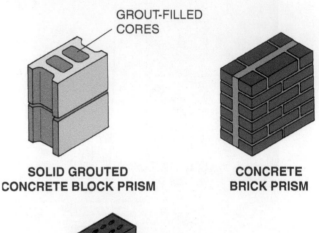

SOLID GROUTED CONCRETE BLOCK PRISM **CONCRETE BRICK PRISM**

PRISMS WITH DIFFERENT SLENDERNESS RATIOS (HEIGHT/THICKNESS RATIO)

28207-14_F02.EPS

Figure 2 Block and concrete brick test prisms.

if uncored masonry units are used. Hollow units should have only face-shell bedding.

Note that the walls may have collar joints mortared or grouted, as in the project specifications. Hollow units may also be grouted, as specified. However, no structural reinforcing rods should be used in the prisms. Metal ties or joint reinforcement can be used for the walls, but vertical or horizontal structural reinforcement is normally not included in the prism. A block prism should be grouted according to the job specifications, but no reinforce-

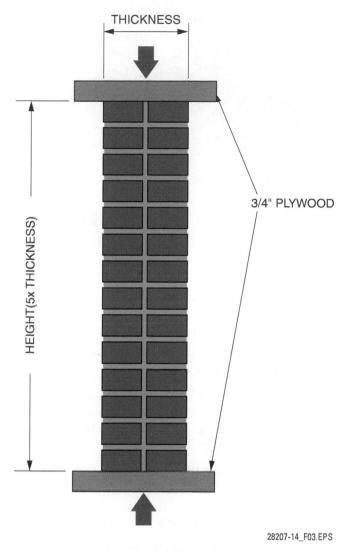

THICKNESS

3/4" PLYWOOD

HEIGHT(5x THICKNESS)

28207-14_F03.EPS

Figure 3 Compressive prism test.

ment should be used. Most prisms are tested by crushing, and steel bars could break the tester.

The prisms must be built on a flat surface where they will not be disturbed for 48 hours. A prism can be built on a foundation of an unattached ¾-inch plywood board. After the prisms are complete, mark each one for identification. After the 48 hours, the prism can be bundled to protect the bond. The prism is topped with another ¾-inch plywood board. The boards are bound tightly with two wires to keep the joints from being disturbed.

Preconstruction prisms are then sent to the laboratory, where they are cured at a temperature between 65°F and 85°F. Quality control prisms are cured at the job site. They must be left in a place where they will be undisturbed but subject to the same weather conditions as the structure on the job itself. Normally, the prisms are cured for 28 days, but some tests are made after only 3 days or 7 days of curing.

2.3.0 Building Grouted Masonry Prisms

Test prisms are also made with grout if the project calls for grouting. The prisms should be grouted only when the actual walls are to be grouted. Do not include any reinforcement in the prism except for the metal ties necessary for bonding the walls. When prisms are to be grouted, follow these steps:

Step 1 Make the block prism using the materials and techniques that will be used on the job. Be sure to place a nonabsorbent board or piece of ¾-inch plywood under the masonry.

Step 2 Allow the mortar joints to cure for 24 hours before pouring the grout.

Step 3 Pour the grout into the cores or collar joints in one layer up to 1½ inches from the top of the prism.

Step 4 Rod the grout 25 times with a 1' × 2" wooden stick, or vibrate the grout with a vibrator, to ensure that the void is completely filled. (If a vibrator is to be used on the job, the prism should be vibrated, not rodded.)

Step 5 Let the prism stand for about 30 minutes to allow the units to absorb some of the water from the grout.

Step 6 Finish filling the void area with grout, and vibrate or rod again at least 25 times. Rod at least 1½ inches into the first layer.

Step 7 Level the top of the prism with a trowel.

After the prism cures for 24 hours, it can be bundled with another piece of wood on top and sent to the laboratory.

2.4.0 Preparing and Testing Mortar and Grout Prisms

Mortar prisms, or test cubes, are normally 2 inches square. They are tested in the laboratory to check the strength characteristics of the mortar to be used on the job. A cube compressive-strength test is shown in *Figure 4*. These cube tests are made only on laboratory-mixed mortar if the mortar design has been specified by proportion according to *ASTM C270*. If the mortar design has been specified by strength according to *ASTM C780*, both lab-mixed and job-mixed mortar will be tested. The procedures for all the mortar tests are specified in *ASTM C1019*.

Some contractors may require you to periodically make mortar prisms so that the field mortar

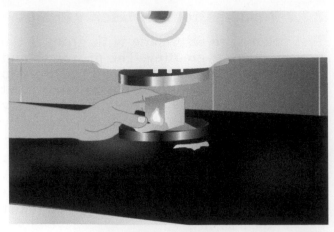

Figure 4 Cube compressive-strength test.

can be checked for uniform batching procedures. However, compressive strength is not specified for field-tested mortar. If you must make test cubes of mortar at the job site, a mortar cube mold or test tray should be used. The mortar should be taken from the latest batch mixed, and the cubes should be labeled.

Grout prisms are also tested. The job specifications will set the number of grout test prisms and the stages of the job at which they are to be made. Typically, grout is sampled whenever there is a change in the mix, materials, or method of mixing. About ½ cubic foot of grout is taken from work in progress to make three prisms. Make the grout test prism rectangles twice as high as wide. The usual size is 3½ inches by 3½ inches by 7 inches high.

These are the steps to follow in preparing grout test prisms, according to *ASTM C1019*:

Step 1   Select a flat location where the prism can harden undisturbed for 2 days.

Step 2   Place a flat piece of nonabsorbent wood, ⅝ inch thick and 3½ inches square, on the flat surface.

Step 3   Tape paper toweling to one face on each of four blocks.

Step 4   Place the block flush to the edges of the wood to make a cavity, with the papered faces on the inside of the cavity as shown in *Figure 5*.

Step 5   Pour grout halfway up into the cavity.

Step 6   Rod the cavity with a wood or metal rod 25 times to eliminate bubbles.

Step 7   Pour grout to fill the cavity completely.

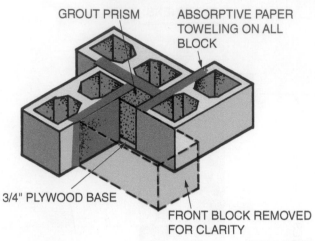

Figure 5 Grout prism mold of block.

Step 8   Rod the cavity 25 times to eliminate bubbles. Be sure the rod penetrates into the lower level also.

Step 9   Trowel the top of the prism flat, and cover it with wet burlap.

Step 10 After 2 days, separate the block to remove the prism. Pack it in sand or sawdust, and send it to the laboratory for further aging and testing.

The grout prisms will be tested for resistance to various stresses.

2.5.0 Testing Masonry Prisms

Masonry prisms are tested to determine the flexural, tensile, and compressive strength of the assembled masonry. The mortar bedding, the thickness and tooling of joints, the grouting, the moisture content of the masonry, and the bonding all affect the strength of the constructed wall. The stress tests discover the breakpoint of the particular prism. A change in the mortar formula and application will probably change the breakpoint more than any other factor. Tests have shown that increasing mortar strength by 160 percent increases the strength of the entire structure by 25 percent.

2.5.1 Impact Test

The impact test measures strength of bond by testing the ability of a masonry prism to withstand impulsive stresses that happen at different points and at random times. The test requires dropping a 60-pound sandbag to hit the middle of the test prism directly on its face, as shown in *Figure 6*. A sandbag is used so the masonry units will not be damaged.

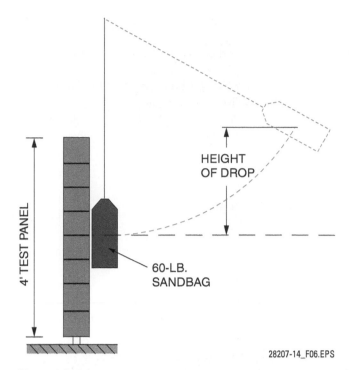

Figure 6 Impact test.

The test prisms are 48 inches high and are supported by a base 6 inches less than the height of the panel; a prism 48 inches high would be supported along 42 inches of its base. The distance from which the bag is dropped is increased until failure occurs. However, the bag is never dropped or swung from above 20 feet. The reported value is the distance of the drop at which the prism breaks.

2.5.2 Flexural Test

The flexural-strength bond test measures the amount of force a masonry beam or wall can resist when the force is applied against the face of the masonry. This test gives information that is important for structures subject to high winds or other types of lateral loads. Under these conditions, walls bend or flex to some degree, which can crack the mortar joints if the strength of bond is not high enough.

There are three methods to test flexural strength: by using a ⅓-point concentrated load, a uniform load, or a uniform transverse load. With the ⅓-point load or the uniform load method, the prism is loaded in the test machine as if it were a simply supported masonry beam. As shown in *Figure 7*, the specimen should be placed with the finish face down and the backup side on the top, or load, side. The load is applied at the two points equidistant from the ends and each other.

With the transverse-load method, the prism is placed upright. A plastic or rubber bag is placed between the wall face and a backboard, and filled

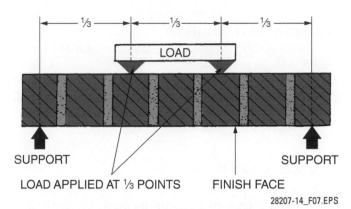

Figure 7 One-third-point flexural-strength test.

with air to give various amounts of pressure against the prism. The amount of pressure on the wall and the degree of movement that results are recorded for use in a detailed design analysis.

2.5.3 Tensile and Other Tests

The tensile-strength bond test measures a prism's resistance to twisting apart. The top bearing block turns the prism in one direction, while the lower bearing block turns the prism in the opposite direction.

The bond wrench test tests prisms of full-sized masonry units rather than cubes. The test machine wrenches the bonded units apart. It measures the flexural tensile strength of the bond.

A diagonal shear test measures the amount of stress a prism will resist along its diagonals. A test machine applies compressive force along the diagonal so that the shear stress is applied to the mortar joints, as shown in *Figure 8*. The prism fails when the applied loads exceed the strength of bond of the mortar and masonry.

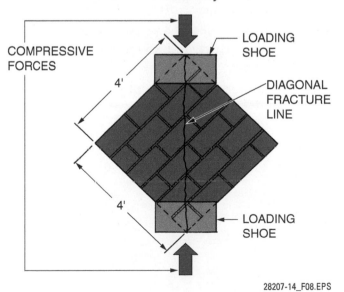

Figure 8 Diagonal tensile shear test.

Additional Resources

ASTM C270, Standard Specification for Mortar for Unit Masonry, Latest Edition. West Conshohocken, PA: ASTM International.

ASTM C780, Standard Test Method for Preconstruction and Construction Evaluation of Mortars for Plain and Reinforced Unit Masonry, Latest Edition. West Conshohocken, PA: ASTM International.

ASTM C1019, Standard Test Method for Sampling and Testing Grout, Latest Edition. West Conshohocken, PA: ASTM International.

2.0.0 Section Review

1. The dimensions of a sample panel are typically _____.

   a. 64 inches long and 48 inches high
   b. 48 inches long and 64 inches high
   c. 32 inches long and 24 inches high
   d. 24 inches long and 32 inches high

2. For a project that will involve the construction of 20,000 square feet of wall area, the typical minimum number of test prisms to be made will be _____.

   a. 12
   b. 9
   c. 6
   d. 3

3. When building a grouted masonry prism, after rodding or vibrating the grout, allow the units to absorb water from the grout by letting the prism stand for approximately _____.

   a. 60 minutes
   b. 45 minutes
   c. 30 minutes
   d. 15 minutes

4. The typical dimensions of mortar test cubes are _____.

   a. 1 inch by 4 inches
   b. 2 inches by 3 inches
   c. 3 inches square
   d. 2 inches square

5. The test that measures the amount of force a masonry beam or wall can resist when the force is applied against the face of the masonry is called the _____.

   a. tensile test
   b. flexural test
   c. bond wrench test
   d. impact test

3.0.0 TESTING MORTAR

Objective

Describe how mortar is tested to ensure quality control on a project.

    a. Describe how to perform sand tests.
    b. Describe how to perform mortar consistency tests.
    c. Describe how to perform brick absorption tests.
    d. Describe how to perform laboratory tests.

Performance Task

Perform a slump test.

Trade Terms

Silt: A sedimentary material consisting of very fine particles intermediate in size between sand and clay.

Slump: A measure (0 to 12 inches) of the consistency or fluidity of a cementitious product such as concrete, masonry, or grout. Lower slump (1 to 4 inches) indicates a stiff mix; higher slump (8 to 12 inches) indicates a fluid mix.

A mason must know the expected properties of each type of mortar and be able to recognize these properties. Mortar tests may be performed to measure the following qualities of mortar:

- Consistency or wetness
- Water retention
- Plasticity
- Strength of bond
- Compressive strength
- Durability
- Flexural strength

Consistency or wetness is determined by the amounts of both water and sand added to the mixture. The consistency of mortar should vary according to the ASTM mortar type. Mortar with a good consistency (*Figure 9*) should be workable, but should not flow freely.

Water retention affects all other qualities. Mortars with high water retention stiffen more slowly. Mortars with low water retention may dry too fast to allow adjusting the masonry units, and may cause weak joints that are not weather tight.

28207-14_F09.EPS

Figure 9 Mortar with good consistency.

Water retention is especially important on hot, dry, and windy construction sites where mortar may quickly lose its workability because of water evaporation.

Plasticity is related to water retention and consistency. The mortar must be plastic enough to stay on a trowel; it must spread easily but hold a uniform thickness. It must stay soft enough to adjust the masonry units before it sets up.

Strength of bond gives the mortar its ability to tie units together. Strength of bond enables masonry to withstand compressive, tensile, and shear stresses.

Compressive strength in mortar increases the masonry unit's resistance to downward force or

Plasticity Test

Testing the plasticity of freshly mixed mortar is simple. Load the trowel with mortar and turn it upside down. The mortar should stay on the trowel. Repeat this test for each new batch of mortar.

28207-14_SA01.EPS

load. The ASTM classifies mortars according to their compressive strength.

Durability is the ability to withstand weathering and freeze-thaw cycles without deterioration.

Flexural strength is the ability to withstand bending forces. Masonry structures may expand or contract due to weather conditions, alternate wetting and drying, or as a result of water loss. Horizontal forces from winds or earthquakes also tend to bend or flex the masonry. These stresses must be resisted by the mortar's flexural strength.

Mortar will have all desired properties only if the proper materials are correctly proportioned and mixed. Some mortar properties can only be determined through laboratory testing, while others can be tested by quick field tests. You should know how to perform the field tests used to check the quality of materials or mortar mixes, and how to prepare samples used in laboratory testing.

3.1.0 Performing Sand Tests

Since the ingredients of mortar directly affect its quality, field tests are used before the mortar is mixed. There are three field tests that can be used to judge the quality of sand:

- *Standard screen sieve test* – This test ensures that the sand is of a grade suitable for mortar.
- *Siltation test* – This test determines the purity of sand.
- *Colorimetric test* – This test indicates the amount of salt or organic materials in the sand.

3.1.1 Sieve Test

The standard screen sieve test checks the gradation of the sand. Sand that is properly graded will result in a mortar that is very uniform and workable. Properly graded sand contains a variety of particle sizes. The small particles will fill the gaps between the larger particles to help maintain the consistency of the mortar. Poorly graded sand has particles all the same size or does not have a variety of sizes. This results in a mixture that has a large proportion of void spaces between the particles, which will result in an inconsistent mortar.

The sieve test determines the percentage of sand in different size classifications. It is performed by passing the sand through a series of sieves or screens with progressively smaller openings. A Size 4 sieve, for example, allows particles ¼ inch and smaller to pass through, while a Size 200 sieve allows the passage of only very fine material. *ASTM C144, Standard Specification for Aggregate for Masonry Mortar*, establishes gradation limits for mortar sand; these should be followed

to ensure a well-graded sand. *Table 2* shows the recommendations for percent of sand by size of the grains.

The proportions of the different sizes of sand can be checked using the specifications set by *ASTM C144*. The process calls for sifting a measured amount of sand through various sizes of sieves and measuring the sand that passes through each. The procedure is as follows:

Step 1 Fill a box with sand, and weigh the box and sand. Record this information as the initial or before weight, or B.

Step 2 Beginning with the largest sieve size, sift the sand through the screen into another container.

Step 3 Place all sand that passes through the sieve back into the first box. Weigh the box and the sand that passed through the sieve. Record this information as the after weight, or A, and record the size of the sieve.

Step 4 Take the next smaller sieve and repeat the procedure and recordings, using each sieve size called for in the job specifications. Different jobs will have different requirements for the fineness of the sand.

Step 5 Determine the percentage passing each sieve size by dividing the after weight, A, by the before weight, B, for each sieve size. The formula is:

$$(\text{weight } A_n \div \text{weight } B_n) \times 100 = \text{percentage passing}$$

For example, if the box and sand weigh 25 pounds before sifting and 20 pounds after sifting, the percentage passing through this sieve is 80 percent:

$$(20 \div 25) \times 100 = 80\%$$

Table 2 Recommended Sand Gradation Limits

	Percent Passing	
Sieve Size	Natural Sand	Manufactured Sand
No. 4	100	100
No. 8	95–100	95–100
No. 16	60–100	60–100
No. 30	35–70	35–70
No. 50	15–35	20–40
No. 100	2–15	10–25
No. 200	–	0–10

The sieve sizes recommended for grading mortar sand are given in *Table 2*. It specifies that the gradation of the material from any one source shall be reasonably uniform and shall not be permitted to vary over the extreme range. It also recommends that sand on any one job be supplied from the same source. Sand from different sources may be very similar in gradation but may have other differences that will result in different mortar qualities.

3.1.2 Siltation Test

The siltation test determines the presence of silt, clay, or other earth particles in the sand. The sand must be clean or the strength of bond, color, and adhesive strength of the mortar will be affected. The following simple process can be performed to measure the presence of loam, soil, silt, clay, mica, or humus:

Step 1 Fill a clear glass quart container with 3 inches of sand.

Step 2 Partially fill the container with potable water (2 or 3 inches above the sand), and shake the container vigorously.

Step 3 Let the mixture stand for at least an hour. The sand will settle to the bottom. Any organic material, such as silt or loam, in the sample will settle just above the sand, as shown in *Figure 10*.

Step 4 Measure the layer of silt or loam. If the layer is more than ⅛ inch thick, the sand is not clean enough for mortar.

If the sand must be used, it must be thoroughly washed with potable water and retested after washing. In this case, it may be easier to replace the sand.

3.1.3 Colorimetric Test

While the siltation test is used to determine the presence of earth contaminants in the sand, the colorimetric test shows the amount of salts and organic material in the sand. Organic material can reduce the strength of bond of the mortar. The following steps can be used for determining the presence of organic material in aggregates:

Step 1 Make a 3 percent solution of sodium hydroxide by dissolving 1 ounce of sodium hydroxide in 1 quart of distilled water. Sodium hydroxide can burn skin or clothes, so take appropriate care to avoid injury.

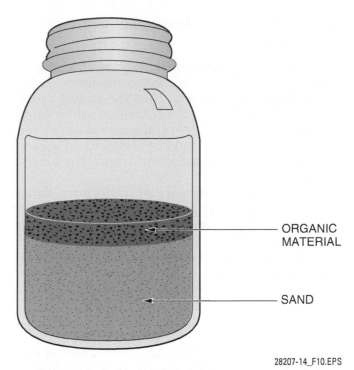

ORGANIC MATERIAL

SAND

28207-14_F10.EPS

Figure 10 Siltation test container.

Step 2 Use a container that has graduated ounce markings. Fill the container with the sand to the 4½-ounce mark.

Step 3 Add the sodium hydroxide solution to the 7-ounce mark. Seal the container tightly, and shake it vigorously.

Step 4 Let the container stand for at least 12 hours. The color of the liquid will show the amount of organic material in the sand.

WARNING!
Sodium hydroxide is caustic and can burn your skin. Always wear hand and eye protection and other personal protection equipment when working with sodium hydroxide or with proprietary cleaners that include it as an ingredient.

See *Figure 11*. If the liquid is clear, the sand is free from impurities and may be used for mortar. If the liquid is slightly yellowed or straw-colored, the sand has some organic content and should be thoroughly washed before it is used. If the liquid is any darker in color, the sand should not be used; the dark color indicates a high degree of foreign matter in the sand.

3.2.0 Performing Mortar Consistency Tests

Immediately after mortar is mixed, it may be tested for plasticity and color consistency. To test mortar plasticity, simply turn the trowel upside down. The mortar should not fall off immediately. Mortar that easily slides off the trowel with a rolling or throwing motion but clings to an upside-down trowel is considered to have the correct plasticity.

To test color consistency or uniformity, flatten some mortar under a trowel. If streaks of color appear, additional mixing is needed. To test color consistency after the mortar has set, make, label, and place test cubes outside at the job site. Make these cubes from the same type of mortar and with the same materials as will be used on the job. Allow the cubes to sit at the job site for at least 5 days before checking the effect that curing and the weather have had on the mortar color.

A slump test may be used to ensure that each batch of mortar has the same consistency. Specifications are normally not established for the slump of mortar, but a slump test will ensure that the mortar consistency is constant throughout the job. Specifications are usually set for the slump of grout, so grout must be tested and records maintained. *Figure 12* shows the relative stiffness of concrete, mortar, and grout. The stiffer concrete and mortar will keep the shape of the cone better than the more fluid grout.

> NOTE
>
> The entire slump test from the start of the filling to the completion of the cone lift should be accomplished in 2½ minutes.

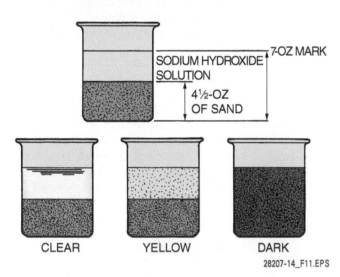

Figure 11 Colorimetric test.

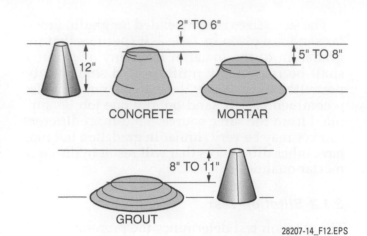

Figure 12 Slump-test comparison.

The slump-test procedure is as follows:

Step 1 A moistened cone mold (*Figure 13*) is placed on a firm, level, moistened, nonabsorbent surface, and is held in place by standing on the foot pieces.

Step 2 The cone is filled with fresh mortar or grout to one-third full by volume, and vibrated or rodded evenly over the surface area 25 times with a standard steel ⅝-inch smooth tamping rod with a rounded end (*Figure 14*).

Step 3 The next layer of fresh mortar or grout is added until the cone is two-thirds full by

Figure 13 Slump cone mold.

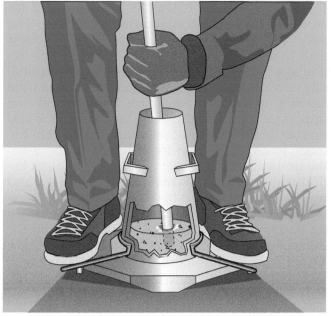

28207-14_F14.EPS

Figure 14 Cone mold one-third full.

volume. This layer is vibrated or rodded evenly 25 times with the rod just penetrating the first layer *(Figure 15)*.

Step 4 The cone is then heaped to overflowing, and this third layer is vibrated or rodded evenly 25 times with the rod just penetrating the second layer *(Figure 16)*. If the mortar or grout sags below the top of the cone, then more mortar or grout is added to keep the mortar or grout above the top of the cone.

28207-14_F15.EPS

Figure 15 Cone mold two-thirds full.

Step 5 After rodding or vibration, the excess mortar or grout is scraped off the top of the cone with the rod *(Figure 17)*. Any spilled mortar or grout is cleaned from around the base of the cone.

Step 6 Using the hand grips, the cone is lifted vertically. This must be done slowly and carefully within 3 to 7 seconds, while avoiding rotational movement or bumping the molded mortar or grout. The

28207-14_F16.EPS

Figure 16 Cone mold full and overflowing.

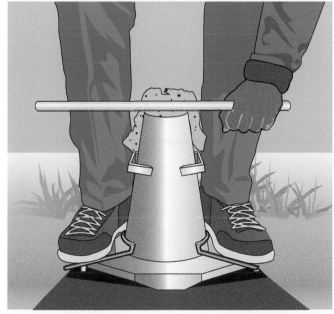

28207-14_F17.EPS

Figure 17 Leveling top of cone mold.

cone is then placed upside down next to, but not touching, the slumped mortar or grout. The rod is laid across the cone over the mortar or grout *(Figure 18)*.

Step 7 The distance from the bottom of the tamping rod to the sagged top of the mortar or grout at the original center of the cone is measured to the nearest ½ inch to determine the slump *(Figure 19)*. Record the slump and discard the test sample.

A 5- to 8-inch fall, or slump, is usually specified for mortar. The more fluid in the mortar, the looser the consistency and the greater the slump. Grout for use in reinforced masonry should have a slump of 8 to 11 inches. The slump for both grout and mortar should be consistent through the span of the job. A minimum of one-half cubic foot of grout or mortar should be taken for slump or strength tests whenever sampling is specified.

3.3.0 Performing Brick Absorption Tests

Concrete masonry units are cured and dried at the manufacturing plant. When delivered to the job site, their moisture content should be within specified limits. Concrete units should never be moistened before or during laying, because they shrink as they dry and the resulting stress will crack the wall.

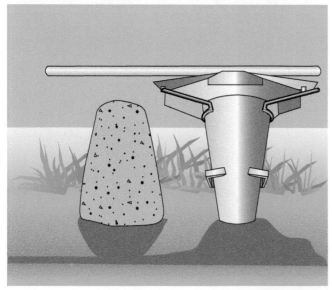

28207-14_F18.EPS

Figure 18 Inverted cone with rod over slumped mortar.

Brick, on the other hand, is not controlled for moisture during manufacturing or curing. The pores (or small openings) in fired-clay products tend to draw or suck water into the unit. This action in a brick is referred to as its suction, or initial rate of absorption. The suction of brick has little bearing on its water resistance, but does have an important effect on the bond between brick and mortar. If the rate of absorption is too high, too much water is diverted from the mortar between the units. Excessive water absorption reduces hydration, changes the strength of the mortar, and results in a weak bond that is not watertight.

To prevent these conditions, brick can be tested for absorption. The method for determining the initial rate of absorption of brick is described in *ASTM C67*. It consists of immersing a dry unit in water to a depth of ⅛ inch for one minute, after which it is removed, weighed, and the final weight compared with its dry weight.

Suctions of brick produced commercially vary from 3 grams per minute to more than 112 grams per minute. There is no consistent relationship between total absorption and suction. Some brick with high total absorptions have low suctions, and vice versa. The suction of the brick when laid is of primary importance and can be controlled at the job site by wetting. Suction is not controlled by product specifications.

ASTM C67 specifications call for wetting the brick at the time of installation if the absorption rate is greater than 30 grams of water per minute over 30 square inches, or 0.035 ounces per square inch per minute at the time of installation. A simple rule-of-thumb absorption test is as follows:

Step 1 Mark a circle 1 inch in diameter on one of the brick to be used in the job. A quarter will serve as a good pattern. Make certain that the surface to be mortared is used for the test.

Step 2 Inside the circle, drop 20 drops of water from a common medicine dropper.

Step 3 Using a watch with a second hand, note the time it takes for the masonry to absorb the water.

If the time for the brick to absorb the water is longer than 1½ minutes, the units need no further preparation before laying. If the absorption time is less than 1½ minutes, the units are very porous and need to be thoroughly wetted before laying. The water can be applied by spray, drip, or soaker hose. In wetting high-suction brick, sprinkling is not enough. A hose stream should be sprayed on the brick pile until water runs from all sides.

Figure 19 Measuring the slump.

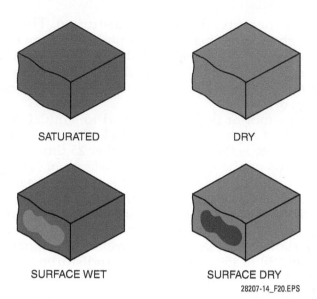

Figure 20 Moisture condition of brick.

The surface of the brick should then be allowed to dry before the brick is laid in the wall. A surface film of water will cause the brick to float on the mortar bed and will prevent proper bonding. Checking a broken unit, as in *Figure 20*, will show whether moisture is evenly distributed.

Masonry units with too much moisture take more time to set because they cannot absorb much water from the mortar. Units with no moisture prevent the mortar from curing properly. Hard-finished brick should be covered on the job so that they do not get too wet. Soft-finished or high-suction brick, on the other hand, must usually be wetted before laying in order to obtain the proper bond.

3.4.0 Performing Laboratory Tests

Laboratory tests are important when mortar is specified by strength rather than by proportions. Mortar is tested when hardened and when plastic. For strength testing, mortar is put into a 2-inch cube mold and cured for 24 hours. It is then unmolded and stored for 27 days to get its 28-day strength. Occasionally, grout is set in a 2-inch cube mold for testing in addition to the larger grout prisms. Grout cubes are moist-cured for 48 hours, then tested.

It is recommended that three mortar cubes be tested for every 100,000 masonry units used. Tests of plastic mortar should also be made on an ongoing basis as the job progresses.

3.4.1 Flow and Flow-After-Suction Tests

The flow-after-suction test measures the ability of plastic mortar to retain water and cohesiveness after the masonry units have absorbed water from it. The test is done in the laboratory using a flow table, flow plate, and mortar made from the same design and materials as that at the job site.

The mortar is put into a mold, rodded, and unmolded on a flow table. The spread of the mor-

Tempering Mortar

Mortar that has become too stiff can be tempered by adding water. The mortar must be thoroughly remixed. This can be done in the wheelbarrow or on a board. Small additions of water will slightly reduce the compressive strength of mortar. Temper colored mortars carefully to avoid color variations. Mixing smaller batches will reduce the need to temper.

tar is measured before it is disturbed. Then, a weighted flow plate is dropped on the mortar from a distance of ½ inch above. The dropping is done 25 times in 15 seconds, and the resulting spread of the mortar is measured. This is the measure of flow.

The mortar is then suctioned to simulate the effect of highly absorbent brick. The weighted flow plate is then dropped another 25 times in 15 seconds on the mortar from a distance of ½ inch. The resulting spread of mortar is measured again. This is the measure of flow after suction.

For laboratory testing, the allowable initial flow ranges from 105 to 115 percent of the undisturbed mortar. Testing in the field gives results ranging from 130 to 150 percent of the undisturbed mortar. After suctioning and the second disturbance, the flow after suction must exceed 75 percent of the original undisturbed diameter.

3.4.2 Cone Penetration Test

The cone penetration test is another method of testing the consistency of plastic mortar. It measures the depth of penetration of a cone-shaped plunger into a mortar sample.

For the test, a metal cup measuring 3 inches in diameter by 3½ inches in depth is filled with the mortar to be tested. The mortar is evenly tamped or spaded so that no air pockets are trapped in the sample. The prepared sample is placed under a cone penetrometer, which drops a metal cone into the mortar and measures the depth the cone penetrates the mortar. The measurement after 30 seconds is the measure of mortar consistency, and is reported in millimeters.

The cone penetration test is also used to measure how long mortar will remain consistent. To measure consistency retention, the cone penetration test is done on several samples at intervals of 15 minutes. These samples are taken directly from the mixer. The laboratory first sets limits on desirable consistency and then measures the time, usually in minutes, during which the mortar keeps that same consistency.

3.4.3 Other Tests

Many other mortar tests are available. There are mortar tests to measure the aggregate-to-cement and aggregate-to-lime ratios (*ASTM C780*), the water content (*ASTM C780*), and the air content (*ASTM C780*; *C231, Standard Test Method for Air Content of Freshly Mixed Concrete by the Pressure Method*; *C173, Standard Test Method for Air Content of Freshly Mixed Concrete by the Volumetric Method*; and *C138, Standard Test Method for Density (Unit Weight), Yield, and Air Content (Gravimetric) of Concrete*). These tests require the use and knowledge of complicated test equipment and procedures, and are used in choosing a mortar for a particular engineering application.

Additional Resources

ASTM C67, Standard Test Methods for Sampling and Testing Brick and Structural Clay Tile, Latest Edition. West Conshohocken, PA: ASTM International.

ASTM C138, Standard Test Method for Density (Unit Weight), Yield, and Air Content (Gravimetric) of Concrete, Latest Edition. West Conshohocken, PA: ASTM International.

ASTM C144, Standard Specification for Aggregate for Masonry Mortar, Latest Edition. West Conshohocken, PA: ASTM International.

ASTM C173, Standard Test Method for Air Content of Freshly Mixed Concrete by the Volumetric Method, Latest Edition. West Conshohocken, PA: ASTM International.

ASTM C231, Standard Test Method for Air Content of Freshly Mixed Concrete by the Pressure Method, Latest Edition. West Conshohocken, PA: ASTM International.

ASTM C780, Standard Test Method for Preconstruction and Construction Evaluation of Mortars for Plain and Reinforced Unit Masonry, Latest Edition. West Conshohocken, PA: ASTM International.

3.0.0 Section Review

1. The test that determines the purity of sand is called the _____.

   a. colorimetric test
   b. standard screen sieve test
   c. siltation test
   d. tensile test

2. When testing the color consistency of mortar, additional mixing is needed if after flattening the mortar under a trowel, the mortar has _____.

   a. inconsistent thickness
   b. uniform thickness
   c. lumps of color
   d. streaks of color

3. The initial rate of absorption of clay brick has an important effect on _____.

   a. mortar bond
   b. water resistance
   c. crack resistance
   d. hardening time

4. The method of testing the consistency of plastic mortar that measures the depth of penetration of a plunger into a mortar sample is called the _____.

   a. volumetric air test
   b. aggregate-to-cement ratio test
   c. flow-after-suction test
   d. cone penetration test

Section Four

4.0.0 Field Inspections and Observations

Objective

Describe how field inspections and observations are used to ensure quality control on a project.

   a. Describe why and how standards and codes inspections are performed.

   b. Describe why and how materials inspections are performed.

   c. Describe the types of observations that are undertaken during construction.

   d. Describe why and how construction tolerances are monitored.

The job contract lists in detail, directly or by incorporation of standards, the specifications for submittals, materials, mixes, testing, preparation, installation, cleaning, and inspections. Field inspections may also include reviewing the level of skill exhibited by the craftworkers. For masonry, especially, the skill and thoroughness of the worker can mean the difference between acceptance or rejection of the finished product. Inspections may include items such as the following:

- Storing and protecting materials
- Mortar mixing, tempering, placing, tooling, and pointing
- Wetting brick with high initial rates of absorption, unit blending, and unit placing
- Keeping CMUs dry
- Installing connectors, joint reinforcement, control joints and/or expansion joints, flashing, and weep holes
- Placing reinforcement and grouting
- Installing temporary bracing and shoring
- Curing and protecting masonry during construction
- Cleaning masonry

Know what inspectors look for, and inspect your own work first. Knowing and previewing what the inspector will check is an excellent way of making sure your own work is up to standard.

4.1.0 Performing Standards and Codes Inspections

Field observation and inspections have become more common and more important with the in-crease in construction litigation. The intention behind field inspection is to ensure that the finished work complies with the contract specifications and that the craftwork meets the required standards.

Good craftwork affects masonry performance over time and is essential to high-quality construction. Masonry construction requires skilled craftworkers working cooperatively with the architect and engineer to execute the design. The goal of quality craftwork is common to all concerned parties.

Responsibility for the quality of a project ultimately lies with the contractor. The architect, engineer, or any independent inspection agencies or testing laboratories are there to verify compliance with standards, designs, and performance criteria; in addition to the contractor, they also may perform field observations and inspections.

Supervisors should see every part of the job at least twice a day. They should continually inspect ongoing work to make sure that the procedures are safe, the work sound, and the progress satisfactory. Federal and state inspectors visit public jobs to check craftwork and progress. County and/or city inspectors visit job sites to check conformance to building codes. Inspectors check for conformance to general construction, electrical, plumbing, fire safety, occupational safety, and public health codes. Architects or project engineers check that the work is in conformance with the specifications.

Many inspectors come unannounced and on an irregular basis. Compliance inspectors usually come when notified of the completion of a stage of work. Routine inspections will take less time if the inspectors or supervisors tracking the work remember the following points:

- Check the specifications for notes of inspections or stages of work that require certificates from officials or insurance companies.
- Check the specifications for the submittals list; note the inspection reports required.
- Contact the agency that granted the building permit for a list of required inspections.
- Ask the mechanical contractors, before they cover their work, if all required inspections are complete. Intermittent subcontractors may forget to tell the supervisor that an inspector is needed before the work can be covered.
- Learn which inspectors insist on seeing the work and which ones allow covering the work without inspection if they are late.
- Call the local fire department to find out what inspections and permits they require.

- Ask the architect or engineer for a list of inspections they have performed as well as those they have scheduled, such as mortgage holders, Federal Housing Administration (FHA) or US Department of Veterans Affairs (VA), tenants, or independent inspecting services.
- Call the health department for regulations about temporary toilets, water supply, sewage, and swimming pools, and ask if any inspections are needed.
- If necessary, ask the permitting agency about permits and inspections for working on public property, obstructing traffic, or connecting to utilities.
- Post a list so that involved supervisors and workers will know the inspection type and schedule.

4.2.0 Performing Materials Inspections

An inspector must be familiar with the project specifications and must verify compliance with standards. Materials inspection is the first part of the process. Manufacturers must supply test certificates showing that materials meet or exceed referenced standards. The inspector must check the certificates, then inspect the materials to make sure they match the certificates and orders.

The inspector checks masonry units for color, texture, size, and match to samples. They are rejected if they are soiled, chipped, or broken in transit. If they do not have certificates, the inspector picks random samples and sends them to a testing laboratory. The inspector checks the moisture condition of clay units and the dryness of CMUs at the time of laying.

The inspector checks mortar and grout ingredients for contamination, and rejects contaminated materials or bagged materials that show signs of water absorption. Packaged materials should be sealed with the manufacturer's labels legible and intact. The inspector checks materials certificates and bills of lading against the specifications and the purchase orders to make sure the material was ordered to specifications and delivered as ordered. Accessories are also checked for design compliance and identified for location within the structure.

The next step is to check that the materials are stored properly. Masonry materials and units must be protected against weather and contamination. CMUs and accessories must be protected against moisture as well.

Acceptable mixing and batching procedures should have been set at the preconstruction conference. The inspector checks that these procedures are being followed. The inspector monitors tempering time to ensure the mortar is fresh and not beginning to set.

Check mortar and grout for contamination, materials for appropriate storage, and accessories for design and location. Use the mortar mixing and batching procedures set in the contract, and monitor tempering time. Inspect and check all of these as you work.

4.3.0 Monitoring Structural Elements during Construction

The project engineer is responsible for structural inspections of loadbearing masonry. Field observation of masonry construction is not intended to certify structural characteristics, but to check the progress of the work.

Foundation beams and other structural elements need to be checked for line and grade be-

Store Masonry Cement Properly

Water will ruin masonry cement. It should be protected from rain either in a storage shed or by being covered with plastic sheets or canvas tarps. Cover any masonry cement that is kept outside. Place bags on wooden pallets to keep them off the ground.

28207-14-SA03.EPS

fore masonry work begins. Areas must be cleaned of dirt, grease, or other materials that could impair the mortar bond. Dimensions and layouts must be checked against drawings and adjustments made to correct discrepancies. Steel reinforcing dowels must be checked for proper location.

Because they are so critical, the inspector checks for proper embedment and coverage of anchors, ties, and joint reinforcements. The inspector monitors vertical coursing and joint widths, placing of reinforcement and ties, and grouting pours. The inspector checks that cavities are clear of mortar droppings and that flashing, weep holes, control joints, expansion joints, lintels, sills, caps, copings, and frames are set properly. The inspector checks that rebar is clean and its coating is intact. Windowsills, door frames, landings, and framing must also be checked to ensure that they are free of mortar.

Inspectors may observe methods or procedures that appear to conflict with the specifications. They will bring these items to the attention of the supervisor or contractor, who will make any adjustments. Masons observing specification conflicts should bring these items to the attention of their supervisor.

4.4.0 Monitoring Construction Tolerances

Inspection of the work in progress is a continuing task. Because so many building components are prefabricated today, tolerances are very important. Different materials have different tolerances due to the nature of their physical properties and manufacturing methods. Masonry has room for adjustment, but it has its own set of tolerances. *Figure 21* shows the tolerances set for mortar joints in block wall construction.

Expansion or control joints filled with elastomeric sealants also have tolerances. A realistic minimum thickness is ⅜ inch, to match the width of the mortar joints.

Concealed concrete and steel structural framing members have greater tolerances than exposed structural members such as masonry panels, veneer, or curtain walls. The AISC sets the tolerance for steel frame, ACI sets the one for concrete frame, and the BIA sets the one for brick veneer.

The adjustment necessary to keep a plumb line across the face must be taken up in the veneer anchors. If adjustments are not made carefully, unanticipated stresses will be put on the anchors, and the veneer may eventually detach as the anchor fails. Improper alignment of anchors with the frame and brick veneer will cause stress on the anchors.

The best solution is to have a variety of anchor lengths on hand to accommodate construction tolerances. Anchors must have a minimum ⅝-inch mortar cover on the outside wall face and a minimum 1½-inch embedment in masonry-unit veneers. If the framing member is out of plumb, choose an anchor size that will safely span the gap.

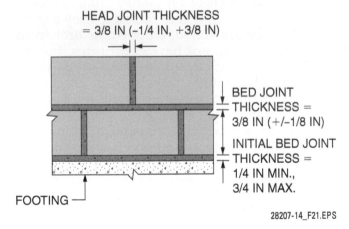

28207-14_F21.EPS

Figure 21 Allowable tolerances for block mortar joints.

OSHA Inspectors

Occupational Safety and Health Administration (OSHA) inspectors regularly visit construction sites unannounced. They can levy fines or stop work if conditions are dangerous. Keep the job site safe and avoid costly fines or work stoppages.

Prevent Materials Waste

Brick can break or chip if not handled carefully. Broken or chipped brick is unusable and will be rejected by an inspector. Brick cubes can break if they are not placed on level ground.

28207-14_SA04.EPS

Additional Resources

ACI 530/ASCE 5/TMS 402, *Building Code Requirements for Masonry Structures*, Latest Edition. Reston, VA: American Society of Civil Engineers.

International Building Code®, Latest edition. Falls Church, VA: International Code Council.

TEK 3-8A, Concrete Masonry Construction. 2001. Herndon, VA: National Concrete Masonry Association.

www.ncma.org

4.0.0 Section Review

1. It is considered a best practice for supervisors to observe every part of the job at least _____.

   a. once a week
   b. three times a week
   c. twice a day
   d. once a day

2. An inspector will reject bagged mortar and grout materials that show signs of _____.

   a. exposure to sunlight
   b. water absorption
   c. settling
   d. stacking too high

3. The inspector is responsible for reviewing each of the following during masonry construction *except* the _____.

   a. proper embedment and coverage of anchors, ties, and joint reinforcements
   b. proper set of weep holes, control joints, expansion joints, lintels, sills, caps, copings, and frames
   c. cleanliness and coating of rebar
   d. results of tests of masonry test panels prisms

4. The minimum embedment that anchors must have in a masonry-unit veneer is _____.

   a. 2¼ inches
   b. 2 inches
   c. 1 ¾ inches
   d. 1½ inches

SUMMARY

This module introduced the concepts of quality assurance and quality control, and reviewed various relevant codes and standards. ASTM specifications for masonry cover clay and concrete masonry units; mortar, grout, and accessories; as well as product testing. Hollow and grouted masonry prisms, mortar prisms, and grout prisms are tested to determine their compressive, flexural, and tensile strength. Sand is tested for cleanliness before it is mixed into mortar. Mortar is tested for consistency and slump. Several mortar tests, such as flow and penetration, are performed in a laboratory.

Field observations and inspections include standards and codes inspections, materials inspections, and observations of structural elements and construction tolerances. However, the responsibility for the quality of the masonry belongs not to the inspector, but to the mason.

1. For most construction projects, quality standards are established _____.
   a. by the general contractor
   b. by the office secretary
   c. in the contract documents and specifications
   d. by the subcontractor

2. Test results, reports, and other documents to be provided for the architect and engineer during the course of a project are called _____.
   a. submittals
   b. periodic updates
   c. project progress documentation
   d. submissions

3. Once model codes are adopted by states or municipalities, they become _____.
   a. subject to interpretation
   b. obsolete after five years
   c. amendable
   d. enforceable laws

4. Documents that name exact properties of materials and methods of installation without using brand names are known as _____.
   a. proprietary specifications
   b. descriptive specifications
   c. performance specifications
   d. reference standard specifications

5. If a type of brick to be used is not specified in a contract, the default is Type _____.
   a. FBA
   b. FBB
   c. FBS
   d. FBX

6. Standards for masonry mortar cover four types: _____.
   a. R, RS, N, O
   b. M, S, N, O
   c. M, S, RN, RO
   d. RM, S, N, RO

7. Masonry prism tests are concerned mainly with the compressive strength of mortar, grout, and masonry _____.
   a. under varying environmental conditions
   b. as individual materials
   c. in varying combinations
   d. performing as a unit

8. Typically, the requirements for constructing a sample panel are set by _____.
   a. testing agencies
   b. the architect
   c. the contractor
   d. local building codes

9. Before construction begins, laboratory requirements may call for the building of up to _____.
   a. three sample panels
   b. five test prisms
   c. eight sample panels
   d. 10 test prisms

10. To provide test results that more nearly reflect the performance of the wall, the height of the prism should be _____.
    a. twice its thickness
    b. three times its thickness
    c. four times its thickness
    d. five times its thickness

11. Laboratory stress tests are conducted to find a masonry prism's _____.
    a. breakpoint
    b. strength limit
    c. failure threshold
    d. index of disintegration

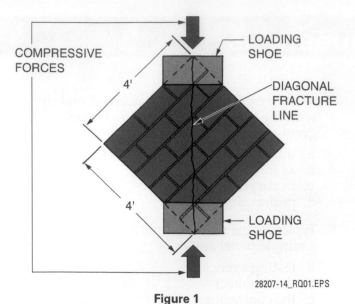

COMPRESSIVE FORCES

LOADING SHOE

DIAGONAL FRACTURE LINE

4'

4'

LOADING SHOE

28207-14_RQ01.EPS

Figure 1

12. The test illustrated in *Review Question Figure 1* is a _____.
   a. bond wrench test
   b. diagonal shear test
   c. diagonal compression test
   d. tensile-strength bond test

13. Mortar's ability to tie units together results from its _____.
   a. compressive strength
   b. plasticity
   c. strength of bond
   d. durability

14. The percentage of sand in different size classifications is determined by a _____.
   a. sieve test
   b. sedimentation test
   c. settling test
   d. screen test

15. Sand is clean enough to use in mortar if the thickness of the siltation test layer is no more than _____.
   a. ¼ inch
   b. ³⁄₁₆ inch
   c. ⅛ inch
   d. ¹⁄₁₆ inch

16. In a colorimetric test, a slightly yellow liquid indicates the presence of _____.
   a. inorganic contaminants
   b. considerable organic matter
   c. some organic matter
   d. no organic matter

17. To check the effects of weather and curing on the color consistency of mortar, make test cubes and expose them to weather at the site for _____.
   a. 24 hours
   b. five days
   c. two weeks
   d. 30 days

18. When testing mortar or grout, slump is measured to the nearest _____.
   a. inch
   b. ¾ inch
   c. ½ inch
   d. ¼ inch

19. Brick's initial rate of absorption is also referred to as its _____.
   a. suction
   b. sponging
   c. wicking
   d. sublimation

20. To reach full strength, a cube of mortar to be strength-tested must be cured for _____.
   a. two days
   b. 14 days
   c. 28 days
   d. 30 days

21. Laboratory testing of mortar's ability to retain water and cohesiveness after masonry units have absorbed moisture from it is done using a _____.
   a. penetrator
   b. flow table
   c. slump cone
   d. hygrometer

22. The results of a cone penetration test are reported in _____.
   a. eighths of an inch
   b. pounds per square inch
   c. kilograms per centimeter
   d. millimeters

23. Inspections for compliance with building codes are performed by _____.
   a. contractor's representatives
   b. city and/or county inspectors
   c. federal inspectors
   d. OSHA inspectors

24. Stored ingredients for mortar and grout are checked by an inspector for _____.

    a. contamination
    b. sunlight exposure
    c. insect infestation
    d. mold or mildew

25. Tolerances have become increasingly important in part because many building materials are _____.

    a. not quality controlled
    b. assembled by robots
    c. damaged in transit
    d. prefabricated

Trade Terms Quiz

Fill in the blank with the correct term that you learned from your study of this module.

1. The totality of features and characteristics of a product or service that bear on its ability to satisfy stated or implied needs is called _____.

2. _____ is a sedimentary material consisting of very fine particles intermediate in size between sand and clay.

3. Procedures for testing, inspecting, checking, and verifying to ensure that the work meets the required standard of quality established by the contract documents is called _____.

4. A(n) _____ is a sample prepared under controlled conditions specifically for testing that uses the same materials, techniques, and craftsmanship as those used on the job site.

5. The measure of the consistency or fluidity of a cementitious product is called _____.

6. _____ includes all planned and systematic actions necessary to provide adequate confidence that an item or a facility will perform satisfactorily in service.

7. Something that is capable of being stretched or extended is said to be _____.

8. The term _____ applies to something that is capable of turning, folding, or bending.

Trade Terms

Flexural
Prism
Quality
Quality assurance

Quality control
Silt
Slump
Tensile

Trade Terms Introduced in This Module

Flexural: Capable of turning, folding, or bending.

Prism: A sample prepared under controlled conditions specifically for testing that uses the same materials, techniques, and craftsmanship as those used on the job site.

Quality: The totality of features and characteristics of a product or service that bear on its ability to satisfy stated or implied needs.

Quality assurance: All planned and systematic actions necessary to provide adequate confidence that an item or a facility will perform satisfactorily in service.

Quality control: Procedures for testing, inspecting, checking, and verifying to ensure that the work meets the required standard of quality established by the contract documents.

Silt: A sedimentary material consisting of very fine particles intermediate in size between sand and clay.

Slump: A measure (0 to 12 inches) of the consistency or fluidity of a cementitious product such as concrete, masonry, or grout. Lower slump (1 to 4 inches) indicates a stiff mix; higher slump (8 to 12 inches) indicates a fluid mix.

Tensile: Capable of being stretched or extended, ductile.

Appendix

APPENDIX: ASTM STANDARDS FOR MASONRY CONSTRUCTION

BRICK

ASTM C27, Standard Classification of Fireclay and High-Alumina Refractory Brick

ASTM C32, Standard Specification for Sewer and Manhole Brick (Made from Clay or Shale)

ASTM C34, Standard Specification for Structural Clay Load-Bearing Wall Tile

ASTM C43, Standard Terminology of Structural Clay Products (note: withdrawn 2009)

ASTM C56, Standard Specification for Structural Clay Nonloadbearing Tile

ASTM C62, Standard Specification for Building Brick (Solid Masonry Units Made from Clay or Shale)

ASTM C106, Specification for Refractories and Incinerators (note: withdrawn 1972)

ASTM C126, Standard Specification for Ceramic Glazed Structural Clay Facing Tile, Facing Brick, and Solid Masonry Units

ASTM C155, Standard Classification of Insulating Firebrick

ASTM C212, Standard Specification for Structural Clay Facing Tile

ASTM C216, Standard Specification for Facing Brick (Solid Masonry Units Made from Clay or Shale)

ASTM C279, Standard Specification for Chemical-Resistant Masonry Units

ASTM C315, Standard Specification for Clay Flue Liners and Chimney Pots

ASTM C410, Standard Specification for Industrial Floor Brick

ASTM C416, Standard Classification of Silica Refractory Brick

ASTM C530, Standard Specification for Structural Clay Nonloadbearing Screen Tile

ASTM C652, Standard Specification for Hollow Brick (Hollow Masonry Units Made from Clay or Shale)

ASTM C902, Standard Specification for Pedestrian and Light Traffic Paving Brick

ASTM C1261, Standard Specification for Firebox Brick for Residential Fireplaces

ASTM C1272, Standard Specification for Heavy Vehicular Paving Brick

CONCRETE MASONRY UNITS

ASTM C55, Standard Specification for Concrete Building Brick

ASTM C73, Standard Specification for Calcium Silicate Brick (Sand-Lime Brick)

ASTM C90, Standard Specification for Loadbearing Concrete Masonry Units

ASTM C129, Standard Specification for Nonloadbearing Concrete Masonry Units

ASTM C139, Standard Specification for Concrete Masonry Units for Construction of Catch Basins and Manholes

ASTM C744, Standard Specification for Prefaced Concrete and Calcium Silicate Masonry Units

ASTM C936, Standard Specification for Solid Concrete Interlocking Paving Units

ASTM C1319, Standard Specification for Concrete Grid Paving Units

NATURAL STONE

ASTM C119, Standard Terminology Relating to Dimension Stone

ASTM C503, Standard Specification for Marble Dimension Stone

ASTM C568, Standard Specification for Limestone Dimension Stone

ASTM C615, Standard Specification for Granite Dimension Stone

ASTM C616, Standard Specification for Quartz-Based Dimension Stone

ASTM C629, Standard Specification for Slate Dimension Stone

MORTAR AND GROUT

ASTM C5, Standard Specification for Quicklime for Structural Purposes

ASTM C33, Standard Specification for Concrete Aggregates

ASTM C91, Standard Specification for Masonry Cement

ASTM C144, Standard Specification for Aggregate for Masonry Mortar

ASTM C150, Standard Specification for Portland Cement

ASTM C199, Standard Test Method for Pier Test for Refractory Mortars

ASTM C207, Standard Specification for Hydrated Lime for Masonry Purposes

ASTM C270, Standard Specification for Mortar for Unit Masonry

ASTM C330, Standard Specification for Lightweight Aggregates for Structural Concrete

ASTM C331, Standard Specification for Lightweight Aggregates for Concrete Masonry Units

ASTM C404, Standard Specification for Aggregates for Masonry Grout

ASTM C476, Standard Specification for Grout for MasonryASTM C595, Standard Specification for Blended Hydraulic Cements

ASTM C658, Standard Specification for Chemical-Resistant Resin Grouts for Brick or Tile

ASTM C887, Standard Specification for Packaged, Dry, Combined Materials for Surface Bonding Mortar

ASTM C1142, Standard Specification for Extended Life Mortar for Unit Masonry

ASTM C1329, Standard Specification for Mortar Cement

REINFORCEMENT AND ACCESSORIES

ASTM A36, Standard Specification for Carbon Structural Steel

ASTM A82, Standard Specification for Steel Wire, Plain, for Concrete Reinforcement

ASTM A153, Standard Specification for Zinc Coating (Hot-Dip) on Iron and Steel Hardware

ASTM A167, Standard Specification for Stainless and Heat-Resisting Chromium-Nickel Steel Plate, Sheet, and Strip

ASTM A185, Standard Specification for Steel Welded Wire Reinforcement, Plain, for Concrete

ASTM A496, Standard Specification for Steel Wire, Deformed, for Concrete Reinforcement

ASTM A615, Standard Specification for Deformed and Plain Carbon-Steel Bars for Concrete Reinforcement

ASTM A641, Standard Specification for Zinc-Coated (Galvanized) Carbon Steel Wire

ASTM A951, Standard Specification for Steel Wire for Masonry Joint Reinforcement

ASTM B227, Standard Specification for Hard-Drawn Copper-Clad Steel Wire

ASTM B766, Standard Specification for Electrodeposited Coatings of Cadmium

ASTM C915, Standard Specification for Precast Reinforced Concrete Crib Wall Members

ASTM C1089, Standard Specification for Spun Cast Prestressed Concrete Poles

ASTM C1242, Standard Guide for Selection, Design, and Installation of Dimension Stone Attachment Systems

SAMPLING AND TESTING

ASTM C67, Standard Test Methods for Sampling and Testing Brick and Structural Clay Tile

ASTM C97, Standard Test Methods for Absorption and Bulk Specific Gravity of Dimension Stone

ASTM C109, Standard Test Method for Compressive Strength of Hydraulic Cement Mortars (Using 2-in. or [50-mm] Cube Specimens)

ASTM C138, Standard Test Method for Density (Unit Weight), Yield, and Air Content (Gravimetric) of Concrete

ASTM C140, Standard Test Methods for Sampling and Testing Concrete Masonry Units and Related Units

ASTM C170, Standard Test Method for Compressive Strength of Dimension Stone

ASTM C173, Standard Test Method for Air Content of Freshly Mixed Concrete by the Volumetric Method

ASTM C231, Standard Test Method for Air Content of Freshly Mixed Concrete by the Pressure Method

ASTM C241, Standard Test Method for Abrasion Resistance of Stone Subjected to Foot Traffic

ASTM C267, Standard Test Methods for Chemical Resistance of Mortars, Grouts, and Monolithic Surfacings and Polymer Concretes

ASTM C426, Standard Test Method for Linear Drying Shrinkage of Concrete Masonry Units

ASTM C780, Standard Test Method for Preconstruction and Construction Evaluation of Mortars for Plain and Reinforced Unit Masonry

ASTM C880, Standard Test Method for Flexural Strength of Dimension Stone

ASTM C952, Standard Test Method for Bond Strength of Mortar to Masonry Units

ASTM C1006, Standard Test Method for Splitting Tensile Strength of Masonry Units

ASTM C1019, Standard Test Method for Sampling and Testing Grout

ASTM C1072, Standard Test Methods for Measurement of Masonry Flexural Bond Strength

ASTM C1093, Standard Practice for Accreditation of Testing Agencies for Masonry

ASTM C1148, Standard Test Method for Measuring the Drying Shrinkage of Masonry Mortar

ASTM C1194, Standard Test Method for Compressive Strength of Architectural Cast Stone

ASTM C1195, Standard Test Method for Absorption of Architectural Cast Stone

ASTM C1196, Standard Test Methods for In Situ Compressive Stress within Solid Unit Masonry Estimated Using Flatjack Measurements

ASTM C1197, Standard Test Method for In Situ Measurement of Masonry Deformability Using the Flatjack Method

ASTM C1262, Standard Test Method for Evaluating the Freeze-Thaw Durability of Dry-Cast Segmental Retaining Wall Units and Related Concrete Units

ASTM C1314, Standard Test Method for Compressive Strength of Masonry Prisms

ASTM C1324, Standard Test Method for Examination and Analysis of Hardened Masonry Mortar

ASTM D75, Standard Practice for Sampling Aggregates

ASTM E72, Standard Test Methods of Conducting Strength Tests of Panels for Building Construction

ASTM E447, Test Methods for Compressive Strength of Laboratory Constructed Masonry Prisms (note: withdrawn 1988)

ASTM E488, Standard Test Methods for Strength of Anchors in Concrete Elements

ASTM E514, Standard Test Method for Water Penetration and Leakage through Masonry

ASTM E518, Standard Test Methods for Flexural Bond Strength of Masonry

ASTM E519, Standard Test Method for Diagonal Tension (Shear) in Masonry Assemblages

ASTM E754, Standard Test Method for Pullout Resistance of Ties and Anchors Embedded in Masonry Mortar Joints

ASSEMBLAGES

ASTM C901, Standard Specification for Prefabricated Masonry Panels

ASTM C946, Standard Practice for Construction of Dry-Stacked, Surface-Bonded Walls

ASTM E835, Standard Guide for Modular Coordination of Clay and Concrete Masonry Units (note: withdrawn 2011)

ASTM C1283, Standard Practice for Installing Clay Flue Lining

ASTM E1602, Standard Guide for Construction of Solid Fuel Burning Masonry Heaters

Additional Resources

This module presents thorough resources for task training. The following resource material is suggested for further study.

ACI 530/ASCE 5/TMS 402, Building Code Requirements for Masonry Structures, Latest Edition. Reston, VA: American Society of Civil Engineers.

ASTM A36, Standard Specification for Carbon Structural Steel, Latest Edition. West Conshohocken, PA: ASTM International.

ASTM A82, Standard Specification for Steel Wire, Plain, for Concrete Reinforcement, Latest Edition. West Conshohocken, PA: ASTM International.

ASTM A153, Standard Specification for Zinc Coating (Hot-Dip) on Iron and Steel Hardware, Latest Edition. West Conshohocken, PA: ASTM International.

ASTM A167, Standard Specification for Stainless and Heat-Resisting Chromium-Nickel Steel Plate, Sheet, and Strip, Latest Edition. West Conshohocken, PA: ASTM International.

ASTM A615, Standard Specification for Deformed and Plain Carbon-Steel Bars for Concrete Reinforcement, Latest Edition. West Conshohocken, PA: ASTM International.

ASTM A951, Standard Specification for Steel Wire for Masonry Joint Reinforcement, Latest Edition. West Conshohocken, PA: ASTM International.

ASTM C5, Standard Specification for Quicklime for Structural Purposes, Latest Edition. West Conshohocken, PA: ASTM International.

ASTM C55, Standard Specification for Concrete Building Brick, Latest Edition. West Conshohocken, PA: ASTM International.

ASTM C62, Standard Specification for Building Brick (Solid Masonry Units Made from Clay or Shale), Latest Edition. West Conshohocken, PA: ASTM International.

ASTM C67, Standard Test Methods for Sampling and Testing Brick and Structural Clay Tile, Latest Edition. West Conshohocken, PA: ASTM International.

ASTM C90, Standard Specification for Loadbearing Concrete Masonry Units, Latest Edition. West Conshohocken, PA: ASTM International.

ASTM C91, Standard Specification for Masonry Cement, Latest Edition. West Conshohocken, PA: ASTM International.

ASTM C126, Standard Specification for Ceramic Glazed Structural Clay Facing Tile, Facing Brick, and Solid Masonry Units, Latest Edition. West Conshohocken, PA: ASTM International.

ASTM C129, Standard Specification for Nonloadbearing Concrete Masonry Units, Latest Edition. West Conshohocken, PA: ASTM International.

ASTM C138, Standard Test Method for Density (Unit Weight), Yield, and Air Content (Gravimetric) of Concrete, Latest Edition. West Conshohocken, PA: ASTM International.

ASTM C140, Standard Test Methods for Sampling and Testing Concrete Masonry Units and Related Units, Latest Edition. West Conshohocken, PA: ASTM International.

ASTM C144, Standard Specification for Aggregate for Masonry Mortar, Latest Edition. West Conshohocken, PA: ASTM International.

ASTM C150, Standard Specification for Portland Cement, Latest Edition. West Conshohocken, PA: ASTM International.

ASTM C173, Standard Test Method for Air Content of Freshly Mixed Concrete by the Volumetric Method, Latest Edition. West Conshohocken, PA: ASTM International.

ASTM C207, Standard Specification for Hydrated Lime for Masonry Purposes, Latest Edition. West Conshohocken, PA: ASTM International.

ASTM C216, Standard Specification for Facing Brick (Solid Masonry Units Made from Clay or Shale), Latest Edition. West Conshohocken, PA: ASTM International.

ASTM C231, Standard Test Method for Air Content of Freshly Mixed Concrete by the Pressure Method, Latest Edition. West Conshohocken, PA: ASTM International.

ASTM C270, Standard Specification for Mortar for Unit Masonry, Latest Edition. West Conshohocken, PA: ASTM International.

ASTM C476, Standard Specification for Grout for Masonry, Latest Edition. West Conshohocken, PA: ASTM International.

ASTM C595, Standard Specification for Blended Hydraulic Cements, Latest Edition. West Conshohocken, PA: ASTM International.

ASTM C652, Standard Specification for Hollow Brick (Hollow Masonry Units Made from Clay or Shale), Latest Edition. West Conshohocken, PA: ASTM International.

ASTM C780, Standard Test Method for Preconstruction and Construction Evaluation of Mortars for Plain and Reinforced Unit Masonry, Latest Edition. West Conshohocken, PA: ASTM International.

ASTM C1019, Standard Test Method for Sampling and Testing Grout, Latest Edition. West Conshohocken, PA: ASTM International.

ASTM C1314, Standard Test Method for Compressive Strength of Masonry Prisms, Latest Edition. West Conshohocken, PA: ASTM International.

ASTM C1329, Standard Specification for Mortar Cement, Latest Edition. West Conshohocken, PA: ASTM International.

International Building Code®, Latest Edition. Falls Church, VA: International Code Council.

NFPA 5000, Building Construction and Safety Code®, Latest edition. Quincy, MA: National Fire Protection Association.

TEK 3-8A, Concrete Masonry Construction. 2001. Herndon, VA: National Concrete Masonry Association. **www.ncma.org**

Figure Credits

Section Review Answers

Answer	Section Reference	Objective
Section One		
1. d	1.1.0	1a
2. b	1.2.0	1b
Section Two		
1. a	2.1.0	2a
2. a	2.2.0	2b
3. c	2.3.0	2c
4. d	2.4.0	2d
5. b	2.5.2	2e
Section Three		
1. c	3.1.0	3a
2. d	3.2.0	3b
3. a	3.3.0	3c
4. d	3.4.2	3d
Section Four		
1. c	4.1.0	4a
2. b	4.2.0	4b
3. d	4.3.0	4c
4. d	4.4.0	4d

NCCER CURRICULA — USER UPDATE

NCCER makes every effort to keep its textbooks up-to-date and free of technical errors. We appreciate your help in this process. If you find an error, a typographical mistake, or an inaccuracy in NCCER's curricula, please fill out this form (or a photocopy), or complete the online form at **www.nccer.org/olf**. Be sure to include the exact module ID number, page number, a detailed description, and your recommended correction. Your input will be brought to the attention of the Authoring Team. Thank you for your assistance.

Instructors – If you have an idea for improving this textbook, or have found that additional materials were necessary to teach this module effectively, please let us know so that we may present your suggestions to the Authoring Team.

NCCER Product Development and Revision
13614 Progress Blvd., Alachua, FL 32615

Email: curriculum@nccer.org
Online: www.nccer.org/olf

❏ Trainee Guide ❏ Lesson Plans ❏ Exam ❏ PowerPoints Other _____

Craft / Level: _____ Copyright Date: _____

Module ID Number / Title: _____

Section Number(s): _____

Description: _____

_____

_____

_____

_____

Recommended Correction: _____

_____

_____

_____

_____

Your Name: _____

Address: _____

_____

Email: _____ Phone: _____

Glossary

Anchor: A metal assembly used to attach masonry to a structural support.

Beam pocket: An opening in a vertical masonry wall that allows an intersecting structural beam to bear on or pass through the wall.

Bearing pressure: The load on a bearing surface divided by its area, expressed in pounds per square inch.

Bedrock: Solid rock that cannot be easily dislodged or removed from the soil. Typically, the rock that forms the outer crust of the earth.

Blowout: The swelling or rupture of a cavity wall from too much pressure caused by pouring liquid grout into the cavity.

Bond beam: A course of masonry units with steel rebar inserted and held in place by a solid fill of grout or mortar; used as a lintel or reinforcement beam to distribute stress.

Breaking the bond: Starting a course with a cut so as to center the header unit over the head joints in the course below.

Bridging: The mounding of grout or cement over an obstruction, creating a void under the obstruction.

Cap: Masonry units laid on top of a finished wall.

Chamfered: Stone block or brick with beveled edges that do not go all the way across the edge or end of the block.

Change order: A document or form used during the construction process to document a change in the construction requirements from the original plans or specifications.

Chase: A continuous recess built into a wall to receive pipes, wires, or heating ducts.

Column: A vertical reinforced masonry element designed to support a load, the width of which never exceeds three times its thickness and the height of which always exceeds four times its thickness.

Coping: The materials or masonry units used to form a cap or finish on top of a wall, pier, chimney, or pilaster to protect the masonry below from water penetration. Coping is usually projected from both sides of the wall to provide a protective covering as well as an ornamental design.

Corbel: The process of laying masonry units to form a shelf or ledge.

Cove: A concave area between two perpendicular planes; the area at the intersection of a wall and floor, or wall and ceiling.

Damper: A metal device used for regulating the draft in the flue of a chimney, usually made of cast iron.

Downdraft: A current of air that moves with force down an opening in a chimney.

Empirically designed: Design based on the application of physical limitations learned from experience or on observations gained through experience, but not based on structural analysis.

Fastener: A metal assembly used to attach building parts to masonry.

Flexural: Capable of turning, folding, or bending.

Footing: An enlargement at the bottom of a wall that distributes the weight of the superstructure over a greater area to prevent settling.

Grouted wall: A hollow masonry wall in which the voids are filled with grout but not reinforcing bar.

Horizontal joint reinforcement: A system of connected steel wire that provides added structural integrity to a masonry wythe by distributing lateral loads evenly and by tying the masonry elements in the structure together mechanically.

Humored: A slang term for gradually adjusting a section of masonry in order to correct alignment issues.

Hydrostatic pressure: The pressure at any point in a liquid at rest, equal to the depth of the liquid multiplied by its density.

Infiltration: The drainage of storm water and other runoff into the ground beneath a paved surface.

Jamb: The side of an opening, or the vertical framing member on the side of the opening, usually for door and window frames.

Key: A recess or groove in one placement of grout or concrete that is later filled with a new placement of grout or concrete so that the two lock together in a tongue-and-groove configuration.

Legend: A listing that explains or defines symbols or special marks placed on plans or drawings. Usually the legend is on the front sheet or index of the plan set.

Lift: One continuous placement of grout or cement without interruption, equivalent to one layer.

Lintel: The horizontal member or beam over an opening that carries the weight of the masonry above the opening.

Loadbearing: Any structure that supports any vertical load in addition to its own weight.

Membrane: A layer of thin, pliable material used to waterproof masonry.

Mortarless paving: The placement of paving brick or block on a horizontal surface in some pattern to form a smooth, flat surface. No mortar is used to bond the units to the surface underneath or to each other.

Panel: A section of wall between control joints, wall ends, or a control joint and wall end.

Parge: To apply a thin coat of mortar or grout on the outside surface of a masonry surface to prepare it for the attachment of veneer or tile, or to waterproof it.

Pavers: Brick, solid concrete block, or patterned concrete block that are used to build smooth, horizontal surfaces. Pavers are manufactured in many different thicknesses and shapes.

Pencil rod: A type of metallic tie that is similar to the shape of a straight wooden pencil; used for control joints in concrete masonry construction.

Pier: A vertical reinforced masonry element that is typically shorter than a column or pilaster. Piers may be designed to carry loads, or may be purely ornamental.

Pilaster: A vertical reinforced masonry element consisting of a thickened section of a wall to which it is structurally bonded; a wall portion projecting from a wall face and serving as a vertical column and/or beam.

Pintle: A type of masonry fastener that allows a horizontal reinforcement to maintain a joint while allowing the wall to have some freedom of vertical movement.

Plasticity: A material's ability to be easily molded into various shapes.

Prism: A sample prepared under controlled conditions specifically for testing that uses the same materials, techniques, and craftsmanship as those used on the job site.

Quality: The totality of features and characteristics of a product or service that bear on its ability to satisfy stated or implied needs.

Quality assurance: All planned and systematic actions necessary to provide adequate confidence that an item or a facility will perform satisfactorily in service.

Quality control: Procedures for testing, inspecting, checking, and verifying to ensure that the work meets the required standard of quality established by the contract documents.

Rebar: Reinforcing bar embedded in concrete, mortar, or grout in such a manner that it acts together with the other components to resist loads.

Reinforced masonry element: A hollow masonry structure other than a wall in which the voids in the masonry units are filled with grout and reinforcing bar.

Reinforced wall: A hollow masonry wall in which the voids in the masonry units are filled with grout and reinforcing bar.

Reveal: The side of an opening in a wall for a window or door. This is the part of the masonry jamb around a window or door frame that can be seen from the frame to the face of the masonry wall; the visible portion of a masonry jamb between the face of the wall and the frame.

Rodding: Poking the grout with a rod in order to consolidate it.

Sectional drawing: A drawing that shows the inside of a component or structure. The view would be as if you cut the item into two pieces and looked at the end of the cut.

Segmental retaining wall (SRW): A wall made of segmental block stacked on top of each other without mortar bonding.

Shop drawing: A drawing that is usually developed by manufacturers, fabricators, or contractors to show specific dimensions and other pertinent information concerning a particular piece of equipment and its installation methods.

Sill: A horizontal member under a door or window. Slip sills fit inside the door or window frame; lug sills extend beyond the frame and into the masonry on the jamb sides of the frame.

Silt: A sedimentary material consisting of very fine particles intermediate in size between sand and clay.

Skew: The condition when two parts come together at an angle that is not 90 degrees, or perpendicular, to each other.

Slump: A measure (0 to 12 inches) of the consistency or fluidity of a cementitious product such as concrete, masonry, or grout. Lower slump (1 to 4 inches) indicates a stiff mix; higher slump (8 to 12 inches) indicates a fluid mix.

Slushing: The process of using a trowel to fill collar joints and other wide openings with mortar.

Tensile: Capable of being stretched or extended, ductile.

Thermal mass: The ability of a building wall to alternately absorb and release heat energy in response to temperature changes.

Toothing: Construction of a temporary end of a wall with the end stretcher of every alternate course projecting. The projecting units are called toothers.

Vibrating: Consolidating grout with the use of a mechanical vibrator.

Waterproofing: The process of treating masonry with a material that will retard penetration of moisture.

Wick: A bundle of fibers that are loosely twisted or braided and woven together to form a cord that can carry water away from an area by capillary action, which occurs as long as the drip end is lower than the absorption end.

Index